T0203030

Communications
in Computer and Information Science   1943

Editorial Board Members

Joaquim Filipe📀, *Polytechnic Institute of Setúbal, Setúbal, Portugal*
Ashish Ghosh📀, *Indian Statistical Institute, Kolkata, India*
Raquel Oliveira Prates📀, *Federal University of Minas Gerais (UFMG),*
Belo Horizonte, Brazil
Lizhu Zhou, *Tsinghua University, Beijing, China*

Rationale

The CCIS series is devoted to the publication of proceedings of computer science conferences. Its aim is to efficiently disseminate original research results in informatics in printed and electronic form. While the focus is on publication of peer-reviewed full papers presenting mature work, inclusion of reviewed short papers reporting on work in progress is welcome, too. Besides globally relevant meetings with internationally representative program committees guaranteeing a strict peer-reviewing and paper selection process, conferences run by societies or of high regional or national relevance are also considered for publication.

Topics

The topical scope of CCIS spans the entire spectrum of informatics ranging from foundational topics in the theory of computing to information and communications science and technology and a broad variety of interdisciplinary application fields.

Information for Volume Editors and Authors

Publication in CCIS is free of charge. No royalties are paid, however, we offer registered conference participants temporary free access to the online version of the conference proceedings on SpringerLink (http://link.springer.com) by means of an http referrer from the conference website and/or a number of complimentary printed copies, as specified in the official acceptance email of the event.

CCIS proceedings can be published in time for distribution at conferences or as post-proceedings, and delivered in the form of printed books and/or electronically as USBs and/or e-content licenses for accessing proceedings at SpringerLink. Furthermore, CCIS proceedings are included in the CCIS electronic book series hosted in the SpringerLink digital library at http://link.springer.com/bookseries/7899. Conferences publishing in CCIS are allowed to use Online Conference Service (OCS) for managing the whole proceedings lifecycle (from submission and reviewing to preparing for publication) free of charge.

Publication process

The language of publication is exclusively English. Authors publishing in CCIS have to sign the Springer CCIS copyright transfer form, however, they are free to use their material published in CCIS for substantially changed, more elaborate subsequent publications elsewhere. For the preparation of the camera-ready papers/files, authors have to strictly adhere to the Springer CCIS Authors' Instructions and are strongly encouraged to use the CCIS LaTeX style files or templates.

Abstracting/Indexing

CCIS is abstracted/indexed in DBLP, Google Scholar, EI-Compendex, Mathematical Reviews, SCImago, Scopus. CCIS volumes are also submitted for the inclusion in ISI Proceedings.

How to start

To start the evaluation of your proposal for inclusion in the CCIS series, please send an e-mail to ccis@springer.com.

Diana Benavides-Prado · Sarah Erfani ·
Philippe Fournier-Viger · Yee Ling Boo ·
Yun Sing Koh
Editors

Data Science and Machine Learning

21st Australasian Conference, AusDM 2023
Auckland, New Zealand, December 11–13, 2023
Proceedings

 Springer

Editors
Diana Benavides-Prado ⓘ
The University of Auckland
Auckland, New Zealand

Philippe Fournier-Viger ⓘ
Shenzhen University
Shenzhen, China

Yun Sing Koh ⓘ
The University of Auckland
Auckland, New Zealand

Sarah Erfani ⓘ
The University of Melbourne
Carlton, VIC, Australia

Yee Ling Boo ⓘ
RMIT University
Melbourne, VIC, Australia

ISSN 1865-0929 ISSN 1865-0937 (electronic)
Communications in Computer and Information Science
ISBN 978-981-99-8695-8 ISBN 978-981-99-8696-5 (eBook)
https://doi.org/10.1007/978-981-99-8696-5

© The Editor(s) (if applicable) and The Author(s), under exclusive license
to Springer Nature Singapore Pte Ltd. 2024

This work is subject to copyright. All rights are reserved by the Publisher, whether the whole or part of the material is concerned, specifically the rights of translation, reprinting, reuse of illustrations, recitation, broadcasting, reproduction on microfilms or in any other physical way, and transmission or information storage and retrieval, electronic adaptation, computer software, or by similar or dissimilar methodology now known or hereafter developed.
The use of general descriptive names, registered names, trademarks, service marks, etc. in this publication does not imply, even in the absence of a specific statement, that such names are exempt from the relevant protective laws and regulations and therefore free for general use.
The publisher, the authors, and the editors are safe to assume that the advice and information in this book are believed to be true and accurate at the date of publication. Neither the publisher nor the authors or the editors give a warranty, expressed or implied, with respect to the material contained herein or for any errors or omissions that may have been made. The publisher remains neutral with regard to jurisdictional claims in published maps and institutional affiliations.

This Springer imprint is published by the registered company Springer Nature Singapore Pte Ltd.
The registered company address is: 152 Beach Road, #21-01/04 Gateway East, Singapore 189721, Singapore

Paper in this product is recyclable.

Preface

It is our great pleasure to present the proceedings of the 21st Australasian Data Science and Machine Learning Conference (formerly known as Australasian Data Mining Conference, AusDM 2023), held at the University of Auckland, Auckland, during 11 to 13 December 2023.

The AusDM conference series first started in 2002 as a workshop initiated by Professor Simeon Simoff (Western Sydney University), Professor Graham Williams (Australian National University), and Emeritius Professor Markus Hegland (Australian National University). Over the years, AusDM has established itself as the premier Australasian meeting for both practitioners and researchers in the area of data mining (or data analytics or data science) and machine learning. AusDM is devoted to the art and science of intelligent analysis of (usually big) data sets for meaningful (and previously unknown) insights. Since AusDM 2002, the conference series has showcased research in data science and machine learning through presentations and discussions on state-of-art research and development. Built on this tradition, AusDM 2023 successfully facilitated the cross-disciplinary exchange of ideas, experiences, and potential research directions, and pushed forward the frontiers of data science and machine learning in academia, government, and industry.

This year, the theme of the conference focused on Data Science and Machine Learning: Now part of everyone everyday. Specifically, Data Science, Machine Learning and AI Innovation Day, Green and Responsible AI Day and Generative AI Day were organized and sparked great discussions. In addition, a journal special issue with Data Science and Engineering by Springer was also planned.

AusDM 2023 received altogether 50 valid submissions, with authors from 16 different countries. The top 5 countries, in terms of the number of authors who submitted papers to AusDM 2023, were Australia (59 authors), New Zealand (46), India (24), Germany (10), and South Korea (8). All submissions went through a double-blind review process, and each paper received at least three peer-reviewed reports. Additional reviewers were considered for a clearer review outcome, if review comments from the initial three reviewers were inconclusive.

Out of these 50 submissions, a total of 20 papers were finally accepted for publication. The overall acceptance rate for AusDM 2023 was 40%. Out of the 34 Research Track submissions, 13 papers (i.e. 38%) were accepted for publication. Out of the 16 submissions in the Application Track, 7 papers (i.e. 44%) were accepted for publication.

The AusDM 2023 Organizing Committee would like to give their special thanks to Albert Bifet, Madeline Newman, and Michael Witbrock for kindly accepting the invitation to give keynote speeches. In addition, the success of three panel discussions themed around Green and Responsible AI, Generative AI and Māori Algorithmic Sovereignty Roundtable was attributed to Kevin Ross, Gabriela Mazorra de Cos, Alvaro Orsi, Kin Lung Chan, Malcolm Fraser, Trevor Kennedy, Christopher Mende, Daniel Wilson, Kiri

West, Paul Brown and Ben Ritchie. The committee would like to express their appreciation to Mingming Gong, Ming Cheuk and Gillian Dobbie for presenting at Spotlight Talks, and Anusha Vidnage, Jessica Moore, Graham Williams, and Giulio Valentino Dalla Riva for presenting at the tutorial sessions. The committee was grateful to have Scott Spencer and Jiamou Liu as the speakers for Industry Talks.

The committee would also like to give their sincere thanks to the University of Auckland for providing admin support and the conference venue. The committee would also like to thank Springer CCIS and the Editorial Board for their acceptance to publish AusDM 2023 papers. This will give excellent exposure of the papers accepted for publication. We would also like to give our heartfelt thanks to all student and staff volunteers at the University of Auckland, who did a tremendous job in ensuring a successful conference event.

Last but not least, we would like to give our sincere thanks to all delegates for attending the conference this year at the University of Auckland. We hope that it was a fruitful experience and you enjoyed AusDM 2023!

December 2023
                                                    Diana Benavides-Prado
                                                    Sarah Erfani
                                                    Philippe Fournier-Viger
                                                    Yee Ling Boo
                                                    Yun Sing Koh

Organization

General Chair

Yun Sing Koh   University of Auckland, New Zealand

Program Chairs (Research Track)

Diana Benavides-Prado   University of Auckland, New Zealand
Sarah Erfani   University of Melbourne, Australia

Program Chair (Application Track)

Philippe Fournier-Viger   Shenzhen University, China

Industry Chair

Shivonne Londt   Amazon Web Services, New Zealand

Special Session Chairs

Jiamou Liu   University of Auckland, New Zealand
Mingming Gong   University of Melbourne, Australia
Amanda J. Williamson   University of Waikato, Deloitte, New Zealand

Publication Chair

Yee Ling Boo   RMIT University, Australia

Diversity, Equity and Inclusion Chair

Richi Nayak   Queensland University of Technology, Australia

Sponsorship Chair

Annelies Tjetjep   AutogenAI, Australia

Web Chair

Bowen Chen   University of Auckland, New Zealand

Publicity Chair

Nick Lim   University of Waikato, New Zealand

Local Organizing Chairs

Nickylee Anderson   University of Auckland, New Zealand
Thomas Lacombe   University of Auckland, New Zealand

Tutorial and Workshop Chair

Andrew Lensen   Victoria University of Wellington, New Zealand

Doctoral Symposium Chairs

Di Zhao   University of Auckland, New Zealand
Asara Senaratne   University of South Australia, Australia

Steering Committee Chairs

Simeon Simoff   Western Sydney University, Australia
Graham Williams   Australian National University, Australia

Steering Committee

Paul Kennedy	University of Technology Sydney, Australia
Jiuyong (John) Li	University of South Australia, Australia
Kok-Leong Ong	RMIT University, Australia
Yanchang Zhao	Data61, CSIRO, Australia
Richi Nayak	Queensland University of Technology, Australia
Ling Chen	University of Technology Sydney, Australia
Dharmendra Sharma	University of Canberra, Australia
Yun Sing Koh	University of Auckland, New Zealand
Lin Liu	University of South Australia, Australia
Warwick Graco	Analytics Shed, Australia
Yee Ling Boo	RMIT University, Australia

Honorary Advisors

John Roddick	Flinders University, Australia
Geoff Webb	Monash University, Australia

Program Committee

Research Track

Alex Peng	University of Auckland, New Zealand
Abhishek Appaji	B.M.S College of Engineering, India
Brad Malin	Vanderbilt University, USA
Dang Nguyen	Deakin University, Australia
Evan Crawford	Western Sydney University, Australia
Gang Li	Deakin University, Australia
Guilherme Weigert Cassales	University of Waikato, New Zealand
Hoa Nguyen	Australian National University, Australia
Jie Yang	University of Wollongong, Australia
Khanh Luong	Queensland University of Technology, Australia
Md Geaur Rahman	Charles Sturt University, Australia
Michal Ptaszynski	Kitami Institute of Technology, Japan
Nuwan Gunasekara	University of Waikato, New Zealand
Quang Vinh Nguyen	Western Sydney University, Australia
Rushit Dave	Minnesota State University at Mankato, USA
Selasi Kwashie	Charles Sturt University, Australia

Sharon Torao Pingi	Queensland University of Technology, Australia
Trung Nguyen	University of Auckland, New Zealand
Vithya Yogarajan	University of Auckland, New Zealand
Warwick Graco	Analytics Shed, Australia
Weijia Zhang	University of Newcastle, Australia
Yang Chen	University of Auckland, New Zealand
Zhaoxu Xi	Upstart, USA

Application Track

Farnoush Falahatraftar	TELUS, Canada
Jing Ma	Auckland University of Technology, New Zealand
Minh Nguyen	Auckland University of Technology, New Zealand
Muhammad Marwan Muhammad Fuad	Coventry University, UK
M. Saqib Nawaz	Shenzhen University, China
Oliver Obst	Western Sydney University, Australia
Samaneh Madanian	Auckland University of Technology, New Zealand
Yue Xu	Queensland University of Technology, Australia
Warwick Graco	Analytics Shed, Australia

Additional Reviewer

Zhonglin Qu

Contents

Research Track

Research Work

Random Padding Data Augmentation

Nan Yang⬡, Laicheng Zhong⬡, Fan Huang⬡, Wei Bao⬡,
and Dong Yuan(✉)⬡

Faculty of Engineering, The University of Sydney, Camperdown, Australia
{n.yang,laicheng.zhong,fan.huang,wei.bao,dong.yuan}@sydney.edu.au

Abstract. The convolutional neural network (CNN) learns the same
object in different positions in images, which can improve the model
recognition accuracy. An implication of this is that CNN may know where
the object is. The usefulness of the features' spatial information in CNNs
has not been well investigated. In this paper, we found that the model's
learning of features' position information hindered the learning of the
features' relationship. Therefore, we introduced Random Padding, a new
type of padding method for training CNNs that impairs the architecture's
capacity to learn position information by adding zero-padding randomly
to half of the border of feature maps. Random Padding is parameter-
free, simple to construct, and compatible with the majority of CNN-
based recognition models. This technique is also complementary to data
augmentation such as random cropping, rotation, flipping, and erasing
and consistently improves the performance of image classification over
strong baselines.

Keywords: Spatial Information · Random Padding · Data
Augmentation

1 Introduction

Convolutional Neural Network (CNN) is an important component in computer
vision and plays a key role in deep learning architecture, which extracts low/mid
/high-level features [24] and classifiers naturally in an end-to-end multilayer, and
the "levels" of features can be evolved by the depth of stacked layers. The idea
of CNN model design is inspired by live organisms' inherent visual perception
process [4]. The shallow layers learn the local features of the image, such as the
color and geometric shape of the image object, while the deep layers learn more
abstract features from the input data, such as contour characteristics and other
high-dimensional properties. The multi-layer structure of the CNN can automatically learn features' spatial information from the input image data. Spatial information is represented by matrices in hierarchical CNN models, which are the conversion between the overall coordinate system of the object and the coordinate
system of each component. Thus, the spatial information of the features learned
by CNN can perform a shift-invariant classification of the input information.

© The Author(s), under exclusive license to Springer Nature Singapore Pte Ltd. 2024
D. Benavides-Prado et al. (Eds.): AusDM 2023, CCIS 1943, pp. 3–18, 2024.
https://doi.org/10.1007/978-981-99-8696-5_1

In fact, what truly provides free translation and deformation invariance for the CNN architecture are the convolution structure and the pooling operation, while the use of sub-sampling and stride will break this special characteristic. Even if the target position is changed, the operations of convolution and pooling can still extract the same information from images and then flatten it to the same feature value in different orders of the following fully connected layer. However, subsampling and stride reduce part of the information of the input image, which leads to the loss of some features, thus breaking the translation invariance of CNN. Recent research on data argumentation has attempted to enhance the invariance by performing the operations of translation, rotation, reflection, and scaling [2, 22] on the input images. However, it cannot really improve the model's learning ability of shift-invariance.

The shift-invariant ability of CNN depends on the learning of features' spatial information, which contains two types of information, i.e., *features relationship* and *position information*. *Features relationship* refers to the relative position among different features, while *position information* represents the absolute position of features in the image.

We deem that feature relationship is helpful in CNN, as if a feature is useful in one image location, the same feature is likely to be useful in other locations. The Capsule Network [10] is designed to learn feature relationships from images, i.e., the spatial relationships between whole objects and their parts. However, it is difficult to implement on complex datasets, e.g., CIFAR-10 and ImageNet. On the other hand, we believe position information is harmful to CNN, as learning it will impede the model's acquisition of feature relationships. Recent evidence suggests that position information is implicitly encoded in the extracted feature maps, thus non-linear readout of position information further augments the readout of absolute position [7]. Additionally, these studies point out the fact that zero-padding and borders serve as an anchor for spatial information that is generated and ultimately transmitted over the whole image during spatial abstraction. Hence, how to reduce the position information introduced by zero-padding has been long-ignored in CNN solutions to vision issues.

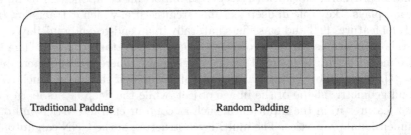

Fig. 1. Traditional Padding and Random Padding.

In this paper, with the purpose of reducing CNN's learning of position information, we proposed the Random Padding operation shown in Fig. 1, that is a variant of traditional padding technology. Random Padding is to add zero-padding to the randomly chosen half boundary of the feature maps, which will weaken the position information and let the CNN model better understand the feature relationship. This technique makes CNN models more robust to the change of an object's absolute position in the images.

The contribution of the paper is summarized as follows:

- We investigate the usefulness of spatial information in CNN and propose a new approach to improving the model accuracy, i.e., through reducing the position information in CNN.
- We propose the Random Padding operation that can be directly added to a variety of CNN models without any changes to the model structure. It is lightweight that requires no additional parameter learning or computing in model training. As the operation does not make changes on the input images, it is complementary to the traditional data augmentations used in CNN training.
- We conducted extensive experiments on popular CNN models. The results show that the Random Padding operation can reduce the extraction of the position information in CNN models and improve the accuracy on image classification.

2 Related Work

2.1 Approaches to Improve Accuracy of CNNs

The evolution of the structure of the convolutional neural networks has gradually improved the accuracy, i.e., Alexnet with ReLu and Dropout [11], VGG with $3 * 3$ kernels [16], Googlenet with inception [19] and Resnet with residual blocks [3]. In addition to upgrading of CNN architectures, the augmentation of input data is also an indispensable part of improving performance.

Data Augmentation. Common demonstrations showing the effectiveness of data augmentation come from simple transformations, such as translation, rotation, flipping, cropping, adding noises, etc., which aim at artificially enlarging the training dataset [15]. The shift invariance of the object is encoded by CNNs to improve the model's learning ability for image recognition tasks. For example, rotation augmentation on source images is performed by randomly rotating clockwise or counterclockwise between 0 and 360°C with the center of the image as the origin, reversing the entire rows and columns of image pixels horizontally or vertically is called flipping augmentation, and random cropping is a method to reduce the size of the input and create random subsets of the original input [11]. Random Erasing [26] is another interesting data augmentation technique that produces training images with varying degrees of occlusion.

Other Approaches. In addition to geometric transformation and Random Erasing, there are many other image manipulations, such as noise injection, kenel filters, color space transformations and mixing images [15]. Noise injection is a method used to help CNNs learn more robust features by injecting a matrix of random values, which are usually drawn from a Gaussian distribution [14]. Kernel filters are a widely used image processing method for sharpening and blurring images [9], whereas color space transformations aims to alter the color values or distribution of images [1,8]. Mixing images appeared in recent years, which has two approaches. The one is cropping images randomly and concatenate the croppings together to form new images [6], the other is to use non-linear methods to combine images to create new training examples [18].

Different from the above approaches, we design the Random Padding operation to improve the performance of CNN models from the perspective of reducing the position information in the network.

2.2   Padding in CNN

The boundary effect is a well-researched phenomenon in biological neural networks [17, 20]. Previous research has addressed the boundary effect for artificial CNNs by using specific convolution filters for the border regions [5]. At some point during the convolution process, the filter kernel will come into contact with the image border [7]. Classic CNNs use zero-padding to enlarge the image for filtering by kernels. The cropped images are filled by paddings to reach the specified size [23]. Guilin Liu and his colleagues proposed a simple and effective padding scheme, called partial convolution-based padding. Convolution results are reweighted near image edges relying on the ratios between the padded region and the convolution sliding window area [12].

Padding is additional pixels which can be added to the border of an image. In the process of convolution, the pixel in the corner of an image will only be covered once by kernels, but the middle pixel will be covered more than once basically, which will cause shrinking outputs and loosing information on corners of the image. Padding works by extending the area in which a convolution neural network processes an image. Padding is added to the frame of the image to enlarge the image size for the kernel to cover better, which assists the kernel with processing the image. Adding padding operations to a CNN helps the model get a more accurate analysis of images.

3   Random Padding for CNN

This section presents the Random Padding operation for training in the convolutional neural network (CNN). Firstly, we introduce the detailed procedure of Random Padding. Next, using comparative experiments to verify that the extraction of position information will be reduced in CNNs with the method of Random Padding. Finally, the implementation of Random Padding in different CNN models is introduced.

3.1   Random Padding Operation

In CNN training, we replace the traditional padding by the technique of Random Padding, which has four types of padding selections shown in Fig. 1. For the feature maps generated in the network, Random Padding will perform the zero-padding operation randomly on the four boundaries according to the required thickness of padding by the feature maps.

When padding thickness equals 1, Random Padding will first randomly select one padding of the four patterns. Assuming that the size of the feature map in the training is $w * w$, the feature map will become $(w + 1) * (w + 1)$ after this new padding method. Then, Random Padding will randomly select one of the four modes again, which will change the size of the feature map to $((w+1)+1) * ((w + 1) + 1)$. In general, Random Padding will perform $2n$ padding selections if the padding thickness is n, where $n = 1, 2, 3$ are the most common in CNN models. The detailed steps of Random Padding are shown in Algorithm 1. In this process, the position of features will be randomly changed by adding Random Padding, hence the learning of object's position information by CNN will be reduced.

Algorithm 1. Random Padding Procedure

Input: Input feature map: I; The thickness of padding: n; The random padding thickness of four boundaries, left, right, top, bottom: l, r, t, b; Padding options: S
Output: Feature map with padding I^*
1: $l, r, t, b \leftarrow 0$
2: $S \leftarrow [[1, 0, 1, 0], [1, 0, 0, 1], [0, 1, 1, 0], [0, 1, 0, 1]]$
3: **for** $i = 1$ to $2n$ **do**
4:     $P_r \leftarrow \text{RandomSelect}(S, 1)$ ▷ select a padding option randomly
5:     $l \leftarrow l + P_r[0]$ ▷ padding_left
6:     $r \leftarrow r + P_r[1]$ ▷ padding_right
7:     $t \leftarrow t + P_r[2]$ ▷ padding_top
8:     $b \leftarrow b + P_r[3]$ ▷ padding_bottom
9: **end for**
10: $I^* \leftarrow \text{ZeroPadding}([l, r, t, b])(I)$

3.2   Validation Method for Position Information Reduction in CNNs

The position information has been proved to be implicitly encoded in the feature map extracted by CNN, which was introduced by the traditional zero padding [7]. In this article, we proposed the hypothesis that the Random Padding operation will reduce the extraction of position information in CNNs. In this sub-section, we prove this hypothesis by comparing position information in an end-to-end manner between the CNN with traditional padding and the CNN with Random Padding.

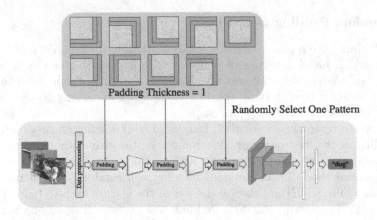

Fig. 2. Random Padding in the CNN.

This validation experiment used the Position Encoding Network [7], which is composed of two critical components: a feedforward convolutional encoder network and a simple position encoding module. For this task, the two encoder networks collected characteristics, from the CNN with traditional padding and the CNN with Random Padding, at different layers of abstraction. After the collection of multi-scale features from the front networks as inputs, the other two position encoding modules outputted the prediction of position information. Due to the fact that position encoding measurement is a novel concept, there is no universal metric. We used Spearman's rank correlation coefficient (SPC) and Mean absolute error (MAE) proposed by [7] to evaluate the position encoding performance to verify that the Random Padding operation reduces the amount of position information extracted by CNNs. The higher the SPC, the higher the correlation between the output and the ground-truth map, while the MAE is the opposite. We will present the detailed setting of the experiment and the results in Sect. 4.

3.3 Construct CNN with Random Padding

The Random Padding operation can be added to different types of backbone networks to construct CNN for image classification.

In order to better analyze the relationship between the way of padding joins CNN and the improvement of model performance, we replaced the traditional padding of the first one, the first two and the first three padding layers with the Random Padding operation to compare the accuracy of image classification in various CNN models, which is shown in Fig. 2. Since the Random Padding operation is complementary to general data augmentation methods, we employ random cropping, random flipping, random rotation, and random Erasing methods to enrich the training datasets. We will present the detailed experiment setting and results in Sect. 5.

4   Evaluation of Position Information in CNNs

This section quantitatively evaluates the impact of CNN using traditional padding and Random Padding on position information extraction. The first part introduces the dataset used by pre-trained models and the evaluation metrics for position information. Secondly, we compare the learning ability of CNN with traditional padding and CNN with Random Padding on position information extraction. In the third part, we analyze the experimental results and justify that CNN with Random Padding reduces the extraction of position information.

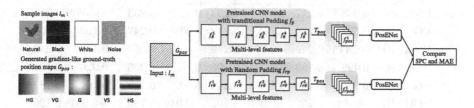

Fig. 3. Compare position information between the CNN with traditional padding and the CNN with Random Padding.

4.1   Dataset and Evaluation Metrics

Dataset. We use the Imagenet dataset to train the basic VGG and VGG with Random Padding as our initialization networks, and then we use the DUT-S dataset [21] as our training set, which contains 10,533 images for training. Following the common training protocol used in [13,25], we train the models on the same training set of DUT-S and evaluate the existence of position information on the synthetic images (white, black and Gaussian noise) and the natural image from the website of News-Leader. Notably, we adhere to the standard-setting used in saliency detection to ensure that the training and test sets do not overlap. Since the position information is largely content independent, any image or model can be used in our studies.

Evaluation. At present, there is no universal standard for measuring position encoding, so we evaluated the performance of position information according to the two different metrics methods (Spearman's rank correlation coefficient (SPC) and Mean Absolute Error (MAE)) previously used by [7]. SPC is a non-parametric measurement of the association between the ground-truth and the predicted position map. We maintain the SPC score within the range [-1, 1] to facilitate understanding. MAE measures the average magnitude of the errors in the predicted position map and the ground-truth gradient position map, without considering their direction.

The lower the SPC value, the less position information the model produces, and the higher the MAE value, the less position information the model outputs.

We expect lower SPC values and higher MAE values after applying the Random Padding operation in CNNs.

Table 1. Comparison of SPC and MAE in CNNs using traditional padding or Random Padding across different image types

H	Model	Nature SPC	Nature MAE	Black SPC	Black MAE	White SPC	White MAED	Noise SPC	Noise MAE
H	PosENet	0.130	0.284	0	0.251	0	0.251	0.015	0.251
	VGG	0.411	0.239	0.216	0.242	0.243	0.242	0.129	0.245
	VGG_RP	−0.116	0.253	0.021	0.252	0.023	0.255	−0.045	0.251
V	PosENet	0.063	0.247	0.063	0.254	0	0.253	−0.052	0.251
	VGG	0.502	0.234	0.334	0.242	0.433	0.247	0.120	0.250
	VGG_RP	−0.174	0.249	−0.174	0.249	−0.027	0.257	0.100	0.249
G	PosENet	0.428	0.189	0	0.196	0	0.206	0.026	0.198
	VGG	0.765	0.14	0.421	0.205	0.399	0.192	0.161	0.187
	VGG_RP	−0.49	0.200	−0.009	0.196	−0.040	0.195	−0.051	0.196
HS	PosENet	0.187	0.306	0	0.308	0	0.308	−0.060	0.308
	VGG	0.234	0.211	0.227	0.297	0.285	0.301	0.253	0.292
	VGG_RP	0.043	0.308	0.049	0.308	0.026	0.308	0.050	0.308

4.2 Architectures and Settings

Architectures. We first build two pre-trained networks based on the basic architecture of VGG with 16 layers. The first network uses traditional padding, and the second one applies the technique of Random Padding on the first two padding layers. The proper number of padding layers adding in the CNN is analyzed in Sect. 5. Meanwhile, we construct a randomization test by using a normalized gradient-like position map as the ground-truth. The generated gradient-like ground-truth position maps contain Horizontal gradient (HG) and vertical gradient (VG) masks, horizontal and vertical stripes (HS, VS), and Gaussian distribution (G). As shown in Fig. 3, the combination of natural images $I_m \in \mathbb{R}^{h \times w \times 3}$ and gradient-like masks $G_{pos} \in \mathbb{R}^{h \times w}$ is used as the input of two pretrained models with fixed weights. We remove the average pooling layer and the layer that assigns categories of the pretrained model to construct an encoder network f_p and f_{rp} for extracting feature maps. The features $(f_\theta^1, f_\theta^2, f_\theta^3, f_\theta^4, f_\theta^5)$ and $(f_{r\theta}^1, f_{r\theta}^2, f_{r\theta}^3, f_{r\theta}^4, f_{r\theta}^5)$ we extract from the two encoder networks respectively come from five different abstraction layers, from shallow to deep. The following is a summary of the major operations:

$$f_\theta^i = \boldsymbol{W}_p * I_m(G_{pos})$$
$$f_{r\theta}^i = \boldsymbol{W}_{rp} * I_m(G_{pos})$$

(1)

where W_p denotes frozen weights from the model using traditional padding, and W_{rp} represents frozen weights from the model using the Random Padding operation. $*$ indicates the model operation.

After multi-scale features collection, the transformation function T_{pos} performs bi-linear interpolation on the extracted feature maps of different sizes to create feature maps with the same spatial dimension. $(f_{pos}^1, f_{pos}^2, f_{pos}^3, f_{pos}^4, f_{pos}^5)$ and $(f_{rpos}^1, f_{rpos}^2, f_{rpos}^3, f_{rpos}^4, f_{rpos}^5)$ can be summarized as:

$$f_{pos}^i = T_{pos}(f_\theta^i)$$
$$f_{rpos}^i = T_{pos}(f_{r\theta}^i)$$

(2)

These resized feature maps should be concatenated together and then send into Position Encoding Module (PosENet) [7], which only has one convolutional layer. The features are delivered to PosENet and trained, where the goal of this training is to generate a pattern that is only related to the position information and has nothing to do with other features. It should be noted that during the training process, the parameters of pretrained networks are fixed. The final stage of this study is to compare the extraction of the amount of position information between the CNN with traditional padding and the CNN with Random Padding.

Settings. The models we choose to compare the position information of feature maps are traditional VGG16 and VGG16 with Random Padding. We initialize the CNN models by pre-training on the ImageNet dataset and keep the weights frozen in our comparison experiment. The size of the input image should be 224×224, which can be a natural picture, a black, a white, or a noise image. We also apply five different ground-truth patterns, HG, VG, G, HS and VS, which represent horizontal and vertical gradients, 2D Gaussian distribution, horizontal and vertical stripes, respectively. All feature maps of five different layers extracted from pre-trained models are resized to a size of 28×28. After the feature map is used as input, Position Encoding Module (PosENet) will be trained with stochastic gradient descent for 15 epochs with a momentum factor of 0.9, and weight decay of 10^{-4}. For this task, PosENet only has one convolutional layer with a kernel size of 3×3 without any padding, which will learn position information directly from the input.

4.3 Comparison and Evaluation

We first use Random Padding on the first two padding layers in one model and then conduct experiments to verify and compare the differences in the position information encoded in the two pre-trained models. Following the same protocol, we train networks, based on traditional VGG16 and VGG16 with Random Padding, on each type of ground-truth, and report the experimental results in Table 1. In addition, we present the result as a reference by training PosENet without using any pre-trained models' feature maps as input. For this task, not only the original image was used, but also pure black, pure white, and Gaussian

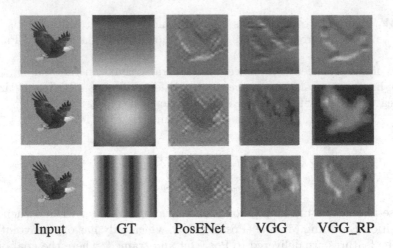

Input   GT   PosENet   VGG   VGG_RP

Fig. 4. Results of PosENet based networks with traditional padding or Random Padding corresponding to different ground-truth (GT) patterns.

noise images were used as inputs of PosENet. This is to verify whether the feature contains position information when there is no semantic information. The structure of PosENet is very simple, which can only read out the input features. If the input feature contains more position information, the output image can better approximate the target object; if the input feature does not contain any position information, the output feature map is similar to random noise and cannot output a regular pattern, which represents that the position information is not derived from prior knowledge of the object. Our experimental results do not evaluate the performance of the model, but compare the impact of different ways of padding on the position information encoded by the CNN model.

Our experiment takes three kinds of features as input, which are the feature maps extracted by VGG16 with traditional padding, VGG16 with Random Padding, and the natural image without any processing, which are recorded as VGG, VGG_RP, and PosENet respectively in the Table 1. The reason why the original image is used as input is that this kind of image does not contain position information, which shows PosENet's own ability to extract position information and plays a comparative role to other results. PosENet can easily extract the position information from the pre-trained VGG models, while it is difficult to extract position information directly from the original image. Only when combined with a deep neural network can this network extract position information that is coupled with the ground-truth position map. Previous studies have noted that traditional zero-padding delivers position information for convolutional neural networks to learn.

According to the results in Table 1, VGG_RP is lower than VGG on the evaluation index SPC value, and almost all higher than VGG on MAE value. Sometimes, VGG_RP is even lower than PosENet's SPC value. The qualitative results for CNNs with traditional padding or Random Padding across different patterns are shown in Fig. 4. The first two columns are the patterns of the input image and the target; the third column is the visualization of directly inputting the source image into PosENet; the fourth and fifth columns are the generation effects of VGG and VGG_RP respectively. We can observe a connection between the predicted and ground-truth position maps for the VG, G, and VS patterns, indicating that CNNs with Random Padding can better learn the object itself that needs to be recognized, which means the technology of Random Padding can indeed reduce CNN's extraction of position information effectively. Therefore, the basic CNN network will learn the position information of the object, while both the PosENet with only one convolution layer and the CNN with Random Padding hardly learn the position information from the input image.

5   Evaluation of Random Padding

This section evaluates the Random Padding operation in CNN for improving the accuracy of image classification.

5.1   Dataset and Evaluation Metrics

Dataset. We use three image datasets to train the CNN model, including the well-known datasets CIFAR-10 and CIFAR-100, and a grayscale clothing dataset Fashion-Mnist. There are 10 classes in the dataset of CIFAR-10, and each has 6000 32×32 color images. The training set has 50000 images and the test set has 10000 images. The CIFAR-100 is just like the CIFAR-10, except it has 100 classes containing 600 images each. Fashion-MNIST is a dataset that contains 28×28 grayscale images of 70,000 fashion items divided into ten categories, each with 7,000 images. The training set has 60,000 pictures, whereas the test set contains 10,000. The image size and data format of Fashion-MNIST are identical to those of the original MNIST.

Evaluation Metrics. The test error assessment is an important part of any classification project, which compares the predicted result of the classified image with its ground-truth label. For image classification, test error is used to calculate the ratio of incorrectly recognized images to the total number of images that need to be recognized.

5.2   Experiment Setting

We use CIFAR-10, CIFAR-100 and Fashion-Mnist to train four CNN architectures, which are Alexnet, VGG, Googlenet and Resnet. We use 16-layer network for VGG and 18-layer network for Resnet. The models with different layers' Random Padding were training for 200 epochs with the learning rate of 10^{-3}. In our first experiment, we compare the different CNN models trained with Random Padding on different layers. For the same deep learning architecture, all the models are trained from the same weight initialization and all the input data has not been augmented. The second experiment is to apply the Random Padding operation and various data augmentations (e.g., flipping, rotation, cropping and erasing) in the CNNs together to justify the complementary performance of the Random Padding operation with data augmentation methods.

Table 2. Test errors (%) with different architectures on CIFAR-10, CIFAR-100 and Fashion-MNIST. **Baseline:** Baseline model, **RP_1:** Random Padding on the first padding layer, **RP_2:** Random Padding on the first two padding layers, **RP_3:** Random Padding on the first three padding layers.

	Model	Alexnet	VGG16	Googlenet	Resnet18
CIFAR-10	Baseline	15.19 ± 0.07	12.41 ± 0.05	11.47 ± 0.08	12.08 ± 0.04
	RP_1	13.39 ± 0.05	11.34 ± 0.07	10.37 ± 0.09	8.21 ± 0.07
	RP_2	12.93 ± 0.12	10.54 ± 0.04	10.25 ± 0.07	–
	RP_3	12.75 ± 0.07	10.61 ± 0.05	10.20 ± 0.13	–
CIFAR-100	Baseline	44.57 ± 0.06	46.93 ± 0.04	34.81 ± 0.07	36.90 ± 0.09
	RP_1	42.36 ± 0.07	42.32 ± 0.06	34.50 ± 0.07	31.29 ± 0.08
	RP_2	41.54 ± 0.06	40.66 ± 0.05	33.12 ± 0.09	–
	RP_3	40.66 ± 0.05	39.66 ± 0.04	32.94 ± 0.06	–
Fashion-MNIST	Baseline	12.74 ± 0.06	7.02 ± 0.08	6.48 ± 0.09	5.80 ± 0.04
	RP_1	12.74 ± 0.06	5.49 ± 0.05	5.88 ± 0.07	5.64 ± 0.07
	RP_2	9.59 ± 0.05	5.49 ± 0.06	5.82 ± 0.05	–
	RP_3	9.40 ± 0.06	5.72 ± 0.08	5.94 ± 0.07	–

5.3   Classification Accuracy on Different CNNs

The experiments in Sect. 4 proved the Random Padding operation can reduce the extraction of position information but did not show how much this method can improve the model performance. So we design a comparative experiment shown in Table 2 to illustrate the results of the Random Padding operation on different padding layers of different CNN models training on the datasets of CIFAR-10, CIFAR-100 and Fashion-MNIST. For each kind of CNN architecture, we apply the Random Padding operation on the first padding layer, the first two padding layers and the first three padding layers. Specially, we only replace the traditional padding with Random Padding on the first padding layer of the Resnet18 due to the unique structure of shortcut in Resnet. Based on the principle of controlling variables, we train and test the basic CNN architectures on the same dataset. All the results are shown in Table 2.

In experiments on the CIFAR-10 dataset, our approach achieved an error rate of 12.75% with Alexnet using Random Padding on the initial three padding layers, 10. 54% in VGG16 with two Random Padding layers, and a state-of-the-art 10.20% with Googlenet after replacing its first three traditional padding layers. On CIFAR-100, Random Padding consistently decreased classification error rates: Alexnet, VGG16, and Googlenet reported rates of 40.66%, 39.66%, and 32.94%, respectively, while Resnet18 reached a significantly low rate of around 31.29%. For Fashion-Mnist, despite its inherently high recognition rate, employing Random Padding still optimized model performance. Specifically, Alexnet improved by 3.34%, whereas VGG16 and Googlenet registered 5.49% and 5.82%, respectively. The Resnet model saw a marginal error reduction of 0.16%.

In general, as the number of layers deepens, the extracted object features gradually become abstracted, and the encoding of position information also becomes more implicit. The addition of the Random Padding operation in subsequent deep layers will jeopardize the models' learning of abstract features. So we can conclude our experiment results that using the technique of Random Padding in the first two padding layers can always improve the performance of various deep learning models.

Table 3. Test errors (%) with different data augmentation methods on CIFAR-10 based on VGG16 with traditional padding and VGG16 with Random Padding. **Baseline:** Baseline model, **RC:** Random Cropping, **RR:** Random Rotation, **RF:** Random Flipping, **RE:** Random Erasing.

VGG16	Random Padding	Test errors (%)
Baseline	–	12.41 ± 0.08
	✓	10.54 ± 0.05
RC	–	10.54 ± 0.14
	✓	10.08 ± 0.09
RR	–	15.12 ± 0.05
	✓	9.82 ± 0.06
RF	–	10.37 ± 0.12
	✓	8.69 ± 0.07
RE	–	11.03 ± 0.08
	✓	8.89 ± 0.09
RF + RE	–	8.75 ± 0.13
	✓	7.75 ± 0.04
RC + RE	–	8.85 ± 0.07
	✓	8.34 ± 0.08
RC + RF	–	8.74 ± 0.06
	✓	8.26 ± 0.07
RC+ RF + RE	–	7.83 ± 0.09
	✓	7.21 ± 0.05

5.4 Classification Accuracy on Different CNNs

In this experiment, we use VGG16 as the benchmark model and apply Random Padding on first two padding layers, and use CIFAR-10 as the test dataset.

We chose four types of data augmentation, which are Random Rotation (RR), Random Cropping (RC), Random Horizontal Flip (RF) and Random Erasing (RE). The test error obtained by CIFAR-10 on the basic VGG16 model is used as the baseline for this task. Then the effectiveness of our approach is evaluated by adding various data augmentations and combining with the Random Padding operation.

As shown in Table 3, the Random Rotation augmentation method is not suitable for the dataset of CIFAR-10 on the VGG16, which means that the use of Random Rotation lowers the accuracy of the model than the baseline. But after adding the Random Padding operation, the model's recognition rate on the CIFAR-10 test set exceeded the baseline, which indicates that the Random Padding operation can help the model learn features better. The model that combines a single data augmentation and the technique of Random Padding has stronger learning ability than the model that only uses data augmentation. Therefore, the Random Padding operation is complementary to the data augmentation methods. In particular, combining all these methods achieves a 7.21% error rate, which has a 5.20% improvement over the baseline.

6   Conclusions

In this paper, by investigating the learning of spatial information in convolutional neural networks (CNN), we propose a new padding approach named "Random Padding" for training CNN. The Random Padding operation reduces the extraction of features' positional information and make the model better understand features' relationships in CNNs. Experiments conducted on CIFAR-10, CIFAR-100 and Fashion-MNIST with various data augmentation methods validate the effectiveness of our method for improving the performances of many CNN models. In the future work, we will apply our approach in large-scale datasets and other CNN recognition tasks, such as, object detection and face recognition.

References

1. Chatfield, K., Simonyan, K., Vedaldi, A., Zisserman, A.: Return of the devil in the details: delving deep into convolutional nets. CoRR abs/1405.3531 (2014). http://arxiv.org/abs/1405.3531
2. Cohen, T., Welling, M.: Group equivariant convolutional networks. In: International Conference on Machine Learning, pp. 2990–2999. PMLR (2016)
3. He, K., Zhang, X., Ren, S., Sun, J.: Deep residual learning for image recognition. In: Proceedings of the IEEE Conference on Computer Vision and Pattern Recognition, pp. 770–778 (2016)
4. Hubel, D.H., Wiesel, T.N.: Receptive fields, binocular interaction and functional architecture in the cat's visual cortex. J. Physiol. **160**(1), 106–154 (1962)
5. Innamorati, C., Ritschel, T., Weyrich, T., Mitra, N.J.: Learning on the edge: investigating boundary filters in CNNs. Int. J. Comput. Vis., 1–10 (2019)
6. Inoue, H.: Data augmentation by pairing samples for images classification. arXiv preprint arXiv:1801.02929 (2018)

7. Islam, M.A., Jia, S., Bruce, N.D.: How much position information do convolutional neural networks encode? In: International Conference on Learning Representations (2019)
8. Jurio, A., Pagola, M., Galar, M., Lopez-Molina, C., Paternain, D.: A comparison study of different color spaces in clustering based image segmentation. In: Hüllermeier, E., Kruse, R., Hoffmann, F. (eds.) IPMU 2010. CCIS, vol. 81, pp. 532–541. Springer, Heidelberg (2010). https://doi.org/10.1007/978-3-642-14058-7_55
9. Kang, G., Dong, X., Zheng, L., Yang, Y.: Patchshuffle regularization. arXiv preprint arXiv:1707.07103 (2017)
10. Kosiorek, A., Sabour, S., Teh, Y.W., Hinton, G.E.: Stacked capsule autoencoders. Adv. Neural. Inf. Process. Syst. **32**, 15512–15522 (2019)
11. Krizhevsky, A., Sutskever, I., Hinton, G.E.: ImageNet classification with deep convolutional neural networks. Adv. Neural. Inf. Process. Syst. **25**, 1097–1105 (2012)
12. Liu, G., et al.: Partial convolution based padding. arXiv preprint arXiv:1811.11718 (2018)
13. Liu, N., Han, J., Yang, M.H.: PiCANet: learning pixel-wise contextual attention for saliency detection. In: Proceedings of the IEEE Conference on Computer Vision and Pattern Recognition, pp. 3089–3098 (2018)
14. Moreno-Barea, F.J., Strazzera, F., Jerez, J.M., Urda, D., Franco, L.: Forward noise adjustment scheme for data augmentation. In: 2018 IEEE Symposium Series on Computational Intelligence (SSCI), pp. 728–734. IEEE (2018)
15. Shorten, C., Khoshgoftaar, T.M.: A survey on image data augmentation for deep learning. J. Big Data **6**(1), 1–48 (2019)
16. Simonyan, K., Zisserman, A.: Very deep convolutional networks for large-scale image recognition. In: Bengio, Y., LeCun, Y. (eds.) 3rd International Conference on Learning Representations, ICLR 2015, San Diego, CA, USA, May 7–9, 2015, Conference Track Proceedings (2015). http://arxiv.org/abs/1409.1556
17. Sirovich, L., Brodie, S.E., Knight, B.: Effect of boundaries on the response of a neural network. Biophys. J . **28**(3), 423–445 (1979)
18. Summers, C., Dinneen, M.J.: Improved mixed-example data augmentation. In: 2019 IEEE Winter Conference on Applications of Computer Vision (WACV), pp. 1262–1270. IEEE (2019)
19. Szegedy, C., et al.: Going deeper with convolutions. In: Proceedings of the IEEE Conference on Computer Vision and Pattern Recognition, pp. 1–9 (2015)
20. Tsotsos, J.K., Culhane, S.M., Wai, W.Y.K., Lai, Y., Davis, N., Nuflo, F.: Modeling visual attention via selective tuning. Artif. Intell. **78**(1–2), 507–545 (1995)
21. Wang, L., et al.: Learning to detect salient objects with image-level supervision. In: Proceedings of the IEEE Conference on Computer Vision and Pattern Recognition, pp. 136–145 (2017)
22. Worrall, D.E., Garbin, S.J., Turmukhambetov, D., Brostow, G.J.: Harmonic networks: Deep translation and rotation equivariance. In: Proceedings of the IEEE Conference on Computer Vision and Pattern Recognition, pp. 5028–5037 (2017)
23. Xie, C., Wang, J., Zhang, Z., Ren, Z., Yuille, A.: Mitigating adversarial effects through randomization. In: International Conference on Learning Representations (2018)
24. Zeiler, M.D., Fergus, R.: Visualizing and understanding convolutional networks. In: Fleet, D., Pajdla, T., Schiele, B., Tuytelaars, T. (eds.) ECCV 2014. LNCS, vol. 8689, pp. 818–833. Springer, Cham (2014). https://doi.org/10.1007/978-3-319-10590-1_53

25. Zhang, P., Wang, D., Lu, H., Wang, H., Ruan, X.: Amulet: aggregating multi-level convolutional features for salient object detection. In: Proceedings of the IEEE International Conference on Computer Vision, pp. 202–211 (2017)
26. Zhong, Z., Zheng, L., Kang, G., Li, S., Yang, Y.: Random erasing data augmentation. In: Proceedings of the AAAI Conference on Artificial Intelligence, vol. 34, pp. 13001–13008 (2020)

Unsupervised Fraud Detection on Sparse Rating Networks

Shaowen Tang[✉] and Raymond Wong

University of New South Wales, Sydney, Australia
{shaowen.tang,ray.wong}@unsw.edu.au

Abstract. Network fraud detection, specifically identifying abnormal users on rating platforms, has attracted considerable interests of researchers due to its wide applicability. However, the performance of existing detection systems suffer from several challenging problems such as class imbalance, lack of annotated data and network sparsity. To address above challenges, in this paper, we propose a novel unsupervised fraud detection algorithm FD-SpaN based on network structure exploration, to effectively rank users based on computed probabilities of being fraudulent and identify abnormal users on sparse networks. Firstly, we model ratings networks as graphs in mathematical manner with introduced metrics. Then, we add variable smoothing terms accordingly when inferring the quality and trustworthiness of each item and rating respectively, to tackle network sparsity on entity level. Meanwhile, for active users, we integrate their rating patterns into our developed formulations as a critical term to avoid overfitting. In addition, our proposed FD-SpaN is scalable to large-scale rating networks in real world due to its linear time complexity with respect to the size of network. Extensive experiments on two real-world datasets show the effectiveness of FD-SpaN under extreme class imbalance and network sparsity, as it outperforms other state-of-the-art baselines in terms of all evaluation metrics.

Keywords: Network Fraud Detection · Unsupervised Learning · Network Sparsity

1 Introduction

User-generated feedback on e-commerce platforms like Amazon and Yelp can provide valuable information for potential customers to help them make decisions about purchase and avoid risks. As a result, malicious users will aim at those rating platforms for different purposes e.g., overrating products for promoting sales or underrating products for defaming competitors [11,15]. Therefore, identifying such abnormal users on rating networks has been a meaningful and popular research topic in recent as the rapid spread of online trading.

Traditional methods [11,17,24] try to extract handcraft features such as length of reviews, rating distributions of users etc., and feed them into classic machine learning models to learn suitable weights. However, conventional

© The Author(s), under exclusive license to Springer Nature Singapore Pte Ltd. 2024
D. Benavides-Prado et al. (Eds.): AusDM 2023, CCIS 1943, pp. 19–33, 2024.
https://doi.org/10.1007/978-981-99-8696-5_2

Table 1. Comparison in terms of algorithmic properties with other baselines.

	Birdnest [9]	BAP [22]	Eagle [1]	Behavior [18]	Rev2 [14]	FD-SpaN
Graph-based models		✓	✓		✓	✓
Address Network Sparsity					✓	✓
Alleviate Overfitting				✓		✓
Linear Scalability		✓		✓	✓	✓

models are bottlenecked in performance and not scalable to larger size of data in real world. Therefore, [9,18] start to investigate the rating patterns of users, examining how fraudsters behave differently from normal users within the rating network. More recently, graph-based approaches [1,3,14,22,29,31] have emerged in the area of network fraud detection and shown promising performance. Generally, they attempt to regard rating networks as graphs and analyze the interactions between nodes to find out anomalies within a graph.

Although the field of network fraud detection is well-investigated by existing graph-based methods, they are still not able to perform consistently well across real-world platforms for many reasons. Firstly, on large trading websites like Amazon, millions of users could potentially conduct shopping activities and post their opinions or ratings on a daily basis. It is impossible to annotate such a huge amount of data manually due to the massive labour and time cost. However, training process without sufficient high-quality labeled data will limit the performance of fraud detectors [19]. Secondly, it is challenging for graph-based detection models to analyze nodes with only a few of inter-connections with others, while this situation is quite common in reality [26,32]. For example, a user entity with only one or two ratings could be easily identified as a spammer by many existing models when its ratings deviate from mainstream opinions. Nevertheless, it still could be a benign user just with a different personal preference on specific products. Besides, when a product only receives unreliable ratings, it is unlikely to infer its inherent quality with interpretation. Lastly, many graph-based models decide the trustworthiness of rating depending on local deviation and ignore overall rating patterns of users, which could lead to overfitting [6,7]. The reason is that both normal and unfair users might occasionally give unusual ratings, which violate their overall rating patterns.

In this paper, we aim to address the challenges mentioned above by proposing an Unsupervised **F**raud **D**etection ranking algorithm on **Spa**rse Rating **N**etworks (**FD-SpaN**) to effectively identify abnormal users. Our FD-SpaN is designed to eliminate the negative influence of network sparsity on entity level by flexibly controlling the weights of global defaults on computation for the quality of items and the trustworthiness of ratings. Moreover, FD-SpaN attempts to further boost performance by including overall rating patterns of active users to avoid overfitting. As shown in Table 1, we compare FD-SpaN with other baselines in terms of four major algorithmic properties and FD-SpaN is the only one to satisfy all of them. The contributions of our work are summarized as follows:

- **Analysis.** We comprehensively analyze the problem of network fraud detection and existing challenges, then model rating networks as graphs to include structural relations.
- **Algorithm.** We propose an efficient graph-based fraud detection algorithm FD-SpaN, in setting of unsupervised learning, to rank and identify fraudulent users on sparse rating networks according to computed fraud scores.
- **Effectiveness.** We conduct extensive experiments on two real-world sparse datasets. Results show the effectiveness of our algorithm FD-SpaN, which outperforms all other baselines in terms of all evaluation metrics.

2   Related Work

In this section, we classify existing works into two categories: general fraud detection and graph-based fraud detection. Comprehensive surveys could be found in [2,19].

General Fraud Detection. [11] is the first work to investigate fraud detection on e-commerce platforms by extracting handcraft features corresponding to reviews, reviewers and products and learning suitable weights. Inspired by it, Behavior [18] expands feature set to detect target-based spamming, regarding multiple reviews from the same user on a single item or brand within a short time period as potential frauds. Co-training [17] builds up two independent models with different feature sets related to reviews and reviewers respectively, then uses a small group of labeled data to train both classifiers and annotate unseen data mutually to create new training samples for boosting each other iteratively. Deceptive [24] and Singleton [26] focus on textual content analysis and measure semantic anomaly scores of reviews by computational linguistic strategies such as Linguistic Inquiry and Word Count (LIWC) and Latent Dirichlet Allocation (LDA). [32] studies the problem of singleton review spam detection by converting it into a temporal pattern discovery problem. Similarly, Birdnest [9] analyzes the distribution and frequency of ratings given by a user in a Bayesian probabilistic model, which measures fraud score depending on how estimated posterior distribution is different from global expectation. SpEagle [25] incorporates behavioral and textual features into FraudEagle [1] to improve the performance.

Graph-Based Fraud Detection. Since PageRank [4] and HITS [12] were introduced to explore relations among website pages based on citing, many studies have been extensively investigating fraud detection using topological information rather than content comparison. BAP [22] values ratings even from biased users if ratings are opposite to the usual patterns of their reviewers. Troll-Trust [31] ranks nodes within signed networks based on calculated trustworthiness using sigmoid-like functions, which provides clear semantic interpretation but is limited in dealing with discrete values i.e., ratings in most of real-world platforms. Trustiness [28,29] formally model network fraud detection with three intrinsic metrics to examine information of nodes and edges in mathematical manner within graphs. Rev2 [14] incorporates behavioral features and sentiment analysis as auxiliary information to improve the performance of Trustiness. FraudEagle [1] regards network fraud detection as a node classification problem and uses loopy belief propagation to update the entire graph. More recently, FdGars [30] and CARE [6] learn the vector representations of user nodes by leveraging the power of Graph Convolutional Network (GCN). For tackling the problem of camouflage, Fraudar [10] proposes a novel metric denoting the suspiciousness of users, which would not decrease as any amount of camouflages added.

3   Preliminaries

In this section, we first model general rating networks as bipartite graphs with mathematical symbols. Then, we formally propose the problem that we will focus in this paper. Lastly, we introduce four basic metrics to mathematically examine information of nodes and edges in the graph for addressing the proposed problem.

3.1   Problem Definition

Firstly, we define a rating network as a bipartite graph $G = (U, P, R)$, where U denotes all user nodes; P denotes all product (interchangeable with item) nodes; and R represents all ratings given by user $u \in U$ on product $p \in P$ within the graph G as edges. In addition, let $R_{u,*}$ be all ratings given by user u and $R_{*,p}$ be all ratings received by product p. We denote the value of rating $r_{u,p}$ as $w(r_{u,p})$ and scale it to the range between -1 and $+1$ for generalization, i.e., $w(r_{u,p}) \in [-1,1] \; \forall r_{u,p} \in R$. Based on above preliminaries, we raise the fraud detection problem as following:

Definition 1 (Problem). *Given a bipartite rating graph $G = (U, P, R)$, what is the fraud score of each user node $u \in U$, also known as the probability of being a fraudster, whose rating behavior is maliciously deviated from normal user nodes?*

3.2   Metrics

To solve this specific problem, we introduce four intrinsic metrics corresponding to bipartite rating graphs: 1) quality of item; 2) deviation of rating; 3) trustworthiness of rating and 4) honesty of user. The definitions of four metrics will be explained as follows:

1. **Quality** measures the inherent goodness of each item, which is also considered as the deserved rating from fair users. Therefore, we use a real number between -1 and $+1$ to indicate the quality of a product p, denoted as $Q(p) \in [-1,1], \; \forall p \in P$. Intuitively, high quality $Q(p)$ (close to $+1$) of product p represents high expected rating from normal users, and vice versa.

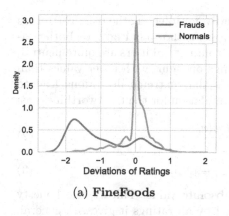

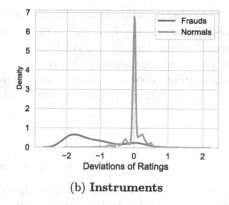

(a) **FineFoods**   (b) **Instruments**

Fig. 1. Rating deviation distributions of fraudulent users and normal users on two real-world datasets.

2. **Deviation** measures the difference between the value $w(r_{u,p})$ of a rating $r_{u,p}$ and the quality $Q(p)$ of its targeted item p, denoted as $dev(r_{u,p})$. Negative deviation suggests lower rating received than what the targeted item deserves, while positive deviation represents higher rating than expectation.
3. **Trustworthiness** measures how reliable is a rating $r_{u,p}$ from user u on item p, denoted as $T(r_{u,p}) \in [0,1] \ \forall r_{u,p} \in R$. The trustworthiness $T(r_{u,p})$ is determined by multiple factors. Intuitively, the trustworthiness scales from 0 to 1, standing for completely untrustworthy and totally reliable, respectively.
4. **Honesty** measures the fairness of a user node u in the graph, denoted as $H(u) \in [0,1], \ \forall u \in U$, where 1 indicates absolute fair users and 0 suggests definite spammers. In particular, the honesty of a user entity is decided by all its ratings given to items within the graph.

4 Methodology

4.1 Unsupervised Learning

After we formulate the bipartite rating networks comprehensively with four basic metrics, we now can explore relations among metrics in mathematical manner. Due to the lack of annotation, we aim to infer true values of metrics in unsupervised setting. Therefore, we design mathematical formulations to calculate values for all metrics interdependently. Firstly, for each item, we use all ratings received by the item to compute the intrinsic quality. Intuitively, each rating weighs variously based on the trustworthiness, as reliable ratings will reflect the real quality of item, while untrustworthy ratings aim to mislead other users. Hence, we develop the formulation for the quality of item as below:

$$Q(p) = \frac{\sum_{r \in R_{*,p}} T(r) \cdot w(r)}{|R_{*,p}|} \tag{1}$$

where $|\cdot|$ is the element count operator. Secondly, for inferring the trustworthiness of a rating, the key factor is the rating deviation. As shown in Fig. 1, on both real-world rating networks, deviations of ratings from normal users are quite neutral, while fraudsters are more likely to give ratings on items with lower values than what items deserve, which generally implies that extreme deviations lead to unreliable ratings. Hence, we develop the formulation for the trustworthiness as:

$$T(r_{u,p}) = 1 - \frac{\|dev(r_{u,p})\|}{2} \tag{2}$$

$$dev(r_{u,p}) = w(r_{u,p}) - Q(p) \tag{3}$$

where $\| \cdot \|$ is an operator for computing absolute value. Thirdly, the honesty of each user node in the graph is determined by all ratings it gives. In general, benign users tend to give trustworthy ratings to items, while fraudsters rarely rate items fairly. Thus, we infer the honesty of each user as:

$$H(u) = \frac{\sum_{r \in R_{u,*}} T(r)}{|R_{u,*}|} \tag{4}$$

Lastly, the fraud score of each user entity in the graph is reversely correlated to its honesty score, denoted as $F(u)$ and expressed as:

$$F(u) = 1 - H(u) \tag{5}$$

4.2   Alleviating Network Sparsity

Normally, introduced formulations will work properly when each node in the graph is densely connected with others. Nevertheless, on real-world rating platforms, the performance of fraud detection systems will suffer from insufficient inter-connections, which is a problem called network sparsity [23]. For example, newly registered users on e-commerce websites will naturally have few ratings on items, which could be easily classified as fraudulent users for first one or two ratings significantly deviating from mainstream [3]. Similarly for item node, it could be hard to infer the real quality of item when most of its received ratings are untrustworthy. Therefore, effective fraud detectors are supposed to address network sparsity, because online platforms will notify suspicious users to verify identification only when feel confident rather than bother normal users frequently in practice, which is harmful to the revenue [16]. Thus, we add a variable Laplace smoothing term in Formula (1) to alleviate network sparsity on entity level as:

$$Q(p) = \frac{\sum_{r \in R_{*,p}} T(r) \cdot w(r) + \sum_{r \in R_{*,p}} (1 - T(r)) * q}{|R_{*,p}| + \sum_{r \in R_{*,p}} (1 - T(r))} \tag{6}$$

where q refers to the default value of quality in the graph, which could be decided as the median value of all items' quality scores. And variable term $\sum_{r \in R_{*,p}} (1 - T(r))$ flexibly controls the proportion of default on calculation for the quality.

Intuitively, the default would account less for the calculation if the item receives more trustworthy ratings. Likewise, we also add a variable smoothing term in Formula (2) as:

$$T(r_{u,p}) = \frac{(1 - \frac{\|dev(r_{u,p})\|}{2}) + C_1 * t}{1 + C_1} \tag{7}$$

$$C_1 = \max(0, \ 1 - \frac{|R_{u,*}|}{K}) \tag{8}$$

where t refers to the default value of trustworthiness in the graph and K is a positive constant number indicating the threshold number of interactions as active users. Intuitively, the calculation will be less dependent on default value as the activity level of user increases. Together, variable smoothing terms in our formulations can automatically adapt to different nodes in the graph, which helps to alleviate network sparsity on entity level.

4.3   Avoiding Overfitting

Even in real-world sparse rating networks, where many user nodes in the graph only give few ratings to item nodes, there still could be some active users who rate items frequently, no matter they are fraudsters or normal users. For an active benign user node, one of its ratings might not be as usual as others due to the specific unsatisfactory shopping experience or unique personal preference. Therefore, it is inaccurate to always decide the trustworthiness of a rating from active users only by local rating deviation [20]. To solve this problem, we introduce a new metric to reflect the rating patterns of users for resisting overfitting when user nodes are active, denoted as $bias(u)$, $u \in U$. Then, we integrate this metric into Formula (7) for computing trustworthiness of rating as:

$$T(r_{u,p}) = \frac{(1 - \frac{\|dev(r_{u,p})\|}{2}) + C_1 * t + C_2 * (1 - \frac{\|bias(u)\|}{2})}{1 + C_1 + C_2} \tag{9}$$

$$bias(u) = \frac{\sum_{r_{u,p} \in R_{u,*}} (w(r_{u,p}) - Q(p))}{|R_{u,*}|} \tag{10}$$

where $C_2 = 1 - C_1$, controlling the importance of term $bias$ on smoothing extreme rating deviation with respect to the activity level of users.

4.4   Proposed Algorithm: FD-SpaN

In this section, we propose algorithm FD-SpaN to effectively rank and identify network frauds as shown in Algorithm 1. Specifically, we iteratively calculate the true values for metrics and adjust default values simultaneously according to the global trend in the graph. In the end, our algorithm will output the likelihood of each user node being abnormal in the graph. In addition, FD-SpaN can also be applied to other fields that information can be organized as bipartite graphs, such as social media [5,13], finance [27] and recommender system [33].

Algorithm 1. The FD-SpaN Algorithm

Input: rating graph $G = (U, P, R)$
Input: constant K, training epochs $epochs$, error threshold ϵ
Output: fraud score $F(u)$ for each user node $u \in U$ in the graph
1: Randomly initialization default values q and t
2: Let $i = 0$
3: **repeat**
4:      $i := i + 1$
5:      Compute $Q(p)$, $\forall p \in P$ using Formula (6) and update q
6:      Compute $T(r)$, $\forall r \in R$ using Formula (9) and update t
7:      $error = \max(mean(G^i(p) - G^{i-1}(p)), \ mean(T^i(r) - T^{i-1}(r)))$
8: **until** $i == epochs$ Or $error < \epsilon$
9: **return** $F(u)$, $\forall u \in \mathcal{U}$, computed by Formula (5)

4.5 Time Complexity Analysis

In each iteration of training, our algorithm FD-SpaN will go through all user nodes once, all item nodes once and all ratings three times to update values of metrics. Hence the time complexity for each iteration is $\mathcal{O}(x + y + 3z)$, where x is the total number user nodes, y is the total number of item nodes and z is the total number of ratings in the graph. Let, n denote the number of training iterations needed for our model FD-SpaN to converge. As a result, the overall time complexity for FD-SpaN would be $\mathcal{O}(n*(x+y+3z))$, where n is a constant number. Thus, theoretically, our proposed model FD-SpaN has a linear time complexity with respect to the size of graph, which is suitable to be applied in larger scale of data of real world.

5   Experiments

In this section, we conduct extensive experiments on two real-world datasets to evaluate FD-SpaN compared with the state-of-the-art baselines to show the effectiveness of our proposed model. Specifically, We aim to answer following research questions:

- **RQ1.** How does FD-SpaN perform on real-world sparse rating networks for identifying frauds compared with other baselines?
- **RQ2.** How does the performance of FD-SpaN and other baselines change under different training percentages?
- **RQ3.** Is FD-SpaN scalable to real-world platforms with larger size of data?
- **RQ4.** How does each different component of FD-SpaN contribute to the overall performance?

5.1   Experiment Setup

Datasets. Following the tradition of previous representative work [6], we also select two real-world datasets for extensive experiments. Table 2 shows the

Table 2. Two real-world datasets used for evaluation.

Datasets	# Users (%normal, %fraud)	# Items	# Reviews	Density
FineFoods	256,059 (0.92%, 0.09%)	74,258	568,454	2.22
Instruments	339,231 (1.37%, 0.12%)	83,046	500,176	1.47

statistics of two real-world datasets, where attribute density refers to the average number of ratings given by each user in the network. Both datasets are relatively sparse and extremely imbalanced to different extents, as we aim to evaluate the performance of FD-SpaN on identifying frauds under actual circumstances.

- **FineFoods** [21] is a bipartite rating network in category of fine foods, with around 65% of user nodes and 41% of item nodes giving and receiving only one rating, respectively. The ground truth is given by the method mentioned in [14] based on provided helpfulness votes.
- **Instruments** [8] refers to the musical instruments trades on Amazon websites. This rating network is even sparser, as roughly 80% of users and 44% of items have only one rating. The ground truth is also given based on the similar method as used in FineFoods dataset.

Baslines. We compare our model with five state-of-the-art baselines on fraud detection. We convert these baselines to calculate the probability of each user being anomaly, as exactly what our model does, for fair comparison.

- **Birdnest** [9] only considers the distribution and timestamps of ratings, and uses Bayesian inference to calculate suspiciousness scores for users.
- **BAP** [22] measures the expected rating for each item and the trend of each user giving high or low ratings. The key of BAP is to value ratings from users even with high bias if ratings are opposite to the usual pattern of reviewers.
- **Eagle** [1] is based on loopy belief propagation and regards the network fraud detection as a node classification problem in graphs.
- **Behavior** [18] designs and extracts multiple behavioral features of users to calculate overall fraud scores for users.
- **Rev2** [14] is based on HITS algorithm [12] and iteratively computes three intrinsic metrics in the graph: the fairness of users, the reliability of ratings and the goodness of products.

Evaluation Metrics. Our task is to infer the probability of each user node being spammer in the graph. Then, we will rank user entities based on the assigned probabilities. However, real-world rating networks are extremely imbalanced and sparse as shown in Table 2. Additionally, effective fraud detectors are supposed to assign as high anomaly scores as possible to real spammers and as low as possible to normal users. Therefore, following by the previous work [14], we extract the most and least suspicious 100 users from the output of model to form the test set and evaluate models by Average Precision (AP), which

Table 3. Performance comparison (%) on two real-world datasets. The best result of each metric is in bold. '-' represents not applicable.

Method	Dataset	FineFoods		Instruments	
	Metric	AP	AUC@200	AP	AUC@200
Baselines	Birdnest	19.09	52.53	1.11	-
	BAP	55.92	65.48	93.29	64.15
	Eagle	47.21	51.55	55.10	43.50
	Behavior	63.39	61.83	93.70	65.98
	Rev2	64.89	60.68	87.81	62.22
Ours	FD-SpaN	**70.87**	**67.03**	**95.34**	**66.85**

measures the capacity of distinguishing fraudsters and normal users. Also, we evaluate models by Area Under ROC Curve (AUC) on the top 200 suspicious user entities, reflecting the ability of each algorithm to identify positive cases on imbalanced dataset.

Implementation Details. For BAP and Rev2, we strictly follow the instructions from their paperwork to implement their algorithms. For Birdnest[1], We use the original code from authors to run experiments. For Eagle[2] and Behavior[3], we use available open-source implementations online. As for our proposed model FD-SpaN, we set parameter $K = 4$ according to the result on validation set, number of training iterations $epochs = 20$ and training error threshold $\epsilon = 10^{-4}$ as many other studies did.

5.2   Effectiveness (RQ1)

In this experiment, we compare the results of proposed model FD-SpaN on both real-world datasets with other state-of-the-art baselines as shown in Table 3. Overall, we can observe that our FD-SpaN consistently outperforms all baselines in terms of all evaluation metrics. More specifically, Birdnest performs poorly across both datasets and has remarkable gaps with other models. This is due to Birdnest only considers the temporal pattern and distribution of ratings provided by users to decide their suspiciousness scores, while it suffers from massive inactive users with only few ratings, which are very common in real-world rating networks but cannot be interpreted by Birdnest.

BAP and Eagle are graph-based models, trying to exploit topological information of graphs to identify anomalies, as what Rev2 and FD-SpaN do. However, they cannot achieve consistently competitive performance because they fail to

---

[1] www.andrew.cmu.edu/user/bhooi/ratings.tar.
[2] https://github.com/rgmining/fraud-eagle.
[3] https://github.com/rgmining/ria.

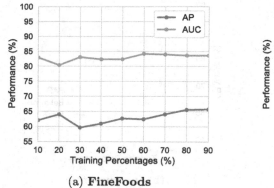

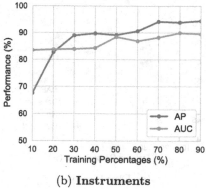

(a) **FineFoods**   (b) **Instruments**

Fig. 2. Performance (%) of model FD-SpaN on two real-world datasets under different training percentages.

alleviate the impacts of network sparsity. Even though Behavior obtains promising results by including multiple designed features regarding to rating behaviors, our FD-SpaN still outperforms Behavior by 7.48% in AP and 5.20% in AUC on FineFoods, which implies the superiority of graph-based methods.

Moreover, Rev2 is a graph-based model and attempts to address network sparsity on graph level. However, even within the same graph, each node is in a different situation. Therefore, applying the same parameters to different nodes in the graph will cause damage to the overall performance. Meanwhile, by alleviating network sparsity on entity level and resisting overfitting through new designed metric, our proposed FD-SpaN achieves a 5.98% improvement in AP value on FineFoods. Also, 91 out of top 100 suspicious user nodes predicted by our model FD-SpaN on dataset Instruments are real spammers, which is the best among all algorithms. In summary, under realistic circumstance of class imbalance and network sparsity, our proposed model FD-SpaN can effectively identify fraudsters within rating networks.

5.3 Training Percentage (RQ2)

In this experiment, we run our model FD-SpaN with different percentages of training data and record the results in terms of AP and AUC values. As we can see from Fig. 2, on FineFoods, FD-SpaN obtains stable performance in both evaluation metrics under different training percentages. On Instruments, even though FD-SpaN scores relatively low in AP with 10% of training data, it rapidly recovers to normal as the training percentage increases. Together, experiment results show the robustness of our FD-SpaN with respect to training percentages.

Moreover, we record the performance variations of our FD-SpaN and other baselines in value of AP under different training percentages, as illustrated in Fig. 3. On FineFoods, FD-SpaN scores the highest when only 10% of training data provided and maintains one of the tops in most choices of training percentages. On Instruments, we can observe FD-SpaN has continuous growth in

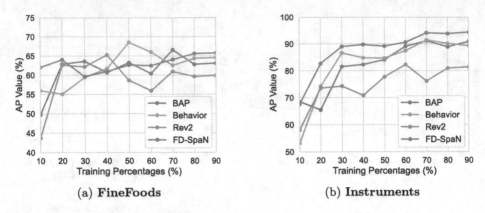

Fig. 3. Average Precision values (%) of different algorithms on two real-world datasets under different training percentages.

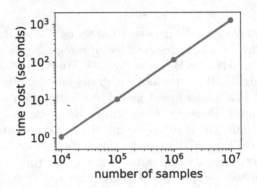

Fig. 4. Training time cost (seconds) of model FD-SpaN, using different number of training samples.

AP value and consistently outperforms other baselines as training percentage increases from 10% to 90%. Together, experiment results suggest our FD-SpaN can effectively analyze topological structure of graphs to identify anomalies.

5.4 Linear Scalability (RQ3)

Alongside with previous theoretical analysis of linear scalability of our proposed FD-SpaN, we also run our algorithm using different numbers of training samples to verify linear scalability. We scale training samples in different magnitudes of amount from 10^4 to 10^7 and record the training time costs for FD-SpaN to converge on the same machine, as illustrated in Fig. 4. Basically, the running time of FD-SpaN is increasing linearly as fed with more training samples. Therefore, experiment results verify our model has linear scalability and is suitable to handle with larger size of data in practice.

Table 4. Performance variations (%) when different component of algorithm FD-SpaN is excluded.

Dataset	FineFoods		Instruments	
	AP	AUC@200	AP	AUC@200
FD-SpaN/AO	62.66	64.67	95.04	65.94
FD-SpaN/AS	63.28	64.62	93.35	63.23
FD-SpaN	70.87	67.03	95.34	66.85

5.5 Ablation Study (RQ4)

In this experiment, we conduct an ablation study on both datasets to validate the effectiveness of designed components of FD-SpaN and demonstrate how each different component contributes to the overall performance of FD-SpaN.

- **FD-SpaN/AO**: it excludes the component for avoiding overfitting.
- **FD-SpaN/AS**: it removes the component for alleviating network sparsity.

As shown in Table 4, on FineFoods, by removing the component of either anti-overfitting or anti-sparsity, the performance of FD-SpaN will drop remarkably by 8.21% and 7.59% in AP and AUC values, respectively. However, we observe a marginal decrease in performance on Instruments when the component of avoiding overfitting is excluded. The reason could be that most of users in Instruments are very inactive so that incorporating their overall rating patterns has very limited effects to resist overfitting. FD-SpaN can achieve the best performance on sparse rating networks only when all designed modules are integrated into it.

6 Conclusion

In this paper, we propose an unsupervised graph-based fraud detection ranking algorithm called FD-SpaN, which targets at effectively identifying anomalies on sparse rating networks in reality. We alleviate the negative impacts of network sparsity on entity level, which leads to finer granularity and better performance than existing methods. We also include the rating patterns of users in our FD-SpaN to avoid overfitting. Our algorithm has a linear time complexity with respect to the size of graph, which makes FD-SpaN is suitable for real-world platforms to deal with large-scale data. More importantly, extensive experiments on two real-world datasets have shown the effectiveness and superiority of FD-SpaN, scoring the highest in all metrics against other state-of-the-art baselines. In the future, we plan to investigate fraud detection on dense rating networks.

References

1. Akoglu, L., Chandy, R., Faloutsos, C.: Opinion fraud detection in online reviews by network effects. In: Proceedings of the International AAAI Conference on Web and Social Media, vol. 7, pp. 2–11 (2013)

2. Akoglu, L., Tong, H., Koutra, D.: Graph based anomaly detection and description: a survey. Data Min. Knowl. Disc. **29**(3), 626–688 (2015)
3. Breuer, A., Eilat, R., Weinsberg, U.: Friend or faux: graph-based early detection of fake accounts on social networks. In: Proceedings of The Web Conference 2020, pp. 1287–1297 (2020)
4. Brin, S., Page, L.: The anatomy of a large-scale hypertextual web search engine. Comput. Netw. ISDN Syst. **30**(1–7), 107–117 (1998)
5. Cheng, L., Guo, R., Shu, K., Liu, H.: Causal understanding of fake news dissemination on social media. In: Proceedings of the 27th ACM SIGKDD Conference on Knowledge Discovery & Data Mining, pp. 148–157 (2021)
6. Dou, Y., Liu, Z., Sun, L., Deng, Y., Peng, H., Yu, P.S.: Enhancing graph neural network-based fraud detectors against camouflaged fraudsters. In: Proceedings of the 29th ACM International Conference on Information & Knowledge Management, pp. 315–324 (2020)
7. Gao, Y., Wang, X., He, X., Liu, Z., Feng, H., Zhang, Y.: Alleviating structural distribution shift in graph anomaly detection. In: Proceedings of the Sixteenth ACM International Conference on Web Search and Data Mining, pp. 357–365 (2023)
8. He, R., McAuley, J.: Ups and downs: modeling the visual evolution of fashion trends with one-class collaborative filtering. In: Proceedings of the 25th International Con World Wide Web, pp. 507–517 (2016)
9. Hooi, B., et al.: BIRDNEST: Bayesian inference for ratings-fraud detection. In: Proceedings of the 2016 SIAM International Conference on Data Mining, pp. 495–503. SIAM (2016)
10. Hooi, B., Song, H.A., Beutel, A., Shah, N., Shin, K., Faloutsos, C.: FRAUDAR: Bounding graph fraud in the face of camouflage. In: Proceedings of the 22nd ACM SIGKDD International Conference on Knowledge Discovery and Data Mining, pp. 895–904 (2016)
11. Jindal, N., Liu, B.: Opinion spam and analysis. In: Proceedings of the 2008 International Conference on Web Search and Data Mining, pp. 219–230 (2008)
12. Kleinberg, J.M.: Authoritative sources in a hyperlinked environment. J. ACM (JACM) **46**(5), 604–632 (1999)
13. Koggalahewa, D., Xu, Y., Foo, E.: A drift aware hierarchical test based approach for combating social spammers in online social networks. In: Xu, Y., et al. (eds.) AusDM 2021. CCIS, vol. 1504, pp. 47–61. Springer, Singapore (2021). https://doi.org/10.1007/978-981-16-8531-6_4
14. Kumar, S., Hooi, B., Makhija, D., Kumar, M., Faloutsos, C., Subrahmanian, V.: REV2: Fraudulent user prediction in rating platforms. In: Proceedings of the Eleventh ACM International Conference on Web Search and Data Mining, pp. 333–341 (2018)
15. Kumar, S., Shah, N.: False information on web and social media: a survey. arXiv preprint arXiv:1804.08559 (2018)
16. Li, A., Qin, Z., Liu, R., Yang, Y., Li, D.: Spam review detection with graph convolutional networks. In: Proceedings of the 28th ACM International Conference on Information and Knowledge Management, pp. 2703–2711 (2019)
17. Li, F.H., Huang, M., Yang, Y., Zhu, X.: Learning to identify review spam. In: Twenty-Second International Joint Conference on Artificial Intelligence (2011)
18. Lim, E.P., Nguyen, V.A., Jindal, N., Liu, B., Lauw, H.W.: Detecting product review spammers using rating behaviors. In: Proceedings of the 19th ACM International Conference on Information and Knowledge Management, pp. 939–948 (2010)

19. Liu, B., Zhang, L.: A survey of opinion mining and sentiment analysis. In: Mining text data, pp. 415–463. Springer, Boston (2012). https://doi.org/10.1007/978-1-4614-3223-4_13
20. Liu, Z., Dou, Y., Yu, P.S., Deng, Y., Peng, H.: Alleviating the inconsistency problem of applying graph neural network to fraud detection. In: Proceedings of the 43rd international ACM SIGIR conference on research and development in information retrieval, pp. 1569–1572 (2020)
21. McAuley, J.J., Leskovec, J.: From amateurs to connoisseurs: modeling the evolution of user expertise through online reviews. In: Proceedings of the 22nd International Conference on World Wide Web, pp. 897–908 (2013)
22. Mishra, A., Bhattacharya, A.: Finding the bias and prestige of nodes in networks based on trust scores. In: Proceedings of the 20th International Conference on World Wide Web, pp. 567–576 (2011)
23. Mukherjee, A., et al.: Spotting opinion spammers using behavioral footprints. In: Proceedings of the 19th ACM SIGKDD International Conference on Knowledge Discovery and Data Mining, pp. 632–640 (2013)
24. Ott, M., Choi, Y., Cardie, C., Hancock, J.T.: Finding deceptive opinion spam by any stretch of the imagination. arXiv preprint arXiv:1107.4557 (2011)
25. Rayana, S., Akoglu, L.: Collective opinion spam detection: bridging review networks and metadata. In: Proceedings of the 21th ACM SIGKDD International Conference on Knowledge Discovery and Data Mining, pp. 985–994 (2015)
26. Sandulescu, V., Ester, M.: Detecting singleton review spammers using semantic similarity. In: Proceedings of the 24th International Conference on World Wide Web, pp. 971–976 (2015)
27. Wang, D., et al.: A semi-supervised graph attentive network for financial fraud detection. In: 2019 IEEE International Conference on Data Mining (ICDM), pp. 598–607. IEEE (2019)
28. Wang, G., Xie, S., Liu, B., Philip, S.Y.: Review graph based online store review spammer detection. In: 2011 IEEE 11th International Conference on Data Mining, pp. 1242–1247. IEEE (2011)
29. Wang, G., Xie, S., Liu, B., Yu, P.S.: Identify online store review spammers via social review graph. ACM Trans. Intell. Syst. Technol. (TIST) 3(4), 1–21 (2012)
30. Wang, J., Wen, R., Wu, C., Huang, Y., Xiong, J.: FdGars: fraudster detection via graph convolutional networks in online app review system. In: Companion Proceedings of the 2019 World Wide Web Conference, pp. 310–316 (2019)
31. Wu, Z., Aggarwal, C.C., Sun, J.: The troll-trust model for ranking in signed networks. In: Proceedings of the Ninth ACM International Conference on Web Search and Data Mining, pp. 447–456 (2016)
32. Xie, S., Wang, G., Lin, S., Yu, P.S.: Review spam detection via temporal pattern discovery. In: Proceedings of the 18th ACM SIGKDD International Conference on Knowledge Discovery and Data Mining, pp. 823–831 (2012)
33. Zhang, S., Yin, H., Chen, T., Hung, Q.V.N., Huang, Z., Cui, L.: GCN-based user representation learning for unifying robust recommendation and fraudster detection. In: Proceedings of the 43rd International ACM SIGIR Conference on Research and Development in Information Retrieval, pp. 689–698 (2020)

Semi-supervised Model-Based Clustering for Ordinal Data

Ying Cui[1,2]($\boxtimes$), Louise McMillan[1,2], and Ivy Liu[1,2]

[1] School of Mathematics and Statistics, Victoria University of Wellington, Wellington, New Zealand
cuiying@myvuw.ac.nz, louise.mcmillan@sms.vuw.ac.nz, ivy.liu@vuw.ac.nz
[2] Centre for Data Science and Artificial Intelligence, Victoria University of Wellington, Wellington, New Zealand

Abstract. This paper introduces a semi-supervised learning technique for model-based clustering. Our research focus is on applying it to matrices of ordered categorical response data, such as those obtained from the surveys with Likert scale responses. We use the proportional odds model, which is popular and widely used for analyzing such data, as the model structure. Our proposed technique is designed for analyzing datasets that contain both labeled and unlabeled observations from multiple clusters. The model fitting is performed using the expectation-maximization (EM) algorithm, incorporating the labeled cluster memberships, to cluster the unlabeled observations.

To evaluate the performance of our proposed model, we conducted a simulation study in which we tested the model from eight different scenarios, each with varying combinations and proportions of known and unknown cluster memberships. The fitted models accurately estimate the parameters in most of the designed scenarios, indicating that our technique is effective in clustering partially-labeled data with ordered categorical response variables.

Keywords: clustering · semi-supervised learning · EM algorithm · ordinal data · Likert scale data · proportional odds model

1 Introduction

A categorical variable is a type of variable with a fixed set of categories. There are two main sub-types of categorical variables: nominal and ordinal variables [1]. Nominal variables have an unordered scale of categories, such as eye colour. In contrast, ordinal variables have an ordered scale of categories, like the Likert scale responses to a survey question with possible categories such as "disagree", "neither agree nor disagree", and "agree".

Many methods treat the data as continuous with equally-spaced categories. This results in significant errors in data interpretation [6,12]. Other methods treat the data as nominal and neglect the ordering of the categories.

© The Author(s), under exclusive license to Springer Nature Singapore Pte Ltd. 2024
D. Benavides-Prado et al. (Eds.): AusDM 2023, CCIS 1943, pp. 34–47, 2024.
https://doi.org/10.1007/978-981-99-8696-5_3

There are methods for analyzing ordinal variables directly, without neglecting the ordering or making assumptions about the spacing. These methods include the proportional odds version of the cumulative logit model, the adjacent-categories logit models, and the multinomial logistic regression model [2]. In particular, the proportional odds model has been popular since McCullagh's survey paper [17] and it is still ubiquitous in many fields such as agriculture [14], medicine [24] and socioeconomic studies [23]. For analyzing ordinal data, the advantage of using the proportional odds model is that it assumes the ordinal response has a latent variable which follows a logistic distribution [4] and that the model's parameters are invariant for all categories of the ordinal response [2].

For clustering data, many methods are available. Common methods include distance-based approaches such as k-means [15,16] and hierarchical agglomerative clustering [13]. However, these methods do not use statistical distributions, which makes statistical model evaluation and model selection techniques invalid [10]. In contrast, model-based clustering approaches describe the clustering process via statistical densities. For example, many publications [5,9,18,20–22] have developed one-mode clustering methods based on finite-mixture densities, fitting models using the Expectation-Maximization (EM) algorithm [8], which is a commonly used strategy.

Most available model-based clustering methods fit the models by assigning observations to specific clusters without prior knowledge of their cluster memberships. We refer to this strategy as "unsupervised" clustering. The term "supervision" is frequently used in the field of machine learning to describe three types of learning strategies: "supervised", "semi-supervised", and "unsupervised". The level of supervision depends on the amount of prior knowledge of cluster membership given [21]. Semi-supervised learning falls between unsupervised and supervised learning, where cluster memberships are available before clustering for some, but not all, observations [26]. For instance, [25] proposed a semi-supervised hybrid clustering algorithm that integrates distance metrics into a Gaussian mixture model.

This paper introduces semi-supervised model-based clustering by applying it to a matrix of ordinal data, using the proportional odds model as the basic model structure. The model is fitted using the EM algorithm, incorporating the known cluster memberships to carry out a probabilistic clustering of the unknown observations. Section 2 presents the model formulation including a clustering form of the proportional odds model. Section 3 presents the model fitting procedure using the EM algorithm. Section 4 presents a simulation study to evaluate the model's performance. Section 5 gives conclusions.

2 Model Formulation

In this section, we define the proportional odds model for ordinal data (Sect. 2.1) and a formulation of it that includes clustering (Sect. 2.2). The likelihood for this model is provided next (Sect. 2.3).

2.1 Proportional Odds Model

We consider the data to be in the form of an $n \times p$ matrix, $\boldsymbol{Y}$, with elements y_{ij}, where each y_{ij} is one of q categories. Let the probabilities for the response categories for y_{ij} be $\theta_{ij1}, \theta_{ij2}, \ldots, \theta_{ijq}$ such that $\sum_{k=1}^{q} \theta_{ijk} = 1$, $\forall i, j$. Each response follows the multinomial distribution:

$$y_{ij} \sim \text{Multinomial}\left(1; \{\theta_{ij1}, \theta_{ij2}, \ldots, \theta_{ijq}\}\right), \ i = 1, 2, \ldots, n \text{ and } j = 1, 2, \ldots, p. \tag{2.1}$$

Initially, we construct a linear predictor for each entry in the data matrix that assumes that each row and each column exhibits different response patterns, and we denote these row and column effects as α_i for $i = 1, 2, \ldots, n$ and β_j for $j = 1, 2, \ldots, p$, respectively. These main effects are the same for all the response categories, i.e., they don't depend on the category index k.

Under the proportional odds model, we also have cut-off points μ_k, for $k = 1, 2, \ldots, q$ with the constraint $\mu_1 < \mu_2 < \cdots < \mu_q = +\infty$. Under this construction, the probabilities θ_{ijk}, $k = 1, 2, \ldots, q$ are formulated as:

$$\theta_{ijk} = \begin{cases} \dfrac{\exp(\mu_k - \alpha_i - \beta_j)}{1 + \exp(\mu_k - \alpha_i - \beta_j)} & k = 1 \\[2ex] \dfrac{\exp(\mu_k - \alpha_i - \beta_j)}{1 + \exp(\mu_k - \alpha_i - \beta_j)} - \dfrac{\exp(\mu_{k-1} - \alpha_i - \beta_j)}{1 + \exp(\mu_{k-1} - \alpha_i - \beta_j)} & 1 < k < q \\[2ex] 1 - \sum_{k=1}^{q-1} \theta_{ijk} & k = q. \end{cases} \tag{2.2}$$

or we can express this model as:

$$\text{logit}\left[P(Y_{ij} \leq k)\right] = \begin{cases} \mu_k - \alpha_i - \beta_j & 1 \leq k < q \\ +\infty & k = q, \end{cases} \tag{2.3}$$

The main effects have the constraints $\sum_{i=1}^{n} \alpha_i = 0$ and $\sum_{j=1}^{p} \beta_j = 0$ and the total number of non-redundant parameters is $\nu = (q-1) + (n-1) + (p-1)$.

2.2 Proportional Odds Model with Clustering

We now consider the situation where clusters are present and where a portion of the data has already been labeled, i.e. has known cluster memberships. For simplicity, we only consider the clustering model in which the rows of the data matrix, representing the observations, are clustered. We assume that there are

R clusters present, though there may be fewer clusters among the labeled observations.

In our semi-supervised strategy, we suppose the rows without given cluster memberships come from a finite-mixture with R components. We suppose that the observations with known clusters have the same model parameters as the mixture components for the observations with unknown clusters. We now drop the effects of the individual rows and assume that all rows belonging to the same row cluster r have the same effect on the response, and these effects are modeled with parameters α_r for $r = 1, 2, \ldots, R$. Therefore, the proportional odds model with row clustering, but retaining the effects of individual columns, can be formulated as follows:

$$\text{logit}[P(Y_{ij} \leq k)] = \mu_k - \alpha_r - \beta_j, \; k < q. \tag{2.4}$$

We assume that both the labeled and unlabeled observations obey this model. In addition, we define $\{\pi_1, \pi_2, \ldots, \pi_R\}$ as the proportion of rows in each row cluster, constrained by $\sum_{r=1}^{R} \pi_r = 1$. Compared with the proportional odds model without clustering, the total number of non-redundant parameters is reduced to $\nu = (q-1) + 2 \times (R-1) + (p-1)$. If we set $R \ll n$, that will ensure that the number of non-redundant parameters is much less than n.

2.3   Likelihood

This section summarises the likelihood and log-likelihood of our proposed model. The complete-data likelihood and log-likelihood, used for the EM algorithm, are also given. We define Ω as the full set of combination of all parameters in the model. The incomplete likelihood is split into the likelihood L_ℓ for observations with known cluster memberships and the likelihood L_u for observations with unknown cluster memberships:

$$
\begin{aligned}
L[\Omega, \pi | Y] &= L_\ell [\Omega | (y_1, \ldots, y_{n_\ell})] \times L_u [\Omega, \pi | (y_{n_\ell+1}, \ldots, y_{n_\ell+n_u})] \\[2mm]
&= \left(\prod_{i=1}^{n_\ell} f_\ell (y_i | \Omega) \right) \left(\prod_{i=n_\ell+1}^{n_\ell+n_u} f_u (y_i | \Omega, \pi) \right) \\[2mm]
&= \left(\prod_{i=1}^{n_\ell} \prod_{r=1}^{R} [f_{r_i} (y_i | \omega_{r_i})]^{I(r_i=r)} \right) \left(\prod_{i=n_\ell+1}^{n_\ell+n_u} \sum_{r=1}^{R} \pi_r f_r (y_i | \omega_r) \right) \\[2mm]
&= \left(\prod_{i=1}^{n_\ell} \prod_{r=1}^{R} \prod_{j=1}^{p} \prod_{k=1}^{q} \theta_{rijk}^{I(y_{ij}=k)I(r_i=r)} \right) \left(\prod_{i=n_\ell+1}^{n_\ell+n_u} \sum_{r=1}^{R} \pi_r \prod_{j=1}^{p} \prod_{k=1}^{q} \theta_{rjk}^{I(y_{ij}=k)} \right),
\end{aligned}
\tag{2.5}
$$

where:

- y_i is i^{th} row of observations, where $i = 1, 2, \ldots, n_\ell + n_u$;
- $I(y_{ij} = k)$ is an indicator variable that is 1 if y_{ij} is in category k, and 0 otherwise;

- n_ℓ is the number of labeled rows with known cluster membership, n_u is the number of unlabeled rows with unknown row cluster membership, and $n = n_\ell + n_u$;
- r_i is the known cluster membership for unlabeled row i;
- ω_r is the set of parameters for component r;
- ω_{r_i} is the set of parameters for component r_i;
- $f_{r_i}(\boldsymbol{y}_i|\omega_{r_i})$ is the component density for the ith row with known row cluster membership r_i;
- $I(r_i = r)$ is an indicator variable that is 1 if row i with known cluster membership r_i belongs to row cluster r, and 0 otherwise;
- $f_r(\boldsymbol{y}_i|\omega_r)$ is the rth component density for row i with unknown row cluster membership;
- $f_\ell(\boldsymbol{y}_i|\boldsymbol{\Omega})$ is the overall mixture density for observation i with known row cluster membership;
- $f_u(\boldsymbol{y}_i|\boldsymbol{\Omega}, \boldsymbol{\pi})$ is the mixture component density for observation i with unknown row cluster membership.

The log-likelihood is as follows:

$$\ell\left[\boldsymbol{\Omega}, \boldsymbol{\pi}|\boldsymbol{Y}\right] = \left(\sum_{i=1}^{n_\ell}\sum_{r=1}^{R}\sum_{j=1}^{p}\sum_{k=1}^{q} I\left(y_{ij} = k\right) I\left(r_i = r\right) \log\left[\theta_{r_ijk}\right]\right)$$
$$+ \sum_{i=n_\ell+1}^{n_\ell+n_u} \log\left(\sum_{r=1}^{R}\pi_r \prod_{j=1}^{p}\prod_{k=1}^{q}\theta_{rjk}^{I(y_{ij}=k)}\right). \qquad (2.6)$$

In order to fit the model using the EM algorithm, we define latent indicator variables Z_{ir} for observations with unknown row cluster memberships:

$$Z_{ir} = \begin{cases} 1 & i \in r \\ 0 & i \notin r \end{cases}$$
$$i = n_\ell + 1, n_\ell + 2, \ldots, n_\ell + n_u,$$
$$r = 1, 2, \ldots, R. \qquad (2.7)$$

These indicator variables obey the constraint $\sum_{r=1}^{R} Z_{ir} = 1$, $n_\ell+1 \le i \le n_\ell+n_u$ and thus:

$$\prod_{r=1}^{R}\alpha_i^{Z_{ir}} = \sum_{r=1}^{R}\alpha_i Z_{ir}, \quad i = n_\ell + 1, n_\ell + 2, \ldots, n_\ell + n_u. \qquad (2.8)$$

The proportions π_r correspond to the prior probabilities of the latent cluster membership indicators, so the prior probability of a vector of latent variables $\boldsymbol{Z}_i$ is:

$$f\left(\boldsymbol{Z}_i = \boldsymbol{z}_i\right) = f\left[\left(Z_{i1} = z_{i1}, Z_{i2} = z_{i2}, \ldots, Z_{iR} = z_{iR}\right) | \left(\pi_1, \pi_2, \ldots, \pi_R\right)\right]$$
$$= \frac{1!}{0!\ldots1!\ldots0!}\pi_1^{z_{i1}}\pi_2^{z_{i2}}\cdots\pi_R^{z_{iR}}$$
$$= \prod_{r=1}^{R}\pi_r^{z_{ir}}. \qquad (2.9)$$

The conditional distribution for the responses in a row i with unknown cluster membership, given the latent row cluster membership indicators $\boldsymbol{z}_i$, can therefore be rewritten as:

$$f_u\left(\boldsymbol{y}_i|\boldsymbol{z}_i,\boldsymbol{\Omega}\right)=\prod_{r=1}^{R}\left[f_r\left(\boldsymbol{y}_i|\boldsymbol{\omega}_r\right)\right]^{z_{ir}}\,,\quad i=n_\ell+1,n_\ell+2,\ldots,n_\ell+n_u.\ (2.10)$$

The joint distribution of the observed responses and the row cluster membership indicators for row i can be written as:

$$
\begin{aligned}
f_u\left(\boldsymbol{y}_i,\boldsymbol{z}_i\right) &= f_u\left(\boldsymbol{y}_i|\boldsymbol{z}_i\right)f\left(\boldsymbol{z}_i\right)\\
&= \left(\prod_{r=1}^{R}\left[f_r\left(\boldsymbol{y}_i|\boldsymbol{\omega}_r\right)\right]^{z_{ir}}\right)\left(\prod_{r=1}^{R}\pi_r^{z_{ir}}\right)\\
&= \prod_{r=1}^{R}\left[\pi_r f_r\left(\boldsymbol{y}_i|\boldsymbol{\omega}_r\right)\right]^{z_{ir}}\,,\quad i=n_\ell+1,n_\ell+2,\ldots,n_\ell+n_u.\ (2.11)
\end{aligned}
$$

The overall complete-data likelihood for the model is:

$$
\begin{aligned}
L_c\left[\boldsymbol{\Omega},\boldsymbol{\pi}|\boldsymbol{Y},\boldsymbol{Z}\right] &= \left(\prod_{i=1}^{n_\ell}f_\ell\left(\boldsymbol{y}_i|\boldsymbol{\Omega}\right)\right)\left(\prod_{i=n_\ell+1}^{n_\ell+n_u}f_u\left(\boldsymbol{y}_i,\boldsymbol{z}_i|\boldsymbol{\Omega}\right)\right)\\
&= \left(\prod_{i=1}^{n_\ell}f_\ell\left(\boldsymbol{y}_i|\boldsymbol{\Omega}\right)\right)\left(\prod_{i=n_\ell+1}^{n_\ell+n_u}f_u\left(\boldsymbol{y}_i|\boldsymbol{z}_i,\boldsymbol{\Omega}\right)f\left(\boldsymbol{z}_i\right)\right)\\
&= \left(\prod_{i=1}^{n_\ell}\prod_{r=1}^{R}\left[f_{r_i}\left(\boldsymbol{y}_i|\boldsymbol{\omega}_{r_i}\right)\right]^{I\left(r_i=r\right)}\right)\left(\prod_{i=n_\ell+1}^{n_\ell+n_u}\prod_{r=1}^{R}\left[\pi_r f_r\left(\boldsymbol{y}_i|\boldsymbol{\omega}_r\right)\right]^{z_{ir}}\right)\\
&= \left(\prod_{i=1}^{n_\ell}\prod_{r=1}^{R}\left[f_{r_i}\left(\boldsymbol{y}_i|\boldsymbol{\omega}_{r_i}\right)\right]^{I\left(r_i=r\right)}\right)\left(\prod_{i=n_\ell+1}^{n_\ell+n_u}\prod_{r=1}^{R}\left[\pi_r\prod_{j=1}^{p}f_r\left(y_{ij}|\boldsymbol{\omega}_r\right)\right]^{z_{ir}}\right)\\
&= \left(\prod_{i=1}^{n_\ell}\prod_{r=1}^{R}\prod_{j=1}^{p}\prod_{k=1}^{q}\theta_{r_ijk}^{I\left(y_{ij}=k\right)I\left(r_i=r\right)}\right)\left(\prod_{i=n_\ell+1}^{n_\ell+n_u}\prod_{r=1}^{R}\left[\pi_r\prod_{j=1}^{p}\prod_{k=1}^{q}\theta_{rjk}^{I\left(y_{ij}=k\right)}\right]^{z_{ir}}\right).
\end{aligned}
$$
$$(2.12)$$

The corresponding complete-data log-likelihood is:

$$
\begin{aligned}
\ell_c\left[\boldsymbol{\Omega},\boldsymbol{\pi}|\boldsymbol{Y},\boldsymbol{Z}\right] &= \left(\sum_{i=1}^{n_\ell}\sum_{r=1}^{R}\sum_{j=1}^{p}\sum_{k=1}^{q}I\left(y_{ij}=k\right)I\left(r_i=r\right)\log\left[\theta_{r_ijk}\right]\right)\\
&\quad+\left(\sum_{i=n_\ell+1}^{n_\ell+n_u}\sum_{r=1}^{R}z_{ir}\log(\pi_r)+\sum_{i=n_\ell+1}^{n_\ell+n_u}\sum_{j=1}^{p}\sum_{k=1}^{q}\sum_{r=1}^{R}z_{ir}I(y_{ij}=k)\log\left[\theta_{rjk}\right]\right).
\end{aligned}
$$
$$(2.13)$$

3   Model Fitting

This section describes the model fitting procedure using the Expectation Maximization (EM) algorithm [8]. This algorithm is an iterative procedure that estimates the maximum likelihood in incomplete-data problems [19]. It iterates over two steps called the *Expectation step* and the *Maximization step*. The Expectation step updates the posterior probabilities of cluster membership expressed by the latent variables. The Maximization step updates the maximum likelihood estimates of the parameters, using the latest estimated values of latent variables from the E-step. There are various possible rules for convergence but we will consider the algorithm to have converged when the incomplete-data log-likelihood is sufficiently similar in two successive iterations. If required, the rows can then be allocated to specific clusters based on the final estimates of the latent cluster memberships $\{Z_i\}$.

3.1   The Expectation Step (E-Step)

In the E-step, the latest estimates of the mixing proportions $\{\pi_r\}$ and parameters Ω are used to calculate the expected values of z_{ir}. The conditional expectation of the complete-data log-likelihood at iteration t is expressed as follows:

$$
E_{\{z_{ir}\}|Y,\Omega^{(t-1)},\pi^{(t-1)}}\left[\ell_c\left(\Omega,\pi|Y,\{z_{ir}\}\right)\right]
$$

$$
= \sum_{i=1}^{n_\ell}\sum_{r=1}^{R}\sum_{j=1}^{p}\sum_{k=1}^{q} I\left(y_{ij}=k\right) I\left(r_i=r\right) \log\left[\theta_{r_ijk}^{(t-1)}\right]
$$

$$
+ \sum_{i=n_\ell+1}^{n_\ell+n_u}\sum_{r=1}^{R} \log\left[\pi_r^{(t-1)}\right] E\left[z_{ir}|\{y_i\};\Omega^{(t-1)},\pi^{(t-1)}\right]
$$

$$
+ \sum_{i=n_\ell+1}^{n_\ell+n_u}\sum_{j=1}^{p}\sum_{k=1}^{q}\sum_{r=1}^{R} I(y_{ij}=k)\log\left[\theta_{rjk}^{(t-1)}\right] E\left[z_{ir}|\{y_i\};\Omega^{(t-1)},\pi^{(t-1)}\right].
$$

$$(3.1)$$

The latent variable Z_{ir} follows a Bernoulli distribution and its expectation is computed as follows:

$$
E\left[z_{ir}|y_i;\Omega^{(t-1)},\pi^{(t-1)}\right]
$$

$$
= \left(0\times P\left[z_{ir}=0|y_i;\Omega^{(t-1)},\pi^{(t-1)}\right]\right) + \left(1\times P\left[z_{ir}=1|y_i;\Omega^{(t-1)},\pi^{(t-1)}\right]\right)
$$

$$
= P\left[z_{ir}=1|y_i;\Omega^{(t-1)},\pi^{(t-1)}\right].
$$

$$(3.2)$$

Then, we apply Bayes' rule to obtain the following expression for the estimated Z_{ir} at iteration t:

$$
\begin{aligned}
\hat{z}_{ir}^{(t)} &= P\left[z_{ir} = 1 | \boldsymbol{y}_i; \boldsymbol{\Omega}^{(t-1)}, \boldsymbol{\pi}^{(t-1)}\right] \\
&= \frac{P\left[\boldsymbol{y}_i | z_{ir} = 1; \boldsymbol{\Omega}^{(t-1)}, \boldsymbol{\pi}^{(t-1)}\right] \times P\left[z_{ir} = 1; \boldsymbol{\Omega}^{(t-1)}, \boldsymbol{\pi}^{(t-1)}\right]}{\sum_{r'=1}^{R} P\left[\boldsymbol{y}_i | z_{ir'} = 1; \boldsymbol{\Omega}^{(t-1)}, \boldsymbol{\pi}^{(t-1)}\right] \times P\left[z_{ir'} = 1; \boldsymbol{\Omega}^{(t-1)}, \boldsymbol{\pi}^{(t-1)}\right]} \\
&= \frac{\hat{\pi}_r^{(t-1)} \prod_{j=1}^{p} \prod_{k=1}^{q} \left[\hat{\theta}_{rjk}^{(t-1)}\right]^{I(y_{ij}=k)}}{\sum_{r'=1}^{R} \left(\hat{\pi}_{r'}^{(t-1)} \prod_{j=1}^{p} \prod_{k=1}^{q} \left[\hat{\theta}_{r'jk}^{(t-1)}\right]^{I(y_{ij}=k)}\right)}.
\end{aligned}
\tag{3.3}
$$

3.2   The Maximization Step (M-Step)

The M-step updates the maximum likelihood estimates for parameters using the estimates $\hat{z}_{ir}^{(t)}$ obtained from the E-step. We apply the Lagrange multiplier λ approach [11] to find an analytical expression for the estimate of π_r:

$$
\hat{\pi}_r^{(t)} = \frac{1}{n_u} \sum_{i=n_\ell+1}^{n_\ell+n_u} \hat{z}_{ir}^{(t)}.
\tag{3.4}
$$

We use numerical optimization to find the latest estimates of the remaining parameters, Ω. In this process we also include the observations with known cluster memberships, because our model assumes that the known clusters have the same parameters as the unknown clusters, so those observations inform the cluster parameter estimation process in the M-step. Each of these observations contributes to the likelihood of the cluster it is known to be in.

Reparameterization of Cut-Off Points μ_k. In the proportional odds model, the ordering of the categories is ensured by imposing a constraint on the parameters μ_k: $\mu_1 < \mu_2 < \cdots < \mu_{k-1}$. It is easier to perform numerical optimization by reparameterizing these parameters in such a way that we optimize unconstrained values. So we introduce parameters $\boldsymbol{w} = (w_2, w_3, \ldots, w_{q-1})$ such that:

$$
\mu_k = \mu_1 + \sum_{\ell=2}^{k} \exp(w_\ell), \quad for \ k = 2, 3, \ldots, q-1,
\tag{3.5}
$$

where $w_2, w_3, \ldots, w_{q-1}$ can take any value from $+\infty$ to $-\infty$, and the parameters μ_k are constrained by construction.

4   Simulation Study

This section gives the results of a simulation study to evaluate the performance of our proposed approach in eight different scenarios. These scenarios vary the percentage of cluster memberships that are known, denoted as $m\%$, and the distribution of memberships within that labeled portion, denoted as $\{g_r\}$.

In the simulated data, the rows are equally distributed among the $R = 3$ clustering groups. Table 1 defines the first four scenarios, in which all the clusters are observed within the portion of observations with known cluster memberships. Table 2 defines the last four scenarios, in which only a subset of the clusters are observed within the portion of observations with known cluster membership.

Table 1. Scenarios where all clusters are observed in the labeled data.

Scenario 1 $m\%= 10\%$	Scenario 2 $m\%= 30\%$	Scenario 3 $m\%= 10\%$	Scenario 4 $m\%= 30\%$
$\pi_1=1/3$	$\pi_1=1/3$	$\pi_1=0.304$	$\pi_1=0.219$
$\pi_2=1/3$	$\pi_2=1/3$	$\pi_2=0.337$	$\pi_2=0.348$
$\pi_3=1/3$	$\pi_3=1/3$	$\pi_3=0.359$	$\pi_3=0.433$
$g_1=1/3$	$g_1=1/3$	$g_1=0.600$	$g_1=0.600$
$g_2=1/3$	$g_2=1/3$	$g_2=0.300$	$g_2=0.300$
$g_3=1/3$	$g_3=1/3$	$g_3=0.100$	$g_3=0.100$

Table 2. Scenarios where only a subset of the clusters is observed in the labeled data.

Scenario 5 $m\%= 10\%$	Scenario 6 $m\%= 30\%$	Scenario 7 $m\%= 10\%$	Scenario 8 $m\%= 30\%$
$\pi_1=0.315$	$\pi_1=0.262$	$\pi_1=0.260$	$\pi_1=0.048$
$\pi_2=0.315$	$\pi_2=0.262$	$\pi_2=0.370$	$\pi_2=0.476$
$\pi_3=0.370$	$\pi_3=0.476$	$\pi_3=0.370$	$\pi_3=0.476$
$g_1=0.500$	$g_1=0.500$	$g_1=1.000$	$g_1=1.000$
$g_2=0.500$	$g_2=0.500$		

We set $n = 300$, $p = 5$, $q = 3$. The responses $\{y_{ij}\}$ are generated from the multinomial distribution with probabilities defined according to (2.4). The true values of the row cluster and column effect parameters are $\{\alpha_1, \alpha_2, \alpha_3\} = \{-2, 0, 2\}$ and $\{\beta_1, \beta_2, \beta_3, \beta_4, \beta_5\} = \{-2, -1.5, 0.3, 1.0, 2.2\}$. The cut-off point values μ_k are obtained such that the response categories have equal probabilities for the baseline row cluster such that $P(y_{ij} = 1) = P(y_{ij} = 2) = \cdots = P(y_{ij} = q)$ when row i belongs to the first row cluster. The true values of the cut-off points μ_k and the reparameterized values w_k are as follows:

$$\mu_1 = w_1 = \log(1/2) \approx -0.693 \tag{4.1}$$
$$\mu_2 = \mu_1 + \exp(w_2) = \log(1/2) + \exp[\log(2)] \approx 1.307.$$

For each scenario, we simulated 100 replicate datasets.

When fitting the models, we apply the k-means algorithm to each simulated dataset to produce the initial clustering groups, and optimised the log-likelihood of the "unsupervised" strategy to generate the initial values for the model's parameters.

All simulation and model-fitting code was written in the R programming language. The EM algorithm converged in all the scenarios and for all the replicate datasets.

Tables 3 and 4 display the average estimates of the parameters and their standard errors obtained from 100 converged replicate datasets in each scenario. We observe that all $\{\beta_j\}$ parameters are estimated exceptionally well, with standard errors below 0.14. The cut-off points $\{\mu_k\}$ are also estimated accurately in most scenarios, but their standard errors are higher in scenarios 5, 6, and 7. Similarly, the clustering parameters $\{\alpha_r\}$ are well-estimated with low standard errors in most scenarios, but higher standard errors in scenarios 5, 6, and 7.

Table 5 and Table 6 give the estimated values of $\{\pi_r\}$ for all scenarios. Generally, we observed that these estimates remain close to the true values in most scenarios, especially in scenarios 1 to 4 where all clusters were observed in the labeled data.

The scatterplots in Fig. 1 show the estimated values of π_1 and π_2 from each of the 100 replicate datasets, split by the eight scenarios. The red triangle in each scatterplot represents the true values of parameters π_1 and π_2. The estimated π_1 and π_2 values are closer to the true values for scenarios 1, 2, 3, 4, and 8 than for scenarios 5, 6, and 7. In general, the estimation of the mixing proportions $\{\pi_r\}$ is accurate in most of the scenarios.

Table 3. The average estimated values of $\{\mu_k\}$, $\{\alpha_r\}$, and $\{\beta_j\}$, along with their standard errors, obtained from 100 replicate datasets. These estimates were obtained for scenarios in which all clusters were observed in the labeled data, and the fitted model had $R = 3$ clusters.

Parameter	True	Scenario1		Scenario2		Scenario3		Scenario4	
		Mean	S.E.	Mean	S.E.	Mean	S.E.	Mean	S.E
μ_1	−0.693	−0.685	0.106	−0.694	0.091	−0.654	0.109	−0.690	0.092
μ_2	1.307	1.269	0.111	1.305	0.097	1.294	0.114	1.309	0.096
α_1	−2.000	−1.895	0.129	−2.007	0.120	−1.887	0.129	−1.998	0.119
α_2	0.000	0.007	0.122	0.003	0.098	0.063	0.133	−0.005	0.099
α_3	2.000	1.888	0.124	2.004	0.115	1.824	0.128	2.003	0.116
β_1	−2.000	−1.961	0.138	−2.008	0.139	−1.951	0.137	−2.007	0.139
β_2	−1.500	−1.469	0.127	−1.503	0.127	−1.464	0.126	−1.503	0.127
β_3	0.300	0.302	0.111	0.306	0.112	0.298	0.111	0.306	0.112
β_4	1.000	0.969	0.115	0.994	0.116	0.964	0.115	0.994	0.116
β_5	2.200	2.159	0.135	2.211	0.136	2.153	0.135	2.210	0.136

Table 4. The average estimated values of $\{\mu_k\}$, $\{\alpha_r\}$, and $\{\beta_j\}$, along with their standard errors, obtained from 100 replicate datasets. These estimates were obtained for scenarios in which a subset of the clusters were observed in the labeled data and the fitted model had $R = 3$ clusters.

Parameter	True	Scenario5		Scenario6		Scenario7		Scenario8	
		Mean	S.E.	Mean	S.E.	Mean	S.E.	Mean	S.E
μ_1	−0.693	−0.339	0.173	−0.189	0.262	−0.240	0.523	−0.696	0.091
μ_2	1.307	1.562	0.181	1.710	0.271	1.709	0.528	1.303	0.095
α_1	−2.000	−1.742	0.287	−1.732	0.475	−2.280	0.996	−2.002	0.118
α_2	0.000	−0.095	0.222	−0.235	0.298	0.010	0.563	−0.013	0.100
α_3	2.000	1.837	0.185	1.967	0.279	2.219	0.538	2.015	0.120
β_1	−2.000	−1.900	0.134	−1.893	0.133	−1.957	0.137	−2.008	0.139
β_2	−1.500	−1.429	0.124	−1.426	0.124	−1.465	0.126	−1.504	0.127
β_3	0.300	0.289	0.109	0.285	0.109	0.298	0.111	0.307	0.112
β_4	1.000	0.936	0.113	0.933	0.113	0.966	0.115	0.994	0.116
β_5	2.200	2.104	0.134	2.100	0.132	2.158	0.136	2.211	0.136

Table 5. The averaged estimated $\{\pi_r\}$ and their standard errors obtained from scenarios where all clusters are known in the labeled data over the 100 converged replicate datasets with fitting row clusters $R = 3$.

Parameter	Scenario1		Scenario2		Scenario3		Scenario4	
	True	Estimated	True	Estimated	True	Estimated	True	Estimated
π_1	0.333	0.357	0.333	0.336	0.304	0.325	0.219	0.221
π_2	0.333	0.292	0.333	0.333	0.337	0.283	0.348	0.345
π_3	0.334	0.351	0.334	0.331	0.359	0.392	0.433	0.435

Table 6. The averaged estimated $\{\pi_r\}$ and their standard errors obtained from scenarios where a subset of the clusters are known in the labeled data over the 100 converged replicate datasets with fitting row clusters $R = 3$.

Parameter	Scenario5		Scenario6		Scenario7		Scenario8	
	True	Estimated	True	Estimated	True	Estimated	True	Estimated
π_1	0.315	0.280	0.262	0.186	0.260	0.203	0.048	0.051
π_2	0.315	0.241	0.262	0.237	0.370	0.364	0.476	0.478
π_3	0.370	0.479	0.476	0.577	0.370	0.432	0.476	0.471

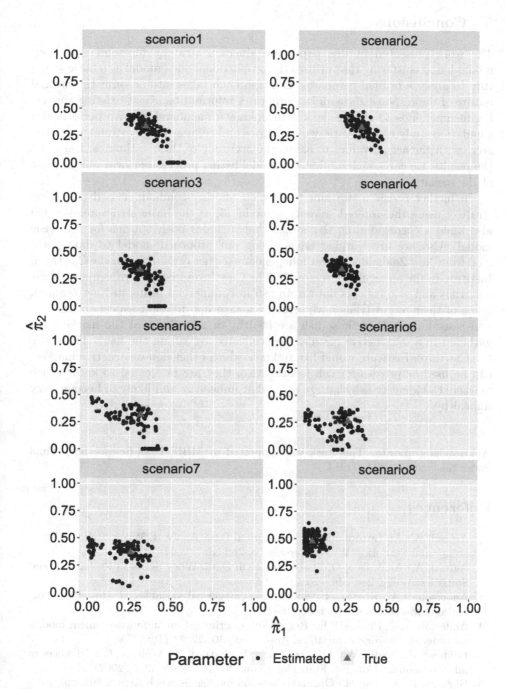

Fig. 1. The scatterplots of estimated π_1 vs π_2 over 100 converged replicate datasets for the semi-supervised model-based clustering model across all the designed the scenarios. The red triangle point represents the true value of π_1 and π_2. (Color figure online)

5   Conclusions

The work presented in this paper introduces a strategy for semi-supervised model-based clustering that utilizes the proportional odds model as a basic structure to analyze ordinal responses. Our approach takes into account the ordinal nature of the response data and incorporates information about existing clustering memberships to cluster data with unknown memberships. We performed a simulation study using a variety of scenarios and compared the estimated parameters with the actual values to assess the accuracy of these estimates. Based on the results, we can conclude that the model fitting process perform well in most of the scenarios.

In further research, we aim to develop another semi-supervised clustering strategy using the ordered stereotype model [3] as the basic structure. We will also build a corresponding R package that includes both options for the basic model. Also, we are working on applying our proposed model to data gathered from New Zealand's largest independent science organization, the Cawthron Institute in the aquaculture sector [7]. The Cawthron Institute runs many different trials and also collects data from fish in commercial farms in New Zealand. Some of their recent trials have involved measuring different aspects of the fish and assessing whether those fish are healthy or not. Some of the markers are gathered in a destructive manner (i.e. collected by killing the fish). Therefore, the Cawthron Institute would like to know which other non-destructive markers can be used as proxies for fish health. Also, they are interested to know which markers can identify fish that are somewhat unhealthy and likely to become very unhealthy.

Acknowledgements. This work was supported by MBIE Data Science SSIF Fund under the contract RTVU1914.

References

1. Agresti, A.: An Introduction to Categorical Data Analysis. Wiley Series in Probability and Statistics. Wiley-Interscience, 2nd edn. (2007)
2. Agresti, A.: Analysis of Ordinal Categorical Data. Wiley Series in Probability and Statistics. Wiley, 2nd edn. (2010)
3. Anderson, J.A.: Regression and ordered categorical variables. J. Roy. Stat. Soc. Series B Methodol. **46**(1), 1–30 (1984)
4. Anderson, J.A., Philips, P.R.: Regression, discrimination and measurement models for ordered categorical variables. Appl. Stat. **30**, 22–31 (1981)
5. Böhning, D., Seidel, W., Alfó, M., Garel, B., Patilea, V., Walther, G.: Advances in mixture models. Comput. Stat. Data Anal. **51**(11), 5205–5210 (2007)
6. Bürkner, P., Vuorre, M.: Ordinal regression models in psychology: a tutorial. Adv. Methods Pract. Psychol. Sci. **2**(1), 77–101 (2019)
7. Cawthron: Cawthron. https://www.cawthron.org.nz/about-us/ (2023)
8. Dempster, A.P., Laird, N.M., Rubin, D.B.: Maximum likelihood from incomplete data via the EM algorithm. J. Roy. Stat. Soc. Ser. B Methodol. **39**(1), 1–38 (1977)

9. Everitt, B.S., Leese, M., Landau, S.: Cluster Analysis. Hodder Arnold Publication, 4th edn. (2001)
10. Fernández, D., Arnold, R., Pledger, S.: Mixture-based clustering for the ordered stereotype model. Comput. Stat. Data Anal. **93**, 46–75 (2014)
11. Grossman, S.I.: Calculus, 3rd edn. Academic Press (1984)
12. Janitza, S., Tutz, G., Boulesteix, A.L.: Random forest for ordinal responses: prediction and variable selection. Comput. Stat. Data Anal. **96**, 57–73 (2016)
13. Johnson, S.C.: Hierarchical clustering schemes. Psychometrika, pp. 241–254 (1967)
14. Lanfranchi, M., Giannetto, C., Zirilli, A.: Analysis of demand determinants of high quality food products through the application of the cumulative proportional odds model. Appl. Math. Sci. **8**, 3297–3305 (2014)
15. Lloyd, S.P.: Least square quantization in PCM. IEEE Trans. Inf. Theory **28**(2), 129–137 (1957)
16. MacQueen, J.: Some methods for classification and analysis of multivariate observations. In: Berkeley Symposium on Mathematical Statistics and Probability 1, pp. 281–297 (1967)
17. McCullagh, P.: Regression models for ordinal data. J. Roy. Stat. Soc. Ser. B, Methodol. **42**(2), 109–142 (1980)
18. McLachlan, G.J., Basford, K.E.: Mixture Models: Inference and Applications to Clustering. Statistics, Textbooks and Monographs, M. Dekker (1988)
19. McLachlan, G.J., Krishnan, T.: The EM Algorithm and Extensions, 2nd edn. John Wiley and Sons Inc. (2015)
20. McLachlan, G.J., Peel, D.: Finite mixture models. Wiley Series in Probability and Statistics (2000)
21. McNicholas, P.D.: Mixture Model-Based Classification. CRC Press, Boca Raton (2017)
22. Melnykov, V., Maitra, R.: Finite mixture models and model-based clustering. Iowa State University Digital Repository (2010)
23. Pechey, R., Monsivais, P., Ng, Y.L., Marteau, T.M.: Why don't poor men eat fruit? Socioeconomic differences in motivations for fruit consumption. Appetite **84**, 271–279 (2015)
24. Skolnick, B.E., et al.: A clinical trial of the progesterone for severe traumatic brain injury. N. Engl. J. Med. **371**, 2467–2476 (2014)
25. Zhang, Y., Wen, J., Wang, X., Jiang, Z.: Semi-supervised hybrid clustering by integrating Gaussian mixture model and distance metric learning. J. Intell. Inf. Syst. **45**(1), 113–130 (2013)
26. Zhu, X., Goldberg, A.B.: Introduction to semi-supervised learning. Synthesis Lectures Artif. Intell. Mach. Learn. **3**(1), 1–130 (2009)

Damage GAN: A Generative Model for Imbalanced Data

Ali Anaissi[1]([⊠])(iD), Yuanzhe Jia[1], Ali Braytee[2], Mohamad Naji[1],
and Widad Alyassine[1]

[1] School of Computer Science, The University of Sydney, Camperdown, Australia
ali.anaissi@sydney.edu.au, yjia5612@uni.sydney.edu.au,
Mohamad.Naji@uts.edu.au
[2] School of Computer Science, The University of Technology Sydney,
Ultimo, Australia
ali.braytee@uts.edu.au

Abstract. This study delves into the application of Generative Adversarial Networks (GANs) within the context of imbalanced datasets. Our primary aim is to enhance the performance and stability of GANs in such datasets. In pursuit of this objective, we introduce a novel network architecture known as Damage GAN, building upon the ContraD GAN framework which seamlessly integrates GANs and contrastive learning. Through the utilization of contrastive learning, the discriminator is trained to develop an unsupervised representation capable of distinguishing all provided samples. Our approach draws inspiration from the straightforward framework for contrastive learning of visual representations (SimCLR), leading to the formulation of a distinctive loss function. We also explore the implementation of self-damaging contrastive learning (SDCLR) to further enhance the optimization of the ContraD GAN model. Comparative evaluations against baseline models including the deep convolutional GAN (DCGAN) and ContraD GAN demonstrate the evident superiority of our proposed model, Damage GAN, in terms of generated image distribution, model stability, and image quality when applied to imbalanced datasets.

Keywords: Damage GAN · ContraD GAN · SimCLR · SDCLR · Imbalanced datasets

1 Introduction

GAN is a popular deep learning architecture composed of a generator and a discriminator. The generator aims to learn the distribution of real samples, while the discriminator evaluates the authenticity of inputs, creating a dynamic "game process". While GAN is extensively used for image generation, their effectiveness in imbalanced datasets has received less attention [1,4,20]. Contrastive learning, a self-supervised training technique that captures augmented image invariants and reduces training effort for image classification, has recently gained prominence [2]. Our objective is to integrate contrastive learning into GAN, leveraging its potential to improve the performance and stability in the context of imbalanced datasets.

© The Author(s), under exclusive license to Springer Nature Singapore Pte Ltd. 2024
D. Benavides-Prado et al. (Eds.): AusDM 2023, CCIS 1943, pp. 48–61, 2024.
https://doi.org/10.1007/978-981-99-8696-5_4

In this paper, we introduce Damage GAN, a novel GAN model constructed by implementing SDCLR as a replacement for the SimCLR module in ContraD GAN, and applying it to the task of training and validation. The imbalanced CIFAR-10 dataset is utilized for both model training and validation. Evaluation metrics such as Fréchet Inception Distance (FID) and Inception Score (IS) are employed to assess the performance of the proposed model. After experiments, we demonstrate that Damage GAN outperforms state-of-the-art models, such as DCGAN and ContraD GAN, when applied to imbalanced datasets, thereby highlighting its potential for improving GAN performance on imbalanced datasets.

This paper is organized as follows. Section 2 is about related work, mainly discusses and analyses the contributions and limitations related to our study in the last ten years. In Sect. 3 we describe our research ideas and model structure in details. In Sect. 4 we clarify the dataset used for experiments, the model training procedure, the evaluation metrics, as well as how to conduct experiments on exploratory data analysis and model comparison to verify the predictive ability, stability and applicability of our model. Finally, we conclude the paper in Sect. 5 and look to the future. Since this paper contains a large number of technical terms, the notations are summarized in Table 1 for the convenience.

2   Related Work

The foundation of our proposed model is built upon the ContraD GAN, which involves training a generative adversarial network through contrastive learning applied to the discriminator. To further improve the model's performance, we recommend incorporating Self-Damaging Contrastive Learning (SDCLR) as a replacement for the SimCLR module within the ContraD GAN framework.

The following section introduces a literature review covering fundamental theories pertinent to our proposed model.

2.1   GAN

The framework of GAN, which is a type of generative algorithm, was proposed in 2014 [10]. The main idea of the generative model is to learn the pattern of the training data and use that knowledge to create new examples [11,29,30]. GAN introduces the concept of adversarial learning to address the limitations of generative algorithms. The basic principle is to produce data that looks very similar to real samples. GAN consists of two modules: a generator and a discriminator, which are typically implemented using neural networks. The generator learns to understand the distribution of real examples and generate new ones, while the discriminator tries to determine if the inputs are genuine or fake. The goal is for the generator to capture the true distribution of real data and generate realistic examples.

In recent years, GANs and their variations have gained widespread use in the fields of Computer Vision (CV) and Natural Language Processing (NLP). These models provide distinct advantages over other generative methods, as

Table 1. Notations used in this paper.

Abbreviation	Description
CIFAR-10	It is an image database containing 60,000 32×32 colour images in 10 classes, with 6,000 images per class
ContraD GAN	Training GANs with stronger augmentations via contrastive discriminator
CNN	Convolutional Neural Network
CV	Computer Vision
Damage GAN	Proposed model in this paper
DCGAN	Deep Convolutional Generative Adversarial Network
FID	Fréchet Inception Distance
ImageNet	It is an image database organized according to the WordNet hierarchy, in which each node of the hierarchy is depicted by hundreds and thousands of images
IS	Inception Score
GAN	Generative Adversarial Network
LeakyReLU	It is a type of activation function based on ReLU. It has a small slope for negative values with which LeakyReLU can produce small, non-zero, and constant gradients with respect to the negative values
MLP	Multi-Layer Perceptron
NLP	Nature Language Processing
ReLU	Rectified Linear Units. It is a non-linear activation function that is widely used in multi-layer neural networks or deep neural networks
SDCLR	Self-damaging Contrastive Learning of Visual Representations
Sigmoid	It is a special form of the logistic function with an S-shaped curve
SimCLR	Simple Framework for Contrastive Learning of Visual Representations
SoftMax	It is a function that turns a vector of K real values into a vector of K real values that sum to 1
Tanh	Hyperbolic Tangent. It is the hyperbolic analogue of the Tan circular function used throughout trigonometry

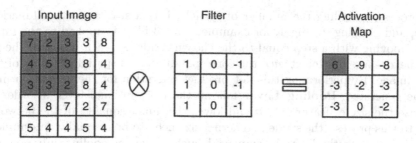

Fig. 1. An example of convolutional layer.

discussed by Goodfellow et al. in their seminal work [9]. GANs offer parallel generation capabilities, a unique feature not present in other generative models. Unlike models like Boltzmann machines, GANs impose fewer restrictions on the generator structure and eliminate the need for Markov chains. Additionally, GANs are recognized for their ability to produce high-quality samples that often outperform those generated by alternative algorithms.

However, GANs also exhibit certain limitations that necessitate consideration. GAN training involves achieving Nash Equilibrium, a complex task that presents challenges. The training process for GANs can be hindered by issues such as oscillation and non-convergence. Partial mode collapses within the generator can result in a limited variety of generated outputs. Imbalances between the generator and discriminator can lead to overfitting concerns.

2.2 CNN

CNNs are a type of neural network that utilizes convolutional filters to capture features in a grid-like manner, resembling the structure of the visual cortex in the human brain. This approach requires less data pre-processing compared to traditional neural networks, as demonstrated by Heaton [13]. Notably, CNNs have become dominant in the field of Computer Vision (CV) with notable works such as LeNet-5 [18], AlexNet [17], VGG [25], Inception [26], and ResNet [12]. Convolutional Neural Networks (CNNs) present a range of advantages compared to traditional neural networks. Firstly, CNNs excel in reducing the number of specified parameters, contributing to enhanced generalization and a reduced risk of overfitting. Additionally, the architecture of CNNs enables them to learn intricate features from input data via convolutional and pooling layers. Simultaneously, they efficiently carry out classification tasks using fully connected layers, ensuring an organized approach to information processing. Furthermore, CNNs simplify the process of implementing large-scale networks, making them a preferred choice for handling complex tasks in various domains.

The CNN architecture comprises different layers, which are outlined below. **Convolutional Layer** effectively captures the features of the input image by using convolutional filters, also known as kernels, to create N-dimensional activation maps. Several hyper-parameters need to be determined and optimized in this process, including the number of kernels, kernel size, activation function, stride, and padding. In Fig. 1, for example, a 3×3 filter is applied to the input image, moving with a step equal to the chosen stride. At each position, the filter's metrics are multiplied with the corresponding 3×3 spatial elements of the input image. By performing this dot product operation for the activation map can be generated. **Pooling Layer** follows the convolutional layer in order to decrease the size of feature maps and lower the dimensionality of the network. Prior to this process, the stride and kernel size need to be manually determined. Several pooling methods can be employed, such as average pooling, min pooling, max pooling and mixed pooling. **Fully Connected Layer** is positioned as the last layer in the CNN structure and functions as a classifier. In this layer, each neuron is connected to every neuron in the preceding layer. Once the convolutional layers extract features and the pooling layers down-sample the outputs, the resulting outputs are passed through the fully connected layer to generate the final outputs of CNN. It is essential to activate the fully connected layer using a non-linear function like SoftMax, Tanh, or ReLU [23].

2.3   DCGAN

Unsupervised learning through CNNs garnered noteworthy attention in 2015. A notable milestone was the inception of DCGAN, which showcased the prowess of CNNs in producing visually captivating outcomes [22]. As depicted in Fig. 2, the structure of DCGAN aligns with the foundational architecture of the original GAN. Nonetheless, DCGAN introduces certain alterations. Notably, the generator in DCGAN generates 100-dimensional noise, subsequently subjecting it to processing and transformation via convolutional layers.

For the effective integration of deep convolutional networks within GANs, a series of architectural principles have been introduced. Firstly, the discriminator replaces pooling layers with stride convolutions, enabling spatial down-sampling. In contrast, the generator employs fractional stride convolutions for spatial up-sampling. Batch normalization is integrated into every layer of both the generator and discriminator, stabilizing training, mitigating initialization issues, and promoting gradient flow to deeper layers. The adoption of deeper architectures in place of fully connected layers accelerates convergence. The output layer employs the Tanh activation function, while the ReLU activation function is used in other

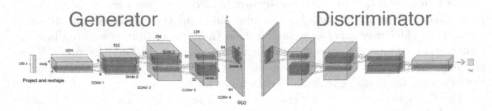

Fig. 2. The structure of DCGAN.

layers, effectively addressing saturation and covering the color space during training. LeakyReLU activation is applied across all discriminator layers, facilitating higher-resolution modeling.

DCGAN has emerged as a more stable GAN training framework, showcasing the ability to learn meaningful image representations in both supervised learning and generative modeling contexts. Despite these advancements, certain challenges persist in practice, including filter collapse and oscillating behavior, which require continued attention.

2.4 Contrastive Learning

Traditional supervised learning techniques heavily rely on annotated data, which can pose challenges when dealing with limited annotations. To address the issue of insufficient labeled data, researchers have explored alternative approaches. Self-supervised learning, a type of unsupervised learning, has gained attention for its ability to create pseudo-labels autonomously through models [19]. This empowers the training of unannotated datasets using a supervised framework, effectively mitigating annotation scarcity.

Among the various methods within self-supervised learning, contrastive learning, introduced in 2021 [14], stands out. It hinges on a multitude of negative samples and compares distinct samples to produce high-quality outcomes. This technique aims to identify both similarities and differences within a dataset, effectively categorizing data based on these attributes. Contrastive learning's primary objective is to map similar sample representations in close proximity within the embedding space, while ensuring that dissimilar representations are distanced from each other [24]. This results in the aggregation of positive sample representations and the separation of negative pairs' representations.

In the subsequent discussion, two prominent contrastive learning frameworks, SimCLR and SDCLR, will be elaborated upon.

SimCLR. Despite showing respectable performance, various self-supervised learning methods consistently fall short of achieving results comparable to supervised learning [7]. However, SimCLR, a straightforward framework that employs self-supervised contrastive learning, manages to outperform supervised learning outcomes, particularly when applied to the ImageNet dataset [6]. SimCLR operates based on three core components:

- Data Augmentation: This process begins by randomly sampling a batch of images, to which two distinct data augmentations are applied.
- Base Encoder: Utilizing ResNet-50, the encoder extracts vectors from the augmented images and generates representations through a pooling layer.
- Projection Head: Introducing a non-linear projection, typically in the form of a single-layer MLP. The loss function integrates two primary elements: cosine similarity and loss calculation.

The noticeable difference between this framework and conventional supervised learning mainly stems from the chosen data augmentation, the incorporation of a non-linear projection, and the design of the loss function. Our proposed model draws inspiration from the principles of SimCLR, as we strive to further optimize it.

SDCLR. To tackle the challenge of data imbalance, specifically the "long tail" problem, a technique known as self-damaging contrastive learning (SDCLR) has emerged to bolster the efficacy of data training [16]. This method employs two networks: the target model, which undergoes training, and the self-competitor, which acts as a pruning mechanism, ensuring consistency between the outputs of both models.

The core principle behind SDCLR is to foster representations capable of capturing nuanced differences among similar data samples. This is achieved by introducing a self-damaging mechanism during the training process. This mechanism penalizes representations that exhibit excessive similarity, even when they pertain to different views of the same data sample.

Through the implementation of SDCLR, we can fortify the robustness of data and effectively confront the complexities linked with data imbalance and the "long tail" phenomenon. Consequently, we can integrate the insights derived from this model into our own approach.

2.5   ContraD GAN

ContraD GAN represents a cutting-edge approach that seamlessly combines contrastive learning and Generative Adversarial Networks (GANs) [15]. This innovative method introduces a unique strategy by integrating the loss function from SimCLR with contrastive learning techniques. This integration empowers the discriminator to undergo training with heightened data augmentation, resulting in improved model stability and a reduced risk of discriminator overfitting.

The ContraD GAN workflow encompasses several distinct steps. Initially, data augmentation is applied to real data, generating two distinct perspectives. Concurrently, data augmentation is executed on the data produced by the GAN's generator. These viewpoints are then presented to the discriminator, yielding corresponding representations. A projection head is subsequently employed to derive corresponding vectors for these representations.

ContraD GAN has demonstrated its prowess as a successful fusion of GANs and contrastive learning, yielding impressive outcomes across a range of widely used public datasets. A primary advantage lies in its ability to train adversarial networks with enhanced data augmentation. However, it's important to note that the original work mainly emphasizes the amalgamation of SimCLR and supervised contrastive learning, with limited focus on GAN architecture design.

Consequently, the ContraD GAN framework offers substantial room for further enhancements. For instance, addressing data imbalance challenges and extending the applicability of ContraD GAN to various dataset types are promising areas that warrant exploration and expansion.

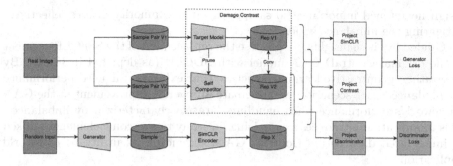

Fig. 3. The structure of Damage GAN.

3 Methodology - Damage GAN

The initial version of ContraD GAN introduced an innovative incorporation of SimCLR into the generator, resulting in a distinctive amalgamation of contrastive learning and GAN. This merger brought forth several benefits, including enhanced performance under rigorous data augmentation and the mitigation of challenges like overfitting. However, the efficacy of ContraD GAN, akin to numerous contrastive learning models, is susceptible to the characteristics of the input dataset.

ContraD GAN has showcased commendable outcomes for classification tasks when operating with sizable, balanced datasets. Nonetheless, complexities arise when confronted with diminutive, highly imbalanced datasets, particularly when classifying items from rare categories characterized by limited representation. Previous research [16] has indicated that while contrastive learning exhibits greater resilience to imbalanced data in comparison to supervised learning, it still encounters hurdles when addressing imbalances within long-tailed datasets. In real-world scenarios, data distributions frequently follow a long-tailed distribution, wherein minority classes are typically inadequately represented. This introduces the notion of "label bias," wherein the classification decision boundary is significantly influenced by the majority classes [28].

To enhance the performance of ContraD GAN in the context of imbalanced data settings, a potential strategy involves refining the discriminator component. Recent advancements in contrastive learning, drawing inspiration from the SimCLR architecture, have demonstrated superior performance compared to the original model. Consequently, exploring alternative contrastive learning architectures by substituting the SimCLR component within ContraD GAN presents a promising avenue.

An illustrative example of a relevant study addressing imbalanced data within contrastive learning is the self-damaging contrastive learning method (SDCLR) [16]. In SDCLR, a branch of the original SimCLR framework is adapted to create a pruned branch, wherein samples from minority classes are treated differently through the assignment of larger losses. This adaptation guides the model to assign heightened importance to samples from the minority classes, effectively mitigating the imbalance issue.

Guided by these insights, we propose the replacement of the SimCLR segment in the original ContraD GAN model with SDCLR (as depicted in Fig. 3). By integrating a contrastive learning structure that exhibits enhanced performance in imbalanced data scenarios, we anticipate an overall improvement in the GAN framework's performance when handling datasets characterized by imbalances. This modification introduces a promising pathway to overcome challenges linked to imbalanced data within ContraD GAN, extending its utility to real-world applications.

4 Experimental Results

4.1 Datasets

The study utilizes the CIFAR-10 dataset, comprising 60,000 entries with dimensions of 32×32 pixels. To comprehensively evaluate the proposed model, the dataset was employed in three distinct configurations:

- Full Dataset: This configuration serves as a baseline for comparison, facilitating an assessment of the proposed model's overall performance.
- Partial Dataset: To delve into the model's efficacy on imbalanced datasets, the original data was reduced to one-fifth of its original size, following the method established by Cui et al. [8]. This smaller dataset not only possesses an imbalance but also serves as the foundation for controlled experiments.
- Imbalanced Dataset: Following the generation of the partial dataset, Cui et al. introduced a balance factor to select examples, thereby constructing an imbalanced dataset. For this study, a balance factor of 100 was implemented, resulting in the largest class comprising 4,500 examples, while the smallest class contains only 45 examples.

The comparative analysis entails evaluating different GANs on the Full, Partial, and Imbalanced datasets, thus yielding insights into their respective performances across varying dataset configurations.

4.2 Evaluation Metrics

This paper employs two primary metrics, namely the Inception Score (IS) and the Fréchet Inception Distance (FID), to gauge the effectiveness of the GANs.

The Inception Score (IS) evaluates the quality of generated images [3]. This assessment involves inputting an image into a neural network, specifically Inception-v3 [27], and acquiring the output layer probabilities for each category, denoted as $p(y \mid x)$. Here, x signifies the data feature, and y represents the label. The distribution of labels is represented by $p(y)$. The IS is calculated using the following formula:

$$IS = \exp\left(\mathbb{E}_{x \sim p_g} D_{KL}(p(y \mid x) \| p(y))\right) \tag{1}$$

The primary aim is for the generator to produce diverse images containing meaningful objects, resulting in a low-entropy distribution $p(y)$ and a high-entropy distribution $p(y \mid x)$. A higher Inception Score (IS) indicates superior performance, showcasing a more substantial KL-divergence [5] between these two distributions.

Fréchet Inception Distance (FID) can be perceived as an advancement of the IS metric, as it takes into account not only the quality of samples generated by GANs but also the influence of real data [21]. FID compares the statistical attributes of generated images with those of real samples by leveraging features from Inception-v3, unlike IS, which directly provides class assignments. The FID is calculated using the subsequent formula:

$$FID = \|\mu_r - \mu_g\|_2^2 + Trace(\sum_r + \sum_g -2(\sum_r \sum_g)^{1/2}) \tag{2}$$

In this context, $x_r \sim N(\mu_r, \sum_r)$ and $x_g \sim N(\mu_g, \sum_g)$ represent the 2,048-dimensional activations of the Inception-v3 pool3 layer for real and generated samples, respectively. A reduced Fréchet Inception Distance (FID) value indicates improved performance, highlighting a higher similarity between real and generated images. This similarity is quantified by measuring the distance between their activation distributions.

4.3   Results

In this section, we present a series of experiments aimed at comparing the performance of three models: DCGAN, ContraD GAN, and our proposed model, Damage GAN. The main objectives of these experiments are as follows:

Reproducing and evaluating results for both DCGAN and ContraD GAN, while simultaneously assessing the performance of our proposed Damage GAN model using the standard CIFAR-10 dataset.

Additionally, generating two subsets from the original CIFAR-10 dataset, each containing 10,000 samples. One subset is balanced, while the other is deliberately imbalanced. By calculating FID and IS metrics across nine different scenarios (combining three GANs and three datasets), we aim to understand the implications of this data modification on performance.

Furthermore, investigating the potential disparities in class distribution when GANs operate on the imbalanced dataset. We also explore whether Damage GAN contributes to improving this distribution.

Lastly, we aim to determine whether Damage GAN has succeeded in enhancing the image quality of the imbalanced dataset. We accomplish this by comparing FID results for the two minor classes (with the largest generated samples) and the two major classes (with the smallest generated samples).

The outcomes of the aforementioned experiments are provided below:

58   A. Anaissi et al.

Table 2. FID and IS results for 3 GANs with 3 datasets.

Dataset	DCGAN			ContraD			Damage		
	FID	IS(mean)	IS(std)	FID	IS(mean)	IS(std)	FID	IS(mean)	IS(std)
Full	25.47	7.40	0.19	10.27	9.02	0.28	11.04	8.66	0.16
Partial	40.57	6.42	0.14	16.72	7.92	0.21	18.56	8.12	0.29
Imbalanced	55.15	5.71	0.16	29.92	7.43	0.16	28.45	7.95	0.15

a) The CIFAR-10 results for DCGAN and ContraD GAN align with previously published findings [15], as observed from Table 2. However, the Damage GAN takes twice as long to run and exhibits slightly inferior performance compared to ContraD GAN.

b) The FID results for all GANs deteriorate when the training sample size is reduced, and the degradation worsens in the case of imbalanced training sets. Among the GANs, the Damage GAN exhibits the lowest rate of deterioration, as depicted in Table 2. Moreover, on the imbalanced dataset, the Damage GAN outperforms the ContraD GAN by 5% in terms of scores.

Table 3. Samples of classes on Partial and Imbalanced datasets.

Dataset	air	car	bird	cat	deer	dog	frog	hrs	ship	truck	Total
Partial	1,116	1,116	1,116	1,116	1,116	1,116	1,116	1,116	1,116	1,116	11,160
Imbalanced	348	969	125	208	1,617	75	4,500	581	2,697	45	11,165

Table 4. Distribution for classes on Partial and Imbalanced datasets.

Dataset	Model	air	car	bird	cat	deer	dog	frog	hrs	ship	truck
Partial	DCGAN	0.94	0.52	1.55	0.75	1.50	0.69	1.74	0.58	1.18	0.53
	ContraD	0.96	1.23	1.10	0.90	0.85	0.74	1.11	1.13	0.93	1.02
	Damage	1.02	1.22	1.14	0.75	0.87	0.79	1.08	1.14	0.93	1.02
Imbalanced	DCGAN	0.94	0.19	4.67	1.04	0.96	0.51	0.98	0.49	0.86	1.73
	ContraD	1.35	1.05	3.32	0.85	0.75	0.89	0.81	1.11	0.83	2.11
	Damage	1.16	0.94	2.85	1.05	0.75	0.64	0.85	1.17	0.85	1.42

c) Table 3, 4 presents the deviation of the generated samples from the training samples for both Partial and Imbalanced datasets after classifying the generated samples into classes using the linear evaluator. As anticipated, the deviation for the balanced set is close to 1, with a mean value of 1. On the other hand, the imbalanced GANs display a mean deviation greater than 1, indicating under-representation of major classes and over-representation of minor classes. The

Table 5. FID scores for minor and major classes in the Imbalance dataset.

Model	Total Classes	Major Classes	Minor Classes
DCGAN	55.15		
ContraD	29.92	31.87	32.65
Damage	28.45	31.15	31.15

Damage GAN generates a distribution that closely resembles the generated distribution.

d) The FID results for the imbalanced dataset show a 5% improvement compared to the ContraD GAN score (see Table 5). The quality of FID data for the majority and minority classes is impacted by the smaller generated sample size of the minority class, but it exhibits a similar 5% improvement. In contrast, the majority class experiences a less than 2% improvement. These results align with the published SDCLR results, which analyze the accuracy of the linear evaluator, the imbalanced dataset, and the minority and majority classes. The two primary categories in this context are referred to as "frogs" and "ships," while the less common categories are "dogs" and "trucks." Due to the limited number of samples available for the less common categories, a subset of 100 samples is used when calculating FID. It has been observed that FID decreases as the sample size increases, until reaching around 5,000 samples. To determine the FID for both the majority and minority classes, 100 samples were randomly selected from each imbalanced dataset and compared with the full imbalanced datasets (ContraD GAN, Damage GAN) to establish a scale factor. This scale factor, approximately 1/16, was then applied to adjust the FID for the minority and majority classes. Upon a visual examination of the majority and minority classes, it appears that the quality of the minority classes is inferior to that of the majority classes. However, this discrepancy is not reflected in the FID scores, which raises the need for further investigation. It is possible that the issue is related to the small sample size and warrants additional exploration.

5 Conclusion

The primary objective of this paper is to enhance the performance of Generative Adversarial Networks (GANs) when confronted with imbalanced datasets that closely resemble real-world data distributions. In contrast to prior researchers who focused on altering the generator, our approach involves modifying the original GAN by replacing the discriminator with Self-Damaging Contrastive Learning (SDCLR). Comparative analyses between the baselines (ContraD GAN and DCGAN) and our Damage GAN model reveal noteworthy enhancements, particularly in terms of Fréchet Inception Distance (FID) and Inception Score (IS), with a distinct emphasis on the standard deviation of IS. This signifies that our model generates more consistent outputs compared to the baselines.

Moreover, the images produced by Damage GAN showcase improved quality for the major classes.

Nonetheless, our study does present certain limitations. Primarily, our experimentation utilized the CIFAR-10 dataset, which comprises relatively small-sized images. Future endeavors should consider evaluating our model on the GTSRB dataset, characterized by larger images. Furthermore, the dataset in our study features mutually exclusive labels, whereas real-world scenarios often involve images with multiple labels. It would be valuable to explore the applicability of our model in generating complex industry images containing diverse elements. Lastly, while Damage GAN effectively balances minor classes and enhances performance for major classes within imbalanced datasets, the FID of the minor classes increases, indicating potentially lower image quality compared to the baselines.

In conclusion, our modification presents improvements for GANs, particularly in the context of addressing imbalanced datasets, and holds promise for future advancements. Further investigations are recommended to gain deeper insights into the factors influencing FID changes across different classes. Additionally, testing the model on datasets featuring larger image sizes and quantities is suggested to validate results obtained from CIFAR-10 and to explore the model's potential in handling high-resolution images.

References

1. Anaissi, A., Suleiman, B.: B 2-fedgan: Balanced bi-directional federated gan. In: International Conference on Computational Science, pp. 380–392. Springer, Cham (2023). https://doi.org/10.1007/978-3-031-35995-8_27
2. Anaissi, A., Zandavi, S.M., Suleiman, B., Naji, M., Braytee, A.: Multi-objective variational autoencoder: an application for smart infrastructure maintenance. Appl. Intell. **53**(10), 12047–12062 (2023)
3. Barratt, S., Sharma, R.: A note on the inception score (2018). arXiv preprint arXiv:1801.01973
4. Borji, A.: Pros and cons of gan evaluation measures: new developments. Comput. Vis. Image Underst. **215**, 103329 (2022)
5. Bu, Y., Zou, S., Liang, Y., Veeravalli, V.V.: Estimation of kl divergence: optimal minimax rate. IEEE Trans. Inf. Theory **64**(4), 2648–2674 (2018)
6. Chen, T., Kornblith, S., Norouzi, M., Hinton, G.: A simple framework for contrastive learning of visual representations. In: International Conference on Machine Learning, pp. 1597–1607. PMLR (2020)
7. Chowdhury, A., Rosenthal, J., Waring, J., Umeton, R.: Applying self-supervised learning to medicine: review of the state of the art and medical implementations. In: Informatics, vol. 8, p. 59. MDPI (2021)
8. Cui, Y., Jia, M., Lin, T.Y., Song, Y., Belongie, S.: Class-balanced loss based on effective number of samples. In: Proceedings of the IEEE/CVF Conference on Computer Vision and Pattern Recognition, pp. 9268–9277 (2019)
9. Goodfellow, I.: Nips 2016 tutorial: Generative adversarial networks (2016). arXiv preprint arXiv:1701.00160
10. Goodfellow, I.J., Shlens, J., Szegedy, C.: Explaining and harnessing adversarial examples. arXiv preprint arXiv:1412.6572 (2014)

11. Gui, J., Sun, Z., Wen, Y., Tao, D., Ye, J.: A review on generative adversarial networks: algorithms, theory, and applications. IEEE Trans. Knowl. Data Eng. (2021)
12. He, K., Zhang, X., Ren, S., Sun, J.: Delving deep into rectifiers: surpassing human-level performance on imagenet classification. In: Proceedings of the IEEE International Conference on Computer Vision, pp. 1026–1034 (2015)
13. Heaton, J., Goodfellow, I., Bengio, Y., Courville, A.: Deep learning: The mit press (2016). 800 pp, isbn: 0262035618. Genetic Programming and Evolvable Machines 19(1-2), 305-307 (2018)
14. Jaiswal, A., Babu, A.R., Zadeh, M.Z., Banerjee, D., Makedon, F.: A survey on contrastive self-supervised learning. Technologies 9(1), 2 (2020)
15. Jeong, J., Shin, J.: Training gans with stronger augmentations via contrastive discriminator (2021). arXiv preprint arXiv:2103.09742
16. Jiang, Z., Chen, T., Mortazavi, B.J., Wang, Z.: Self-damaging contrastive learning. In: International Conference on Machine Learning, pp. 4927–4939. PMLR (2021)
17. Krizhevsky, A., Sutskever, I., Hinton, G.E.: Imagenet classification with deep convolutional neural networks. In: Advances in Neural Information Processing Systems 25 (2012)
18. LeCun, Y., Bottou, L., Bengio, Y., Haffner, P.: Gradient-based learning applied to document recognition. Proc. IEEE 86(11), 2278–2324 (1998)
19. Liu, R.: Understand and improve contrastive learning methods for visual representation: areview (2021). arXiv preprint arXiv:2106.03259
20. Naji, M., Anaissi, A., Braytee, A., Goyal, M.: Anomaly detection in x-ray security imaging: a tensor-based learning approach, p. 1–8 (2021)
21. Obukhov, A., Krasnyanskiy, M.: Quality assessment method for gan based on modified metrics inception score and fréchet inception distance. In: Software Engineering Perspectives in Intelligent Systems: Proceedings of 4th Computational Methods in Systems and Software 2020, vol. 1 4, pp. 102–114. Springer (2020)
22. Radford, A., Metz, L., Chintala, S.: Unsupervised representation learning with deep convolutional generative adversarial networks (2015). arXiv preprint arXiv:1511.06434
23. Sharma, S., Sharma, S., Athaiya, A.: Activation functions in neural networks. Towards Data Sci. 6(12), 310–316 (2017)
24. Shim, J., Kang, S., Cho, S.: Active cluster annotation for wafer map pattern classification in semiconductor manufacturing. Expert Syst. Appl. 183, 115429 (2021)
25. Simonyan, K., Zisserman, A.: Very deep convolutional networks for large-scale image recognition (2014). arXiv preprint arXiv:1409.1556
26. Szegedy, C., Reed, S., Erhan, D., Anguelov, D., Ioffe, S.: Scalable, high-quality object detection (2014). arXiv preprint arXiv:1412.1441
27. Xia, X., Xu, C., Nan, B.: Inception-v3 for flower classification. In: 2017 2nd International Conference on Image, Vision and Computing (ICIVC), pp. 783–787. IEEE (2017)
28. Yang, Y., Xu, Z.: Rethinking the value of labels for improving class-imbalanced learning. Adv. Neural. Inf. Process. Syst. 33, 19290–19301 (2020)
29. Yao, Y., et al.: Conditional variational autoencoder with balanced pre-training for generative adversarial networks. In: 2022 IEEE 9th International Conference on Data Science and Advanced Analytics (DSAA), pp. 1–10. IEEE (2022)
30. Zhou, Y., et al.: Vgg-fusionnet: a feature fusion framework from CT scan and chest x-ray images based deep learning for covid-19 detection. In: 2022 IEEE International Conference on Data Mining Workshops (ICDMW), pp. 1–9. IEEE (2022)

Text-Conditioned Graph Generation Using Discrete Graph Variational Autoencoders

Michael Longland[1,3,4]($\boxtimes$), David Liebowitz[1,2], Kristen Moore[3,4], and Salil Kanhere[1]

[1] The University of New South Wales, Sydney, Australia
{m.longland,salil.kanhere}@unsw.edu.au
[2] Penten, Canberra, Australia
david.liebowitz@penten.com
[3] CSIRO's Data61, Eveleigh, Australia
kristen.moore@data61.csiro.au
[4] Cyber Security Cooperative Research Centre, Joondalup, Australia

Abstract. Inspired by recent progress in text-conditioned image generation, we propose a model for the problem of text-conditioned graph generation. We introduce the Vector Quantized Text to Graph generator (VQ-T2G), a discrete graph variational autoencoder and autoregressive transformer for generating general graphs conditioned on text. We curate two multimodal datasets of graph-text pairs, a real-world dataset of subgraphs from the Wikipedia link network and a dataset of diverse synthetic graphs. Experimental results on these datasets demonstrate that VQ-T2G synthesises novel graphs with structure aligned with the text conditioning. Additional experiments in the unconditioned graph generation setting show VQ-T2G is competitive with existing unconditioned graph generation methods across a range of metrics.

Keywords: Graph generation · Generative modelling · Graph neural networks · Multimodal modelling

1 Introduction

Graphs are a natural way of representing relational and structural data such as molecules, social networks and knowledge graphs. The range of data types they may represent makes the problem of learning and generating graphs important, with broad applications, for example in drug [5,25] or protein [15] design and network science [17,21]. The study of graph generation dates back to the 1950s with the Erdős-Rényi random graph model [9]. This and similar early work in network science [1,12,28] focused on models that generate a single class of graphs with known statistical properties, with few parameters to control the structure. Because they are simple and not learned from data, these models have limited

© The Author(s), under exclusive license to Springer Nature Singapore Pte Ltd. 2024
D. Benavides-Prado et al. (Eds.): AusDM 2023, CCIS 1943, pp. 62–74, 2024.
https://doi.org/10.1007/978-981-99-8696-5_5

capacity to mimic the complex dependencies and structures frequently observed in real graphs [2].

Modern approaches to graph generation use neural networks to learn a model over a dataset of graphs, and then sample novel graphs from the model that mimic those in the dataset. Such models demonstrate far stronger ability to synthesise graphs from areas including molecules, citation networks, 3D point clouds, and varieties of synthetic graphs [14,15,23,25,31]. While these approaches have proven powerful, many of these generators lack capacity for controllable generation of general graphs. This limits their potential use in real-world applications where it may be desirable to have samples aligned with some provided context or conditioning, such as for molecule design [8].

Inspired by progress in conditional image generation from text [19,20,22,32], we propose the novel task of conditional *graph generation* from text. Replicating the success of such multi-modal generative systems in the graph domain requires that the essential computational elements of those architectures be redesigned such that it can use graph data. This is not a trivial task due to the non-Euclidean nature of graphs. No existing work to date has studied multimodal modelling for graphs and text in in this context. We address the gap in this paper. Specifically, we investigate the novel problem of generating graphs conditioned on text, *text-to-graph generation*. That is, given a dataset of paired graphs G_i and text samples T_i, $\mathbf{G} = \{(G_1, T_1), (G_2, T_2), \ldots (G_m, T_m)\}$, jointly learning over both modalities to generate novel and realistic graphs that are aligned with given conditioning text. This alignment can refer to two settings. The first is where the graph structure is explicitly described by the caption. For example, the text specifies one or more attributes such as the graph's size, type, or structural metrics. The second setting is where the graph is associated or semantically related to the caption in some manner, but does not include direct information about the graph structure.

There are three key challenges in this task. First, graphs have irregular structure. Graphs in a dataset may contain differing numbers of nodes or edges, along with varied topological features at both the local and global level. The varied input dimension poses issues for many approaches, including the VQ-VAEs that we employ. Second, large-scale datasets of graphs and text that are appropriate for this graph synthesis task do not yet exist, so we must curate them. Finally, there are no robust evaluation methods for text-to-graph generation. We must modify existing methods for evaluating unconditioned graph generative models to be accurate and sufficient methods in this text conditioned setting.

Our proposed model for this problem is named the Vector-Quantized Text-to-Graph Generator (VQ-T2G). It is a generative model for graphs that outputs diverse and realistic graphs aligned with conditioning text. The model utilises an adaptation of the vector-quantized variational autoencoder (VQ-VAE) [26] to encode and reconstruct graph-structured data, followed by an autoregressive prior trained over both the discrete latents and text captions for text-conditioned sampling. In summary, our contributions are:

- An adaptation of the VQ-VAE framework for graph structured data that learns to encode then reconstruct graphs through a discrete bottleneck. In particular, it is able to handle graphs with varying sizes.
- VQ-T2G, a model for the proposed text-to-graph generation task based on this VQ-VAE for graphs, along with a multimodal autoregressive prior learned over graphs and text to allow text conditioning.
- Curation of two datasets of graph-text pairs.

Datasets and code are available at https://github.com/longland-m/vqt2g.

2   Related Work

2.1   Deep Learning on Graphs

Graph neural networks (GNN) have proven powerful in a variety of applications such as recommender systems and social network community detection. While many GNN architectures exist in the literature, the most relevant to our work is VQ-GNN [7], which, to date, is the only model that uses vector quantization in a GNN. It is designed for node representation learning in large graphs, introducing discrete representations in an approximated message passing scheme to improve memory efficiency and scalability. As the design targets large-graph node-level learning and the approximated message passing doesn't carry over to graph-level learning or reconstruction, VQ-GNN cannot be used in graph generation.

2.2   Unconditioned Graph Generation

In the unconditioned setting, graph generators aim to learn a distribution $p(G)$ over graphs from a dataset of observed graphs, then sample from this distribution to synthesise novel graphs similar to those in the training set. Autoregressive models such as GraphRNN [31] and GRAN [14] approach graph generation as an autoregressive decision process, adding nodes and/or edges to build graphs sequentially. Models such as GraphVAE [25] instead generate the entire graph adjacency matrix in a single step.

2.3   Conditioned Graph Generation

Including some type of conditioning in the graph generation process may be useful in applications where more model control is desirable. VQ-T2G falls into this category of graph generator, with the type of conditioning being natural language text. Many other varieties of conditioning have been explored in prior work. SPECTRE [15] generates graphs in two stages. Part of the generated graph spectrum is first sampled, with the model learning to sample realistic spectra during training. The graph structure is then generated conditioned on this sampled spectrum. The MolT5 [8] model allows for controllable generation of molecules through conditioning on a natural language description of a molecule's desired characteristics, such as its physical properties. While apparently similar

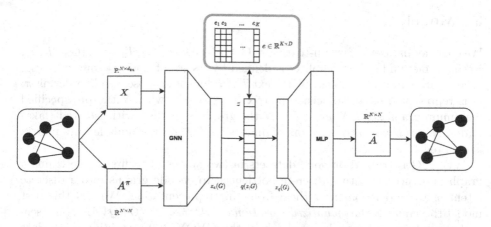

Fig. 1. Architecture of the GVQVAE model trained during the first stage of VQ-T2G training. The input graph (left) is represented under node ordering π as adjacency matrix A^π. This adjacency matrix and the graph's node feature matrix X are the input to the encoder. The graph is encoded then the output mapped to the nearest codebook vectors before being input to the decoder, to construct probabilistic adjacency matrix $\tilde{A}$. The graph is then reconstructed from this probabilistic matrix.

to VQ-T2G in the sense that the conditioning is text-based, MolT5 only operates on a single data modality through the SMILES representation. Since it does not consider graph structure, it remains fundamentally different to our model. Condgen [29] addresses the most similar setting to VQ-T2G. This model uses a conditional generative adversarial network (GAN) architecture to condition graph generation on a real-valued vector of graph level features. This is done through concatenating a continuous graph embedding with one or more conditioning vectors associated with the graph. For example, in a dataset of citation graphs, generated graphs could be conditioned on the publication venue and other citation metadata. The metadata is represented as a real-valued vector by conversion to an appropriate numeric value or one-hot encoding. In contrast with our setting of natural language conditioning, these features used in Condgen are relatively simple.

2.4 Text-to-Graph Problems

Problems known as *text to graph generation* exist in two other domains. These are knowledge graphs [10,27], and scene graphs (in settings where text is used in the graph construction) [3,30]. In contrast with our goal of generating novel graphs, these tasks perform graph *conversion* [10] or graph *extraction* [16] from text. In other words, models for these tasks aim to parse and infer relationships between entities then construct a graph reflecting those relationships. Our task is fundamentally different in that we study the generation of *novel* graphs from text.

3   Model

We consider datasets of graph-text pairs $\mathbf{G} = \{(G_1, T_1), (G_2, T_2), \ldots, (G_m, T_m)\}$. Each simple, undirected graph G is defined by its set of nodes V and edges E. Graphs are paired (captioned) with text T. Under some chosen node ordering π, G is represented as an adjacency matrix $A^\pi \in \mathbb{R}^{N \times N}$ with N the pre-specified maximum graph size. When $|V| < N$ the graph is padded with isolated (fake) nodes. Nodes have feature vectors of dimension d_{in} and the node feature matrix is denoted $X \in \mathbb{R}^{N \times d_{in}}$.

VQ-T2G is trained on such datasets in two stages. The first stage trains a graph autoencoder with a discrete bottleneck to encode graphs into a discrete latent space and reconstruct them from this representation. We call this first model the *graph vector-quantized variational autoencoder (GVQVAE)*. The second stage learns to sample graphs from the GVQVAE by training a separate model, an autoregressive transformer, over the discrete latent space learned in the GVQVAE. This second stage model may also learn to use text as conditioning for graphs. Figure 1 shows the architecture of the GVQVAE model. It takes as input a graph adjacency matrix A^π and feature matrix X and learns to reconstruct the graph structure. The decoder outputs a probabilistic adjacency matrix $\tilde{A}$ representing the predicted graph.

3.1   Graph Vector-Quantized Variational Autoencoder

We use a VAE in the GVQVAE architecture, with a discrete bottleneck to encode G into a discrete latent vector z, then reconstruct the original graph from this representation.

The encoder is defined by a variational posterior $q_\phi(z|G)$ and the decoder by a generative distribution $p_\theta(G|z)$. Each element of z maps to a continuous vector in the model's codebook of size $K \in \mathbb{N}$. The codebook is an embedding space $e \in \mathbb{R}^{K \times D}$, with K the size of the codebook, D the dimension of each codebook vector, and e_i denoting the i'th embedding vector. As in [26], encoder outputs $z_e(G)$ are mapped to their nearest embedding vector to produce a vector of codebook indices. During the forward pass the decoder takes as input the encoding of the graph mapped to the nearest codebook vectors, $z_q(G)$. However the operation mapping encoder outputs to codebook vectors has no defined gradient, so in the backwards pass the gradient skips the codebook and is passed directly to the encoder.

The GVQVAE loss function $\mathcal{L}_G$ is similar to the original VQ-VAE loss with the first term replaced with a graph reconstruction loss:

$$\mathcal{L}_G = \log p(G|z_q(G)) + \|\mathrm{sg}[z_e(G)] - e\|_2^2 + \beta\|z_e(G) - \mathrm{sg}[e]\|_2^2 \tag{1}$$

Here, sg denotes a stopgradient operator. This operator is defined as identity during forward pass. During the backward pass it blocks gradients from passing through its argument, as such its argument is treated as a constant. The first term is the graph reconstruction loss, measured with binary cross-entropy loss. The second term is the VQ loss. This encourages embedding vectors to move closer to encoder outputs, as measured by l_2 distance. The final term is a commitment loss which encourages the encoder outputs to not stray too far from codebook vectors. Since backpropagation bypasses the discrete bottleneck, only the second term optimises the codebook. The first term is optimised by both the encoder and decoder, and the final term is optimised by the encoder.

We now describe the graph specific architectures of the encoder and decoder, which both differ substantially from their image domain counterparts.

Encoder. The encoder begins with message passing GraphSAGE [11] layers with ReLU activation and BatchNorm to obtain an embedding for each node. The node embeddings are then concatenated to a single vector and projected using a linear layer to produce a graph-level latent representation. This representation is then reshaped to N_Z vectors of dimension D which is then mapped to the nearest codebook codes. The hyperparameter N_Z denotes the number of codes in the latent representation; the use of linear layers means N_Z need not be equal to the number of nodes in the original graph. This assists scalability to larger graphs as, for example, a dataset with $N = 500$ may use $\|z\| = 100$ codes to represent the graph.

The initial node feature vector for the i-th node of a graph G is the concatenation of three features. The first is the i-th row of the graph adjacency matrix A^π; graphs smaller than the maximum size N have this vector padded with -1's. Second is the node degree, and the final features are randomly sampled from a Gaussian distribution and are resampled during each forward pass of the encoder.

Decoder. The decoder learns to reconstruct the graph adjacency matrix from the discrete representations. It uses a multi-layer perceptron (MLP) architecture with four layers, having Tanh activation and dropout after each layer. The output of the final layer is reshaped to a tensor $\mathbf{R} \in \mathbb{R}^{N \times d_{out}}$, corresponding to individual node representations of dimension d_{out}. An inner product followed by sigmoid is applied to this tensor to obtain a probabilistic adjacency matrix $\hat{\mathbf{A}} = \sigma\left(\mathbf{R}\mathbf{R}^\top\right) \in \mathbb{R}^{N \times N}$ from which the final adjacency matrix $\mathbf{A}$ is assembled.

VQ Bottleneck. The bottleneck is simply the codebook, or embedding space, $e \in \mathbb{R}^{K \times D}$. These are latent vectors the encoder outputs are mapped to in the forward pass before being passed to the decoder. The indices of these vectors will correspond to tokens in the vocabulary of the transformer during the second stage of training. The *codebook collapse* problem in vector-quantized models is a common [6,13] issue and refers to the tendency of these models to end up utilising only one or few codebook vectors, thereby limiting the model's overall representational capacity. To alleviate this issue we follow [13] and batchnorm the output of the encoder before assignment to codebook vectors, and set the learning rate of the codebook to be ten times that of the encoder and decoder parameters. Empirically we find this successful in promoting good codebook usage.

3.2 Transformer Decoder

The autoregressive transformer learns the prior $p(z)$ over the discrete latents from encoded graphs, with associated text captions as conditioning. It uses a combined vocabulary of text tokens and graph tokens (i.e. indices of the codebook vectors). For a *(graph, text)* pair, the text is byte-pair encoded and concatenated with the vector of graph tokens from the GVQVAE encoder to form a single sequence. The transformer uses a decoder-only architecture similar to GPT-2 [18]. We train the tokenizer and transformer from scratch, rather than using a pretrained language model checkpoint, to keep model size small. Text is lowercased prior to byte-pair encoding. The number of text token positions in the model is set to the maximum encoded length in any caption; encoded texts shorter than this are padded to the max length with a text padding token.

To generate text-conditioned graphs following training, text captions are input to the transformer following byte-pair encoding and padding to the maximum text token length. All graph tokens are then generated autoregressively using this transformer, and these are mapped back to their corresponding codebook vectors from the GVQVAE. This is used as input for the GVQVAE decoder, from which the graph is generated.

4 Datasets

We have curated two graph-text datasets to evaluate our model. Both are significantly larger than existing datasets used to evaluate general graph generative models. The first is a real-world dataset of egocentric networks from the English Wikipedia page link network, the other is a dataset of synthetic graphs. Texts in the synthetic dataset contain explicit descriptions of the graph's structure, while the Wikipedia dataset has text associated with the central node. These each correspond to a distinct graph-text alignment setting outlined in Sect. 1.

4.1 Graph-Text Paired Datasets

Wikipedia Ego Nets. Graphs in the Wikipedia dataset consist of two-hop ego networks from the English Wikipedia inter-page link network. Nodes correspond to articles and an edge indicates an in-text link between pages. We use the cleaned 2018 link network from WikiLinkGraphs [4] and do not consider edge direction. The graphs are paired with the concatenation of the ego node's article title and first sentence. There are 8000 graph-text pairs with graph sizes $60 \leq |V| \leq 160$, and text length $26 \leq |V| \leq 571$. To handle the neighbor explosion problem common in large graph sampling, we construct eight subgraphs of the full link network to instead sample from. We traverse sections of the Wikipedia category hierarchy, ensuring little to no overlap, take the induced subgraph of pages in these visited categories and sample 1000 graphs from each. The topology of the link structure is known to differ across Wikipedia, such as links in *Mathematics* being sparse compared to a dense core observed in *Physics* [24]. This structural diversity implies some relationship between the semantic content of articles and the link network from which we may learn.

Synthetic Graphs. The synthetic dataset consists of sixteen distinct varieties of common synthetic graphs. Both graphs and text are constructed methodically. The dataset has a total of 3186 graph-text pairs, with graph sizes $20 \leq |V| \leq 160$ and text length $50 \leq |V| \leq 228$. Texts include a description of the graph variety along with specific attributes such as the parameters used to generate it or the number of nodes. While rule-based methods are sufficient to construct the graphs in this dataset, it is useful for understanding our model. The explicit relationships between texts and graphs allows easy visual evaluation of the text and graph alignment.

4.2   Graph Datasets

To compare with existing graph generation models we also experiment in the unconditioned setting with three graph-only datasets. These are: (1) A two-community graph dataset [15,31] with $60 \leq |V| \leq 160$, (2) The Wikipedia ego graph dataset introduced above (excluding the texts), (3) A dataset of protein graphs with $100 \leq |V| \leq 500$.

5   Experiments

Table 1. Comparison with Condgen in the text-conditioned setting on the (graph, text) datasets. Smaller MMD scores are better. Degree: degree distribution, Clust.: clustering coefficient, Orbit: 4-node orbit counts, Spect.: graph Laplacian spectrum.

	Synthetic				Wiki ego			
	Degree	Clust	Orbit	Spect	Degree	Clust	Orbit	Spect
Condgen	0.33	$9.0e^{-2}$	$8.9e^{-4}$	0.32	0.16	0.19	$9.0e^{-2}$	0.10
VQ-T2G	$8.5e^{-2}$	$4.5e^{-2}$	$4.5e^{-2}$	$2.3e^{-2}$	$4.7e^{-2}$	$3.2e^{-2}$	$8.0e^{-2}$	$1.2e^{-2}$

We empirically verify the ability of VQ-T2G to generate graphs from text captions, as well as its performance in unconditioned generation. For text-conditioned experiments we evaluate with distributions of graph statistics between generated graphs and the test set. We also perform visual inspection for generated graphs in the synthetic dataset. When training VQ-T2G we use a data split of 90% of data for training and 10% for testing, sharing the split in both stages of training. For text-conditioned experiments we use Condgen [29] as a baseline to compare with. For Condgen we encode the texts as a continuous vector and use this as the conditioning vector. Other existing graph generation models cannot readily incorporate text conditioning.

Table 2. Results of unconditioned graph generation of VQ-T2G against existing graph generator models. Smaller MMD scores are better. Degree: degree distribution, Clust.: clustering coefficient, Orbit: 4-node orbit counts, Spect.: graph Laplacian spectrum.

	Two-community				Wiki ego (graphs only)				Proteins			
	Degree	Clust	Orbit	Spect.	Degree	Clust	Orbit	Spect.	Degree	Clust	Orbit	Spect
GRAN	0.23	**$3.9e^{-2}$**	$8.5e^{-2}$	**$2.4e^{-2}$**	**$6.7e^{-3}$**	$2.7e^{-2}$	$1.3e^{-2}$	$3.3e^{-2}$	$2.0e^{-3}$	**$4.7e^{-2}$**	0.13	$5.1e^{-3}$
SPECTRE	0.29	$5.6e^{-2}$	**$2.5e^{-2}$**	0.23	0.44	$4.0e^{-2}$	$2.0e^{-2}$	0.23	**$1.3e^{-3}$**	**$4.7e^{-2}$**	**$2.9e^{-2}$**	**$2.0e^{-3}$**
Condgen	**0.19**	$6.5e^{-2}$	$4.5e^{-2}$	0.1	$5.0e^{-2}$	$1.6e^{-2}$	$1.1e^{-2}$	$5.9e^{-2}$	-	-	-	-
VQ-T2G	0.23	$6.0e^{-2}$	$3.4e^{-2}$	$8.1e^{-2}$	$1.5e^{-2}$	**$1.4e^{-2}$**	**$7.1e^{-3}$**	**$1.6e^{-2}$**	0.14	0.12	0.36	$4.5e^{-2}$

5.1 Evaluation Metrics

Our metrics follow those commonly used in the evaluation of unconditional graph generators [14,15,23,31]. That is, the maximum mean discrepancy (MMD) over the: degree distribution, clustering coefficient distribution, count of all orbits with 4 nodes, and eigenvalues of the normalised graph Laplacian. We use the total variation (TV) distance kernel in the MMD due to its speed and scalability. We compute these statistics between graphs in the test set (the ground-truth) and graphs generated by the model, one generated from each of the test set texts. This setup will demonstrate the ability of a model to align generated graphs with unseen texts. Low MMD scores indicate a model has learned well.

5.2 Results

Text-Conditioned Graph Generation. Results in experiments on the graph-text datasets, in the text-conditioned setting, are reported in Table 1. VQ-T2G strongly outperforms Condgen on all but one measured MMD metric. These low scores imply our model successfully learns to align generated graphs with the text conditioning, significantly better than the baseline.

Unconditioned Graph Generation. We next evaluate the model in the unconditioned graph generation setting. We again compare against Condgen [29] along with the unconditioned models GRAN [14] and SPECTRE [15]. To train VQ-T2G in the unconditioned setting we train the GVQVAE as normal, then train the transformer over only graph tokens (i.e. a text length and text vocabulary of zero). For the unconditioned Wiki ego dataset we reuse the GVQVAE from the conditioned experiments and only train the transformer separately. For Condgen we simply leave feature vectors empty. Note we were unable to get Condgen to work with the proteins dataset. Of most note are the two-community and the Wiki ego datasets; we include the proteins dataset to highlight current limitations of the VQ-T2G model. Table 2 lists the unconditioned results. We again generate the same number of graphs as there are in the test set and compute MMD scores between these. Notably, VQ-T2G performs well on the Wiki ego dataset but fails to beat other models in the two-community and proteins datasets. We expect this is due to the smaller dataset sizes. It is difficult to train transformer models with small datasets, it was found to be easy to overfit. Ensuring sufficient graph diversity from the unconditioned sampling was also difficult. However it did perform well for the unconditioned Wiki ego dataset, which is more evidence that dataset size helps VQ-T2G performance significantly. For the proteins dataset we point out the much lower performance of our model compared to baselines. The current architecture of VQ-T2G has difficulty scaling to large graphs, and this, combined with the small size of the proteins dataset, leads to poor generalisation to unseen graphs.

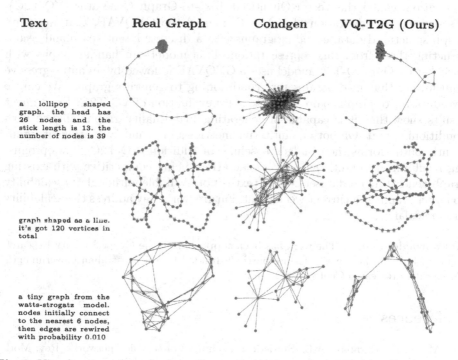

Text	Real Graph	Condgen	VQ-T2G (Ours)

a lollipop shaped graph. the head has 26 nodes and the stick length is 13. the number of nodes is 39

graph shaped as a line. it's got 120 vertices in total

a tiny graph from the watts-strogatz model. nodes initially connect to the nearest 6 nodes, then edges are rewired with probability 0.010

Fig. 2. Visualization of graphs from the synthetic dataset generated by Condgen and VQ-T2G, along with the text used to generate the graph and the real graph paired with that text.

Visual Evaluation. Examples of generated graphs and their conditioning text from the synthetic dataset are in Fig. 2. We compare samples from our model against those generated by Condgen, along with the ground-truth graph matching the caption. While it is clear that neither of the models generate graphs that exactly match the text, the graphs generated by VQ-T2G are far closer to the real graph than those generated by Condgen. It is unsurprising that Condgen's results are not as accurate as the model was not designed for text conditioning. VQ-T2G is able to generate graphs that are approximately the correct topology but still has much room to improve. Longer and more precise text descriptions could assist with this. Also of note is that the conditioning text contains the number of nodes of the graph as an integer. This is likely a non-optimal part to include in the dataset and may be another partial reason for VQ-T2G's performance.

6 Conclusion

In this paper we propose a model for the novel problem of text-conditioned graph generation, called the Vector-Quantized Text-to-Graph Generator (VQ-T2G). We develop an adaptation of VQ-VAE, named the GVQVAE, that works on graph structured data, encoding graphs into a discrete latent space and reconstructing them from this representation. This model can handle graphs with varied sizes. Our VQ-T2G model uses a GVQVAE followed by an autoregressive transformer that incorporates text conditioning to generate graphs. We curate two datasets of graphs paired with text for evaluation of VQ-T2G. Experiment results show that it is capable of generating high-quality graphs aligned with conditioning text. On both quantitative metrics and visual evaluations it significantly outperforms the Condgen baseline. In addition, VQ-T2G shows promising results in unconditioned graph generation, being competitive with existing graph generators on datasets where text is not available, although its scalability to larger graphs is a current weak point. Future work may address this scalability in particular.

Acknowledgements. The work has been supported by the Cyber Security Research Centre Limited whose activities are partially funded by the Australian Government's Cooperative Research Centres Programme.

References

1. Albert, R., Barabási, A.L.: Statistical mechanics of complex networks. Rev. Mod. Phys. **74**(1), 47 (2002)
2. Broido, A.D., Clauset, A.: Scale-free networks are rare. Nat. Commun. **10**(1), 1–10 (2019)
3. Chang, X., Ren, P., Xu, P., Li, Z., Chen, X., Hauptmann, A.: Scene graphs: a survey of generations and applications. arXiv preprint arXiv:2104.01111 (2021)
4. Consonni, C., Laniado, D., Montresor, A.: Wikilinkgraphs: a complete, longitudinal and multi-language dataset of the Wikipedia link networks. In: Proceedings of the International AAAI Conference on Web and Social Media, vol. 13, pp. 598–607 (2019)
5. De Cao, N., Kipf, T.: MolGAN: an implicit generative model for small molecular graphs. arXiv preprint arXiv:1805.11973 (2018)
6. Dhariwal, P., Jun, H., Payne, C., Kim, J.W., Radford, A., Sutskever, I.: Jukebox: a generative model for music. arXiv preprint arXiv:2005.00341 (2020)
7. Ding, M., Kong, K., Li, J., Zhu, C., Dickerson, J., Huang, F., Goldstein, T.: VQ-GNN: a universal framework to scale up graph neural networks using vector quantization. In: Advances in Neural Information Processing Systems, vol. 34 (2021)

8. Edwards, C., Lai, T., Ros, K., Honke, G., Ji, H.: Translation between molecules and natural language. arXiv preprint arXiv:2204.11817 (2022)
9. Erdős, P., Rényi, A., et al.: On the evolution of random graphs. Publ. Math. Inst. Hung. Acad. Sci **5**(1), 17–60 (1960)
10. Guo, Q., Jin, Z., Qiu, X., Zhang, W., Wipf, D., Zhang, Z.: Cyclegt: unsupervised graph-to-text and text-to-graph generation via cycle training. arXiv preprint arXiv:2006.04702 (2020)
11. Hamilton, W., Ying, Z., Leskovec, J.: Inductive representation learning on large graphs. In: Advances in Neural Information Processing Systems, vol. 30 (2017)
12. Holland, P.W., Laskey, K.B., Leinhardt, S.: Stochastic blockmodels: first steps. Soc. Netw. **5**(2), 109–137 (1983)
13. Łańcucki, A., et al.: Robust training of vector quantized bottleneck models. In: 2020 International Joint Conference on Neural Networks (IJCNN), pp. 1–7. IEEE (2020)
14. Liao, R., et al.: Efficient graph generation with graph recurrent attention networks. In: Advances in Neural Information Processing Systems, pp. 4257–4267 (2019)
15. Martinkus, K., Loukas, A., Perraudin, N., Wattenhofer, R.: Spectre: spectral conditioning helps to overcome the expressivity limits of one-shot graph generators. arXiv preprint arXiv:2204.01613 (2022)
16. Melnyk, I., Dognin, P., Das, P.: Grapher: multi-stage knowledge graph construction using pretrained language models. In: NeurIPS 2021 Workshop on Deep Generative Models and Downstream Applications (2021)
17. Newman, M.: Networks. Oxford University Press, Oxford (2018)
18. Radford, A., Wu, J., Child, R., Luan, D., Amodei, D., Sutskever, I.: Language models are unsupervised multitask learners. OpenAI Blog **1**(8), 9 (2019)
19. Ramesh, A., Dhariwal, P., Nichol, A., Chu, C., Chen, M.: Hierarchical text-conditional image generation with clip latents. arXiv preprint arXiv:2204.06125 (2022)
20. Ramesh, A., et al.: Zero-shot text-to-image generation. In: International Conference on Machine Learning, pp. 8821–8831. PMLR (2021)
21. Robins, G., Pattison, P., Kalish, Y., Lusher, D.: An introduction to exponential random graph (p*) models for social networks. Soc. Netw. **29**(2), 173–191 (2007)
22. Rombach, R., Blattmann, A., Lorenz, D., Esser, P., Ommer, B.: High-resolution image synthesis with latent diffusion models (2021)
23. Shirzad, H., Hajimirsadeghi, H., Abdi, A.H., Mori, G.: TD-GEN: graph generation using tree decomposition. In: International Conference on Artificial Intelligence and Statistics, pp. 5518–5537. PMLR (2022)
24. Silva, F.N., Viana, M.P., Travençolo, B.A.N., Costa, L.d.F.: Investigating relationships within and between category networks in Wikipedia. J. Informetrics **5**(3), 431–438 (2011)
25. Simonovsky, M., Komodakis, N.: GraphVAE: towards generation of small graphs using variational autoencoders. In: Kůrková, V., Manolopoulos, Y., Hammer, B., Iliadis, L., Maglogiannis, I. (eds.) ICANN 2018. LNCS, vol. 11139, pp. 412–422. Springer, Cham (2018). https://doi.org/10.1007/978-3-030-01418-6_41
26. Van Den Oord, A., Vinyals, O., et al.: Neural discrete representation learning. In: Advances in Neural Information Processing Systems, vol. 30 (2017)
27. Wang, L., Li, Y., Aslan, O., Vinyals, O.: Wikigraphs: a Wikipedia text-knowledge graph paired dataset. arXiv preprint arXiv:2107.09556 (2021)
28. Watts, D.J., Strogatz, S.H.: Collective dynamics of 'small-world' networks. Nature **393**(6684), 440–442 (1998)

29. Yang, C., Zhuang, P., Shi, W., Luu, A., Li, P.: Conditional structure generation through graph variational generative adversarial nets. In: NeurIPS, pp. 1338–1349 (2019)
30. Yang, X., Tang, K., Zhang, H., Cai, J.: Auto-encoding scene graphs for image captioning. In: Proceedings of the IEEE/CVF Conference on Computer Vision and Pattern Recognition, pp. 10685–10694 (2019)
31. You, J., Ying, R., Ren, X., Hamilton, W.L., Leskovec, J.: GraphRNN: generating realistic graphs with deep auto-regressive models. arXiv preprint arXiv:1802.08773 (2018)
32. Yu, J., et al.: Scaling autoregressive models for content-rich text-to-image generation. arXiv preprint arXiv:2206.10789 (2022)

Boosting QA Performance Through SA-Net and AA-Net with the Read+Verify Framework

Liang Tang[1], Qianqian Qi[2], Qinghua Shang[1(✉)], Yuguang Cai[1], Jiamou Liu[2], Michael Witbrock[2], and Kaokao lv[1]

[1] Qi An Xin Technology, Beijing, China
{shangqinghua,caiyuguang}@qianxin.com
[2] The University of Auckland, Auckland, New Zealand
qqi518@aucklanduni.ac.nz, {jiamou.liu,m.witbrock}@auckland.ac.nz

Abstract. This paper proposes an ensemble model for the Stanford Question Answering Dataset (SQuAD) with the aim of improving performance compared to baseline models such as Albert, and Electra. The proposed ensemble model incorporates Sentence Attention (SA-Net) and Answer Attention (AA-Net) components, which leverage attention mechanisms to emphasize important information in sentences and answers, respectively. Additionally, the model adopts a read+verify architecture. In the Read stage, the model's focus is on accurately predicting answer text, while in the Verify stage, it emphasizes the ability to determine the presence or absence of an answer, providing a probability for the existence of an answer. To enhance the training data, techniques for data augmentation are utilized, including Synonyms Replacement and Random Insertion. The experiment results demonstrate significant improvements on the Albert and Electra baseline models, highlighting the effectiveness of the proposed ensemble model for SQuAD.

Keywords: answer attention (AA) · sentence attention (SA) · SQuAD

1 Introduction

Natural Language Processing (NLP) has witnessed remarkable advancements in recent years, with various deep learning models achieving impressive results in tasks like question answering. Among these tasks, the Stanford Question Answering Dataset (SQuAD) [7] has served as a benchmark for evaluating the performance of question-answering systems. The success of models like Bert, Albert, and Electra has demonstrated the power of large-scale language models in capturing complex linguistic patterns and understanding contextual information.

L. Tang and Q. Qi—These authors contributed equally to this work.

© The Author(s), under exclusive license to Springer Nature Singapore Pte Ltd. 2024
D. Benavides-Prado et al. (Eds.): AusDM 2023, CCIS 1943, pp. 75–89, 2024.
https://doi.org/10.1007/978-981-99-8696-5_6

Ensemble models have emerged as a promising technique to further enhance the performance of NLP models. By combining multiple components or techniques, ensemble models can effectively leverage the strengths of individual models and mitigate their weaknesses, leading to better overall performance. In this context, this paper proposes a novel ensemble model for the SQuAD task, aiming to outperform existing baseline models and push the boundaries of question answering accuracy.

The primary goal of this paper is to demonstrate the effectiveness of the proposed ensemble model compared to existing baseline models (RoBERTa, Albert, Electra, etc.) on the SQuAD 2.0 [6]. We aim to showcase significant improvements in question answering accuracy by adopting the Read+Verify approach, leveraging data augmentation techniques, and incorporating attention mechanisms. The results of this study provide valuable insights into the potential of ensemble models for enhancing NLP applications, particularly in question answering scenarios.

This research makes significant contributions to the field of NLP and QA. Firstly, it proposes an ensemble model for SQuAD 2.0 that outperforms existing baseline models. The ensemble model utilizes Sentence Attention (SA-Net) and Answer Attention (AA-Net) modules, which emphasize important context and relevant cues, leading to a deeper understanding of the input and more accurate predictions. Moreover, the model leverages the Read+Verify two-stage approach, where the Read stage focuses on predicting answer text accurately, while the Verify stage determines the presence or absence of an answer. This two-stage process enhances the model's performance and robustness, making it more practical for real-world applications. Additionally, the incorporation of data augmentation techniques, such as Synonyms Replacement (SR) and Random Insertion (RI), improves the model's generalization capabilities by introducing linguistic variations in the training data.

Secondly, the evaluation results demonstrate substantial performance improvements over baseline models, showcasing the effectiveness of the proposed ensemble approach. The ensemble model's enhanced accuracy and robustness hold practical implications for various NLP applications, including chatbots, virtual assistants, and information retrieval systems. Furthermore, the research provides valuable insights into the benefits of ensemble modeling in NLP tasks, offering guidance to researchers and practitioners seeking to design and implement effective ensemble models for other applications. In conclusion, the contributions of this study advance the state-of-the-art in question answering by introducing an innovative ensemble model that demonstrates superior performance, practicality, and potential for further advancements in NLP.

2   Related Work

In the field of NLP, researchers have made significant progress in question answering tasks, particularly using SQuAD 2.0 [6] as a benchmark. Many state-of-the-art models have been proposed to tackle this task. Some notable ones include BERT [3], ALBERT [4], and ELECTRA [2] which have demonstrated the effectiveness of models in SQuAD 2.0.

BERT is a former Google-released model, and ALBERT is a variant of BERT. Both BERT and ALBERT have been empirically proven to excel in various language processing tasks, attesting to their effectiveness and performance. ALBERT, in particular, has demonstrated remarkable prowess across a diverse spectrum of tasks, showcasing its ability to achieve exceptional results in challenging endeavors like the SQuAD 2.0 challenge, which is the focal task under examination in this study. Moreover, the research extends the investigation to ELECTRA model. By conducting further training on ELECTRA-Large, a more powerful model is obtained, outperforming ALBERT in terms of performance on the General Language Understanding Evaluation (GLUE) benchmark [8]. Furthermore, ELECTRA-Large establishes a new state-of-the-art benchmark on the SQuAD 2.0 dataset.

Syntax-Guided Network (SG-Net) is introduced in [11] that enhances machine reading comprehension by integrating explicit syntactic constraints into a self-attention network (SAN) based Transformer encoder. Experimental results on benchmarks like SQuAD 2.0 and RACE showcase SG-Net's effectiveness in improving language representation and comprehension of complex passages. [12] addresses the challenge of Machine Reading Comprehension (MRC) with a focus on effectively handling unanswerable questions. It introduces the Retro-Reader, which combines two stages: sketchy reading for initial judgments and intensive reading for answer verification, achieving state-of-the-art performance on benchmark MRC datasets like SQuAD 2.0 and NewsQA.

Data augmentation techniques have also been employed to address the limited training data challenge in NLP. Techniques like word replacement, back-translation, and paraphrasing have been shown to enhance model robustness and generalization capabilities. For instance, Easy Data Augmentation (EDA) is introduced in [9], which involves random word replacement, insertion, deletion, and synonym replacement, leading to improved performance in various NLP tasks.

Building on these prior works, this research introduces a novel ensemble model tailored for the SQuAD 2.0 task, leveraging the substantial performance and robustness improvements offered by ensemble techniques in question answering systems [1]. The model adopts a two-stage approach, called Read+Verify, focusing first on predicting answers accurately and then determining the probability of answer existence. Moreover, the model incorporates Synonyms Replacement and Random Insertion to diversify the training data. The performance of the proposed ensemble model is evaluated on the SQuAD 2.0 dataset, demonstrating promising results compared to baseline models like ALBERT and ELECTRA. The study contributes valuable insights into improving question answering accuracy and practicality through ensemble models and data augmentation techniques.

3 Problem Definition

3.1 QA Task

Given a passage P and a question Q, the goal is to find the answer span $A \subseteq P$ that correctly answers the question.

Mathematically, we can represent the passage P as a sequence of tokens:

$$P = \{p_1, p_2, \ldots, p_N\},$$

where p_i represents the i-th token in the passage.

Similarly, the question Q can be represented as a sequence of tokens:

$$Q = \{q_1, q_2, \ldots, q_M\},$$

where q_i represents the i-th token in the question.

The answer span A is represented by the start index a_s and the end index a_e in the passage P:

$$A = \{p_{a_s}, p_{a_s+1}, \ldots, p_{a_e}\}.$$

The goal is to find the values of a_s and a_e that maximize the probability $P(a_s, a_e | P, Q)$, where the answer span is a valid response to the question Q.

An illustrative example is provided below to demonstrate this problem.

Consider the following passage and question, the task is to find the answer span that correctly answers the question based on the information in the passage.

Passage:

The capital of France is Paris. It is known for its art, culture, and history.

Question:

What is the capital of France?

The correct answer span is:

$$A = \{\text{Paris}\}.$$

3.2 Identifying the Presence or Absence of Answers

Besides, in SQuAD 2.0, there are two different types of questions:

1. **No-Answer Questions**: These are questions for which the given context does not contain the information necessary to provide a valid answer. These questions challenge models to recognize when an answer is not present in the provided context and to appropriately indicate that fact.

2. **Answerable Questions**: These are questions for which the context does contain relevant information that can be used to extract a specific answer. These questions test the model's ability to locate and understand information within the context and produce an accurate response.

To model the probability of an answer being present or absent, we can use a binary random variable $\mathcal{Y}$:

$$\mathcal{Y} = \begin{cases} 1 & \text{if an answer is present} \\ 0 & \text{if an answer is absent} \end{cases}$$

The SQuAD 2.0 dataset is designed to encompass both types of questions, mimicking real-world scenarios where some questions have answers within the provided context, while others do not. This diversity of question types encourages the development of models that can accurately identify the presence or absence of answers and provide meaningful responses accordingly.

4 Method

The proposed ensemble model adopts a two-stage approach known as Read+Verify. In the Read stage, the model's primary objective is to predict the answer text with high accuracy. By maximizing the model's ability to predict the correct answer, it lays the foundation for improved performance in question answering. However, predicting answers alone may not be sufficient in real-world scenarios, as some questions may not have any valid answers. Therefore, the second stage, Verify, becomes crucial, where the model focuses on determining the probability of the presence or absence of an answer to a given question. This stage introduces a new dimension of robustness and practicality to the ensemble model.

To address the challenge of limited training data, data augmentation techniques are incorporated into the ensemble model. Specifically, Synonyms Replacement (SR) randomly selects words from sentences and replaces them with synonyms from a synonym dictionary. This process introduces additional linguistic variations into the training data, allowing the model to generalize better to unseen examples. Moreover, the model leverages Random Insertion (RI), where it randomly selects a word from a sentence and inserts a synonym from the word's synonym set at a random position. This repeated process further diversifies the data and enhances the model's ability to handle various linguistic structures.

Furthermore, the proposed ensemble model incorporates Sentence Attention Layer (SA-Net) and Answer Attention Layer (AA-Net) components. These attention mechanisms enable the model to focus on critical information in both sentences and answers, respectively. By emphasizing important context and relevant cues, the ensemble model gains a deeper understanding of the input, leading to more accurate predictions.

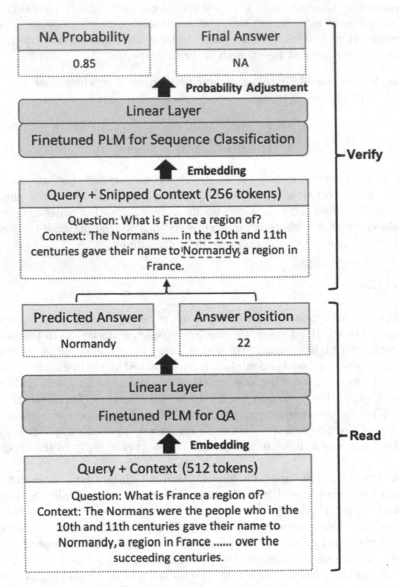

Fig. 1. In our approach, we begin by inputting the **Query + Context** pair into a pretrained language model like BERT. The output goes through a linear layer, predicting both the answer ('Normandy') and its position ('22'). This prediction guides the extraction of the relevant context portion. Initially, the **Query + Context** contains 384 words, which reduces to 192 words after a 'snip' process. The resulting **Query + Snipped Context** is then embedded. This embedding moves through a linear layer before being fed into a PLM for sequence classification. The none-answer probability obtained helps us infer the final answer.

4.1   Model Architecture

In Fig. 1, we show the proposed ensemble model follows the Read+Verify two-stage approach include Read phase and Verify Phase which are shown in Sects. 4.1 and 4.1 respectively.

We illustrate the approach using an example from the SQuAD 2.0 dataset:

> **Question:** What is France a region of?
> **Context:** The Normans were the people who in the 10th and 11th centuries gave their name to Normandy, a region in France ... over the succeeding centuries.

In our process, the **Query + Context** pair is fed into a pretrained language model, such as BERT. The resulting text representation is then passed through a linear layer to predict the answer as 'Normandy' and its position as '22'. This information is used to extract the relevant portion of the context. Prior to the **snipping** procedure, there exist 384 words within the **Query + Context**, which reduces to 192 words subsequently. Following this, we obtain the embedding for **Query + Snipped Context**:

> **Question:** What is France a region of?
> **Context:** The Normans ... in the 10th and 11th centuries gave their name to Normandy, a region in France.

This embedding is forwarded to a sequence classification pretrained language model (PLM) after being passed through a 1-layer CNN classifier. The result is the none-answer probability, from which we can infer the final answer.

Read Stage. The Read Stage is a part of a system that helps answer questions using PLMs. The stage starts by taking two pieces of information: a passage of text that gives some background (context) and a question. Next, the passage and question are tokenized and embedded. Then, a group of different models (such as BERT) work together. Each model looks at the organized context and question and makes a representation of the context and question. This linear layer is applied to predict the answer to the question and where in the context the answer might be. The predicted answer and position from all the models are collected and given as the output of this stage. The detailed work flow of Read Stage is shown in Algorithm 1.

Verify Stage. The process involves obtaining an embedding for a combination of a query and a snipped of context. This embedding is then sent to a sequence classification PLM after undergoing a 1-layer CNN classifier. The outcome is the probability of there being no answer, which can help infer the final answer.

Algorithm 1. Read Stage Model

1: **function** READSTAGEMODEL(*contextPassage, question*)
2:     *inputIds,   attentionMask*   ←   TOKENIZEANDFORMAT(*contextPassage,*
    question)
3:     *readModelOutputs* ← {}
4:     **for** *readModel* **in** *ensembleReadModels* **do**
5:         *output* ← READMODEL(*inputIds, attentionMask*)
6:         *readModelOutputs*.append(*output*)
7:     **end for**
8:     *averagedPredictions* ← AVERAGEPREDICTIONS(*readModelOutputs*)
9:     **return** *averagedPredictions*
10: **end function**

Algorithm 2. Verify Stage Model

1: **function** VERIFYSTAGEMODEL(*contextPassage, question*)
2:     *inputIds,   attentionMask*   =   TOKENIZEANDFORMAT(*contextPassage,*
    question)
3:     *verifyModelOutputs* = {}
4:     **for** *verifyModel* **in** *ensembleVerifyModels* **do**
5:         *output* = VERIFYMODEL(*inputIds, attentionMask*)
6:         *verifyModelOutputs*.append(*output*)
7:     **end for**
8:     *averagedPredictions* = AVERAGEPREDICTIONS(*verifyModelOutputs*)
9:     **return** *averagedPredictions*
10: **end function**

4.2   SA-Net and AA-Net

The process of calculating the final context embedding for a context and question
representation using the attention mechanism is shown in Fig. 2. The input to
the PLM consists of a question and context, separated by '[SEP]' and preceded
by '[CLS]'. This input's representation is denoted as h_i. Through the utilization of specialized masked attention layers, referred to as Sentence Attention
and Answer Attention, a transformed representation, h_i', is formulated. Employing a weighted summation of both h_i and h_i', we get the combined representation $\overline{h}_i$. Subsequently, upon passing through a task-specific layer, the final
question+context embedding is obtained.

    We elaborate on each stage, breaking down the process step by step with the
aid of Eqs. 1, 2, 3 and 4:

1. Equation 1 defines the Question+Context representation H as a matrix. It
   consists of n hidden states $(h_1, h_2, ..., h_n)$, which are likely the output of some
   previous layers in a neural network or pre-trained embeddings representing
   the context and question.
2. Equation 2 introduces the attention mechanism, which calculates attention
   weights, denoted as A_i', for each hidden state h_i in H. The attention weights
   are computed using the dot product between the query vector Q_i' associated

with h_i and the key vectors K_i'. The result of the dot product is then divided by the square root of d_k, which likely represents the dimensionality of the key vectors. The division by the square root is used to scale the dot product and prevent large values that could lead to unstable gradients during training. The result is then passed through a softmax function to obtain normalized attention weights. We add the M as mask matrix in our Sentence Attention and Answer Attention layers. The process of Sentence Attention involves identifying the tokens in a sentence that are relevant to the given question and marking them as true. Additionally, Answer Attention operates on top of Sentence Attention to identify and mark the sentences that are related to the answer.

3. Equation 3 calculates the weighted sum of the value vectors V_i' based on the attention weights A_i'. This step combines the relevant information from the value vectors according to the attention distribution obtained in Eq. 2. The resulting weighted sum is denoted as W_i'.

4. Equation 4 defines how to compute the final context embedding $\overline{h}_i$ using a weighted combination of the original hidden state h_i and the attention-based state h_i' (obtained in Eq. 3). The parameter α controls the balance between the two representations. If α is 1, only the original state h_i is used, and if α is 0, only the attention-based state h_i' is used. The combination of the two representations allows the model to leverage both local and global information in the final question+context embedding.

In summary, the given equations depict the process of applying attention to the question+context representations h_i to calculate the final question+context embedding $\overline{h}_i$.

$$H = [h_1, h_2, \ldots, h_n] \tag{1}$$

$$A_i' = \text{Softmax}\left(\frac{M \cdot (Q_i' K_i'^T)}{\sqrt{d_k}} \right) \tag{2}$$

$$W_i' = A_i' V_i' \tag{3}$$

$$\overline{h}_i = \alpha h_i + (1 - \alpha) h_i' \tag{4}$$

4.3   Ensemble Model Inference

The ensemble strategy is utilized during both the Read and Verify stages to arrive at a final answer prediction. In the Read stage, the prediction pertains to the answer text, while in the Verify stage, it determines the probability of answer existence.

In Fig. 3, the process of attaining the predicted answer, along with its corresponding position and probability, is illustrated for the read phase. With a total of N Read Models, we acquire projected answers, answer positions, and

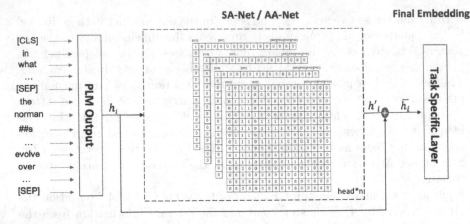

Fig. 2. This outlines the process for calculating the question+context embedding. The tokenized question+context, separated by '[SEP]' and introduced by '[CLS]', are input to the PLM to obtain the representation h_i. This representation is improved through masked attention layers to create a modified version, h'_i, and is subsequently refined by a weighted sum to yield $\bar{h}_i$, producing the final question+context embedding after being processed by a task-specific layer.

their associated probabilities. Subsequently, we employ the algorithm, as demonstrated by Eqs. 5 and 6, utilizing the predictions and their respective probabilities to generate the collective forecast. $Z_1, Z_2, \ldots, Z_N$ represent the outcomes of these forecasts, while the corresponding probabilities are denoted as $u_1, u_2, \ldots, u_N$. Initially, we address scenarios where g models predict the same result (g being any integer greater than 1, signifying multiple models predicting identically). The combined probability U_{combined} is articulated as follows:

$$U_{\text{combined}} = \begin{cases} u_1 \cdot (1.5)^g, & \text{if } Z_1 = Z_2 = \ldots = Z_g \\ u_1, & \text{otherwise} \end{cases} \tag{5}$$

Next, we calculate the normalized probabilities $U_{\text{normalized}}$ for all N models, ensuring that their sum is 1:

$$U_{\text{normalized}} = \frac{U_{\text{combined}}}{\sum_{l=1}^{N} U_{\text{combined},l}} \tag{6}$$

Note that in the above equations, Z_l represents the prediction outcomes of the models, u_l represents the corresponding probabilities, and g represents the number of same predictions.

Differently, the Verify phase adopts XGBoost as the chosen ensemble strategy. In addition to the probabilities of answer existence from M verify models, the feature representation obtained from the Read phase is also incorporated as a feature. The conclusive answer is derived from the output of XGBoost. A visual representation of this process is shown in Fig. 4.

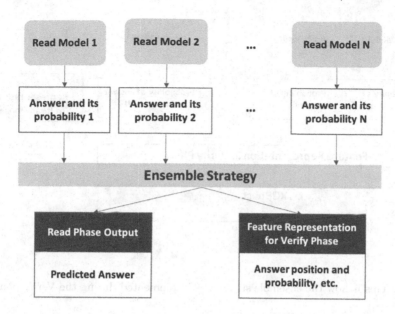

Fig. 3. This is how the ensemble strategy is implemented during the Read phase.

4.4 Data Augmentation

Synonym Replacement (SR) and Random Insertion (RI) techniques are utilized to enhance the training data. The procedures for these techniques are outlined in Algorithms 3 and 4 respectively.

In the Read phase, a total of 5 sentences, randomly selected, are inserted. These sentences constitute 20% of the overall content. Additionally, 1 sentence, chosen at random, is inserted, making up 33% of the text. These sentences are extracted from Wikipedia. During the Verify phase, questions along with their corresponding answers are employed. Subsequently, sentences related to the context of these questions are omitted, generating samples devoid of answers.

5   Experiments

During our experiments, we chose the most recent models available at the time of our research, which include BERT [3], XLNet [10], RoBERTa [5], SG-Net [11], ALBERT [4], and ELECTRA [2].

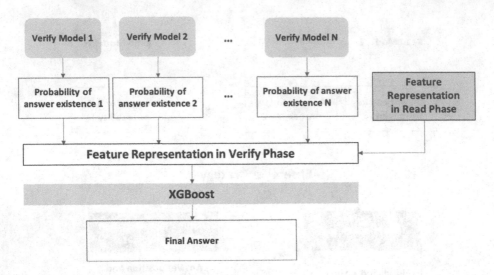

Fig. 4. This is how the ensemble strategy is implemented during the Verify phase.

Algorithm 3. Synonyms Replacement (SR)

1: **function** SYNONYMSREPLACEMENT(*text*, *n*)
2:    *words* = TOKENIZE(*text*)
3:    *nonStopwords* = REMOVESTOPWORDS(*words*)
4:    *selectedWords* = RANDOMLYSELECTNWORDS(*nonStopwords*, *n*)
5:    *replacedText* = REPLACESELECTEDWORDSWITHSYNONYMS(*text*, *selectedWords*)
6:    **return** *replacedText*
7: **end function**

5.1 Evaluation

The SQuAD evaluation method is used to assess the performance of question-answering systems on the SQuAD 2.0 dataset. The dataset consists of a set of questions and corresponding passages, and the task is to predict the answer span within the passage for each question. The evaluation is based on two main metrics:

Exact Match (EM). EM measures the percentage of predicted answers that exactly match the ground-truth (gold standard) answers. If the predicted answer matches the exact span of the gold standard answer, the prediction is considered correct and contributes to the EM score.

Algorithm 4. Random Insertion (RI)

```
1: function RANDOMINSERTION(text, n)
2:     words = TOKENIZE(text)
3:     nonStopwords = REMOVESTOPWORDS(words)
4:     insertedText = text
5:     for i in RANGE(n) do
6:         wordToInsert = RANDOMLYSELECTWORD(nonStopwords)
7:         synonymToInsert = RANDOMLYSELECTSYNONYM(wordToInsert)
8:         insertedText = RANDOMLYINSERTSYNONYM(insertedText, wordToInsert,
       synonymToInsert)
9:     end for
10:    return insertedText
11: end function
```

F1 Score. The F1 score is a measure of the overlap between the predicted answer span and the ground-truth answer span. It considers both precision and recall. Precision is the ratio of the number of correctly predicted tokens to the total number of predicted tokens, while recall is the ratio of the number of correctly predicted tokens to the total number of ground-truth tokens. The F1 score is the harmonic mean of precision and recall.

The EM and F1 scores are commonly used to evaluate the performance of question-answering models on the SQuAD 2.0 dataset. A higher EM score indicates that the model's predictions exactly match the ground-truth answers more often, while a higher F1 score indicates that the model's predictions have a higher overlap with the ground-truth answers in terms of token-level accuracy. During evaluation, each question's predicted answer span is compared to the corresponding ground-truth answer span, and the EM and F1 scores are computed over the entire dataset by averaging the scores for individual questions.

5.2 Experiment Results

Snip is a commonly used method in SQuAD 2.0 dataset. This approach involves selecting the shortest snippet within a passage that contains the answer to a given question. By focusing on the most concise and relevant context, the snip method forces the model to prioritize crucial information for accurate answers while reducing computational overhead associated with processing longer passages.

Based on the results presented in Table 1, it is evident that the Snip technique has a positive impact on the performance of the ensemble model. Comparing the 'Ensemble without Snip' and 'Ensemble with Snip', we can see that the inclusion of the Snip technique results in a marginal improvement in both the Exact Match (EM) score and the F1 score. With the 'Ensemble with Snip', the EM score increased from 90.516% to 90.525%, and the F1 score increased from 92.657% to 92.665%. Even though the improvements are small, they still indicate that the Snip technique enhances the ensemble's ability to generate more accurate and precise answers.

Therefore, it is reasonable to conclude that the Snip technique is good and beneficial for improving the ensemble model's performance on the given task.

Table 1. The Snip technique in the ensemble model, as shown in this table, yields a marginal yet positive enhancement in both Exact Match and F1 scores (EM: 90.516% to 90.525%, F1: 92.657% to 92.665%) on SQuAD 2.0 dev. These improvements suggest the Snip technique's value in refining the ensemble's precision and accuracy for the task."

	EM	F1
Ensemble without Snip	90.516	92.657
Ensemble with Snip	90.525	92.665

Table 2. SA-Net-enhanced models, like ALBERT+SA-Net and ELECTRA+SA-Net, demonstrate improved performance over their base models, achieving higher Exact Match and F1 scores on SQuAD 2.0's development and test sets. This enhancement signifies the efficacy of the SA-Net approach in enhancing QA capabilities.

Model	SQuAD 2.0 dev		SQuAD 2.0 test	
	EM	F1	EM	F1
BERT [3]	–	–	82.1	84.8
SG-Net [11]	–	–	87.2	90.1
XLNet [10]	86.1	88.8	86.4	89.1
RoBERTa [5]	86.5	89.4	86.8	89.8
ALBERT [4]	87.4	90.2	88.1	90.9
ALBERT+SA-Net	87.92	90.53	88.4	90.92
ELECTRA [2]	88.0	90.6	88.7	91.4
ELECTRA+SA-Net	**88.59**	**91.03**	**88.85**	**91.49**

As the experiment results in Table 2, it appears that SA-Net (presumably referring to the model variants with "+SA-Net" in their names) performs slightly better than the base models. Both ALBERT+SA-Net and ELECTRA+SA-Net achieve higher Exact Match (EM) and F1 scores on both the development and test sets of SQuAD 2.0 compared to their respective base models (ALERT and ELECTRA). The improvement in performance indicates that the SA-Net approach is effective in enhancing the question-answering capabilities of these models. We will update the our model result on the leaderboard once it's been accepted.

6   Conclusion

This research presents significant contributions to NLP and QA by proposing an ensemble model for SQuAD 2.0 that surpasses existing baseline models (Bert, Albert, and Electra). The model adopts a Read+Verify two-stage approach, improving performance and robustness for real-world applications. Data augmentation techniques, such as Synonyms Replacement (SR) and Random Insertion (RI), enhance the model's generalization. Integration of Sentence Attention Layer (SA-Net) and Answer Attention Layer (AA-Net) components provides a deeper understanding of input and more accurate predictions. Evaluation results demonstrate substantial performance improvements, with practical implications for NLP applications like chatbots and virtual assistants. The study offers valuable insights into ensemble modeling in NLP tasks, advancing the state-of-the-art in question answering with its innovative and practical approach.

References

1. Aniol, A., Pietron, M., Duda, J.: Ensemble approach for natural language question answering problem. In: 2019 Seventh International Symposium on Computing and Networking Workshops (CANDARW), pp. 180–183. IEEE (2019)
2. Clark, K., Luong, M.T., Le, Q.V., Manning, C.D.: Electra: pre-training text encoders as discriminators rather than generators. arXiv preprint arXiv:2003.10555 (2020)
3. Devlin, J., Chang, M.W., Lee, K., Toutanova, K.: BERT: pre-training of deep bidirectional transformers for language understanding. arXiv preprint arXiv:1810.04805 (2018)
4. Lan, Z., Chen, M., Goodman, S., Gimpel, K., Sharma, P., Soricut, R.: ALBERT: a lite BERT for self-supervised learning of language representations. arXiv preprint arXiv:1909.11942 (2019)
5. Liu, Y., et al.: RoBERTa: a robustly optimized BERT pretraining approach. arXiv preprint arXiv:1907.11692 (2019)
6. Rajpurkar, P., Jia, R., Liang, P.: Know what you don't know: unanswerable questions for squad. arXiv preprint arXiv:1806.03822 (2018)
7. Rajpurkar, P., Zhang, J., Lopyrev, K., Liang, P.: Squad: 100,000+ questions for machine comprehension of text. arXiv preprint arXiv:1606.05250 (2016)
8. Wang, A., Singh, A., Michael, J., Hill, F., Levy, O., Bowman, S.R.: GLUE: a multi-task benchmark and analysis platform for natural language understanding. arXiv preprint arXiv:1804.07461 (2018)
9. Wei, J., Zou, K.: EDA: easy data augmentation techniques for boosting performance on text classification tasks. arXiv preprint arXiv:1901.11196 (2019)
10. Yang, Z., Dai, Z., Yang, Y., Carbonell, J., Salakhutdinov, R.R., Le, Q.V.: XLNet: generalized autoregressive pretraining for language understanding. In: Advances in Neural Information Processing Systems, vol. 32 (2019)
11. Zhang, Z., Wu, Y., Zhou, J., Duan, S., Zhao, H., Wang, R.: SG-Net: syntax-guided machine reading comprehension. In: Proceedings of the AAAI Conference on Artificial Intelligence, vol. 34, pp. 9636–9643 (2020)
12. Zhang, Z., Yang, J., Zhao, H.: Retrospective reader for machine reading comprehension. In: Proceedings of the AAAI Conference on Artificial Intelligence, vol. 35, pp. 14506–14514 (2021)

Anomaly Detection Algorithms: Comparative Analysis and Explainability Perspectives

Sadeq Darrab[1($\boxtimes$)], Harshitha Allipilli[1], Sana Ghani[1],
Harikrishnan Changaramkulath[1], Sricharan Koneru[1], David Broneske[2],
and Gunter Saake[1]

[1] University of Magdeburg, Magdeburg, Germany
sadeq.darrab@ovgu.de,
{harshitha.allipilli,harikrishnan.changaramkulath}@st.ovgu.de,
{sana.ghani,sricharan.koneru}@ovgu.de, gunter.saake@ovgu.de
[2] German Centre for Higher Education Research and Science Studies,
Hanover, Germany
broneske@dzhw.eu

Abstract. In order to detect outliers and potential anomalies in datasets, anomaly detection plays a pivotal role in identifying infrequent and irregular occurrences. The purpose of this paper is to examine and compare the effectiveness of prominent anomaly detection algorithms, including Isolation Forest, Local Outlier Factor (LOF), and One-Class Support Vector Machines (SVM). A variety of datasets are used in our assessment to evaluate key metrics such as precision, recall, F1-score, and overall accuracy. We also introduce innovative techniques that enhance the interpretability of these algorithms, shedding light on the underlying factors that contribute to anomaly detection. By providing insights into the attributes and behaviors associated with anomalies, our research empowers decision-makers to cultivate a profound comprehension of the identified anomalies, subsequently facilitating well-informed decisions grounded in the outcomes of anomaly detection. Through our meticulous comparative analysis and our dedication to unraveling the elements of explainability, we provide invaluable perspectives and pragmatic suggestions to facilitate effective anomaly detection in real-world scenarios.

Keywords: Anomaly Detection · Unsupervised Learning · Explainability

1 Introduction

Anomalies are outliers, noise, exceptions, and deviations from the real behavior of the system. Detecting anomalies involves identifying objects, patterns, occurrences, and observations that do not follow an anticipated pattern [6]. Anomaly detection plays a crucial role in various domains, including Cybersecurity, fraud detection, industrial monitoring, and healthcare. It is possible to identify outliers and potential anomalies that require special attention by identifying and

© The Author(s), under exclusive license to Springer Nature Singapore Pte Ltd. 2024
D. Benavides-Prado et al. (Eds.): AusDM 2023, CCIS 1943, pp. 90–104, 2024.
https://doi.org/10.1007/978-981-99-8696-5_7

flagging rare and irregular instances within datasets. The traditional supervised learning approach relies on labeled data, making it less suitable for detecting anomalies in unsupervised datasets. Due to their ability to uncover unknown anomalies without relying on pre-labeled data, unsupervised anomaly detection algorithms have gained increasing attention [15].

Anomaly detection methods have been found to be useful in a variety of fields, but they are not without their challenges. Since anomalies are often rare and difficult to identify in real-world scenarios, obtaining labeled data for anomalies can be challenging [12]. Furthermore, conventional anomaly detection methods may face difficulties in explaining the anomalies detected in the data. Many existing anomaly detection methods, often operate as "black-box" models, providing little to no insight into how they arrive at their anomaly detection decisions. This lack of transparency and interpretability hinders the adoption of these algorithms in critical domains where explanations for detected anomalies are essential for decision-making [4].

In our paper, we investigate unsupervised anomaly detection algorithms and explore their effectiveness in identifying anomalies within diverse datasets, as well as their limitations. In this paper, we focus on three widely used algorithms: isolation forest, local outlier factor (LOF), and one-class support vector machines (SVM). In terms of computational efficiency, scalability, and interpretability, these algorithms leverage different techniques to detect anomalies.

The main contribution of this paper can be summarized as follows.

- We provide a comprehensive analysis and comparison of these algorithms, evaluating their performance metrics such as precision, recall, F1-score, and overall accuracy on various datasets.
- Our study is focused on benchmarking a variety of algorithms against each other within the realm of anomaly detection. The primary objective is to discern and highlight the unique strengths and weaknesses exhibited by these algorithms across diverse anomaly detection scenarios.
- Through an exploration of novel techniques, we explore how to make algorithms more interpretable. By uncovering the underlying factors contributing to detected anomalies, we empower decision-makers to gain deeper insights into the anomalies detected and make informed decisions based on the anomaly detection results.
- We bridge the gap between algorithmic efficiency and human interpretability by combining performance evaluation with explainability.

This paper has the following structure. The background of anomaly detection models is discussed in Sect. 2. The methodology used in this paper is described in Sect. 3. The detailed implementation of data preprocessing, model training and evaluation, and explainability is presented in Sect. 4. In Sect. 5, we present the evaluation results of these models and datasets whereas in Sect. 6, we present a review of the exiting work. Our final section concludes our paper and outlines our future plans.

2   Background

Data mining and machine learning use anomaly detection to identify occurrences that are significantly different from the norm. Since machine learning has gained popularity in anomaly identification, both supervised and unsupervised learning methods have been used to capture complicated patterns and identify anomalies. For tasks, where anomalies are identified in the training data, supervised learning techniques such as Naive Bayes, Support Vector Machines (SVM) and K-Nearest Neighbors have been investigated and compared again, and One Class SVM, an unsupervised anomaly detection algorithm that outperforms these supervised algorithms [7]. These algorithms may face difficulties with labeled data availability in real-world applications where anomalies are uncommon and difficult to obtain. Hence, we focus on anomaly detection methods based on unsupervised learning techniques that do not require labeled training data. Unsupervised methods such as Local Outlier Factor (LOF), Isolation Forest, and One-Class SVM are used.

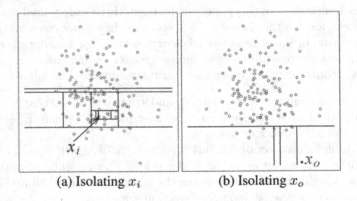

(a) Isolating x_i (b) Isolating x_o

Fig. 1. Isolation Forest [9]

2.1   Isolation Forest

Isolation Forest is a well-known ensemble-based anomaly detection algorithm lauded for its efficiency and capacity to handle high-dimensional data. This algorithm operates by isolating anomalies into partitions in a random forest-like manner. Decision trees are constructed by recursively selecting random features and splitting randomly, which isolates the anomalies effectively.

In Fig. 1(a), a relatively small number of anomalies is visible. This results in the formation of smaller partitions and shorter paths within the tree structure. Contrarily, Fig. 1(b) demonstrates instances characterized by distinct attribute values, leading to early partitioning. Notably, anomalies are likely to be swiftly isolated within a few steps, thereby setting them apart from the larger cluster of data points, which necessitate more partitions for segregation. The path length traversed by an instance within the tree serves as an anomaly score, where a shorter path corresponds to higher anomaly levels. This scoring mechanism enables efficient differentiation between anomalies and normal data points.

2.2 Local Outlier Factor (LOF)

LOF is a density-based anomaly detection algorithm that assesses the local density of instances relative to their neighbors. It calculates the local reachability density for each data point by comparing its distance to its k-nearest neighbors. The LOF score indicates how much an instance's density deviates from that of its neighbors. Low LOF scores correspond to points with significantly lower densities than their neighbors, indicating anomalies [3]. Figure 2 illustrates the LOF mechanism: for data point o2, its local density is computed using its k-nearest neighbors; if o2 has lower density than its neighbors, it is labeled as an anomaly.

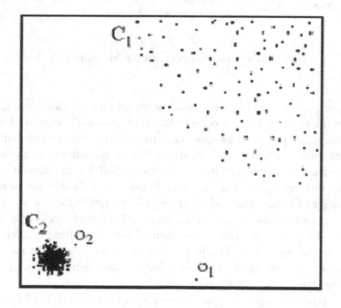

Fig. 2. Local Outlier Factor [1]

2.3 One-Class Support Vector Machine (One-Class SVM)

One-Class SVM is a widely-used algorithm for unsupervised anomaly detection. It is based on the principles of Support Vector Data Description (SVDD). Its primary objective is to learn a hyper-sphere to characterize a single class of data points. Instances outside the decision boundary are classified as anomalies [2]. In Fig. 3, the working principle of a One Class Support Vector Machine is depicted.

Points near the origin are classified as anomalies (-1), while all other points are considered normal.

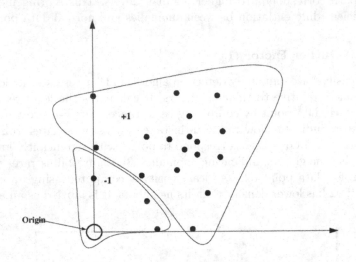

Fig. 3. One Class Support Vector Machine [10]

This paper evaluates model performance based on precision, recall, F1-score, and accuracy metrics. Accuracy reflects the correctness of normal and anomaly classifications, while precision gauges accurate positive predictions, emphasizing correct identification without false positives. Recall measures the model's ability to identify actual positives within the dataset, vital for imbalanced scenarios, and F1-score balances precision and recall, proving valuable for uneven data. Given the rarity of anomalies and their significant deviations, a focus on recall is crucial to minimize false negatives, ensuring robust real-world anomaly detection. The use of F1-score addresses imbalanced data's challenges, accounting for false positives and negatives. The implementation leverages Anchors, an interpretable rule extraction method, chosen for its simplicity and high precision, enhancing the model's trustworthiness.

The research encompasses cardiovascular disease and credit card fraud datasets, addressing class imbalances and distinct data distributions. Comparative analysis sheds light on algorithm effectiveness, offering insights into various contextual advantages and disadvantages. Additionally, the paper highlights explainability's pivotal role in anomaly detection, striving to demystify algorithm decisions through interpretable methodologies. By developing unsupervised anomaly detection algorithms and providing practical guidance, the research empowers users to select optimal approaches tailored to their unique scenarios.

3   Methodology

The main purpose of this research paper is to identify anomalies using different models and explain why these models made this decision. Figure 4 shows a the workflow of the proposed model. This method consists of the following steps.

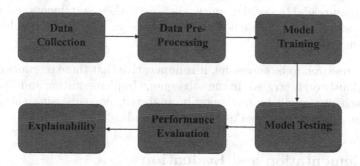

Fig. 4. Graphical Representation of Proposed Methodology

- **Data Collection:** To begin with, we have carefully selected datasets that contain anomalies, including Heart Disease Prediction and Credit Card Fraud detection. These datasets are both expansive and diverse, providing a comprehensive testing ground for the subsequent steps.
- **Data Pre-processing:** The second step involves data pre-processing. A judicious selection of pertinent features becomes imperative for optimizing model performance when recognizing the presence of both useful and extraneous features in the dataset. Due to the heterogeneous nature of the data, consisting of both categorical and continuous attributes, adept handling is required. In order to overcome this, a two-pronged approach is adopted: discretization of continuous values to impart discreteness, followed by feature encoding, such as one-hot encoding and label encoding, for categorical attributes. In order to ensure homogeneity and comparability, normalization or standardization techniques are used.
- **Model Training:** This step involves training the model. As outlined in Sect. 2, the three models are trained here for practical implementation. To pinpoint the optimal hyperparameters for each dataset-model pair, meticulous experiments are conducted.
- **Model Testing:** Testing the model takes place in the fourth step. In order to assess the models, a distinct test dataset is used, which was not used during training. As part of the performance evaluation, the models are scrutinized against the defined evaluation metrics - precision, recall, F1-score, and accuracy.
- **Performance Evaluation:** In this step, the efficacy of the model is comprehensively evaluated. Leveraging the outcomes of the previous phase, the models are subjected to rigorous evaluation across different test datasets. Performance metrics, including precision, recall, F1-score, and accuracy, form the

bedrock of assessment, interlacing both the individual model strengths and trade-offs. Based on these evaluations, along with training times and recall scores, informed comparisons can be made.
- **Interpretability:** In the final phase of this research journey, we will examine the interpretability of model outputs. After training the models, we utilize rule-generating methodologies, such as Anchors, to generate intelligible and human-readable decision rules, such as if-then statements. Our ability to understand the rationale behind the models' classifications is enhanced by this transparency.

For our research to be successful, it is imperative that these interlocking steps are meticulously orchestrated. In the subsequent Implementation and Evaluation section, each phase will be meticulously analyzed, encompassing the intricate details that collectively propel us toward our goals.

4   Implementation and Evaluation

This section outlines the implementation of the mentioned unsupervised machine learning algorithms for anomaly detection on diverse datasets: Heart Disease and Credit Card Fraud from Kaggle. We conduct a comparative analysis to assess Isolation Forest, Local Outlier Factor, and One-class SVM's anomaly detection performance and generalization across real-world scenarios. It encompasses data pre-processing, hyper-parameter tuning, model evaluation, and comparative effectiveness analysis with other algorithms.

4.1   Datasets

The Heart Disease dataset [16] contains 70000 instances and 11 features. The second dataset [5], the Credit Card Fraud, contains 1296675 instances and 22 features. We balanced the dataset by sampling 20,000 instances from the majority class (normal) and 5,000 instances from the minority class (fraud).

4.2   Data Pre-processing

Several data preprocessing steps were performed in this section to ensure the data's quality and suitability for modeling:

- Data Cleaning: In this stage, we focused on identifying and resolving errors, inconsistencies, and missing values present in the dataset. We took necessary steps to handle duplicate values, null values, and inconsistent data entries. Fortunately, there were no null values in the provided datasets. However, we did encounter an inconsistency in the heart disease dataset, specifically in the features ap_hi and ap_lo, (since systolic and diastolic blood pressure cannot be negative) where some values were negative. We promptly addressed and corrected such discrepancies to ensure data integrity.

For the heart disease dataset, we converted the age feature from days to years for better interpretation and the height feature from centimeters to feet for standardization as part of the data pre-processing step. Additionally, we dropped the 'id' column as it does not contribute to the anomaly detection task.

– Data Transformation: We used the following techniques for data transformation:

- Log Transformation: To address data skewness and reduce the impact of extreme values, we employed the natural logarithm of the data values. Specifically, we applied this technique to the *amt* and *city_pop* features in the Credit Card Fraud Dataset. Prior to the transformation, the *amt* feature had minimum and maximum values of 1 and 28,949, respectively, while the *city_pop* feature ranged from 23 to 2,906,700. After applying the Log Transformation, the minimum and maximum values of *amt* were adjusted to 0 and 10, respectively, and the *city_pop* feature ranged from 3 to 15. This transformation helped to normalize the data and mitigate the effects of extreme values, thereby improving the analysis.

- Scaling: To bring all features to a common range and standardize the data, we utilized min-max scaling, also known as normalization. Specifically, we applied this operation to four features in the Credit Card Fraud Dataset: *lat*, *long*, *merch_lat*, and *merch_long*. Before scaling, the mean values for *lat* and *merch_lat* were 39, while the means for *long* and *merch_long* were -90. After applying the min-max scaler, the means of these features were adjusted to 0, effectively standardizing the data and ensuring that they fall within the same range. This normalization process facilitates more consistent and reliable analyses across the dataset.

- Encoding Categorical Variables: To facilitate effective interpretation by the models, we converted categorical variables into numerical representation using label encoding. Within the credit card fraud dataset, we identified 14 unique categories of transactions, 2 genders, and 50 states, all represented with strings. Through label encoding, we transformed the features *category*, *gender* and *state* into corresponding integers, enabling the models to handle them efficiently during training and prediction processes.

– Feature Engineering: As part of this crucial step, we focused on extracting relevant information from the raw data and creating new features to improve the model's learning capabilities. Our objective was to provide more meaningful and informative input for the models to better understand and process the data. For instance, in the heart disease dataset, we converted the *age* feature from days to years for better interpretability. Additionally, we transformed the *height* feature from centimeters to feet to ensure consistency and ease of understanding. Similarly, in the credit card fraud dataset, we performed feature engineering on the *trans_date_trans_time* feature by splitting it into separate components such as hour, day, and month. This transformation enabled the models to capture temporal patterns more effectively and gain deeper insights into the data's temporal dynamics.

4.3   Algorithm Implementations

- Isolation Forest: The python implementation of isolation forest from sci-kit learn was utilized in our experiments. Among the nine available parameters, we focused on tuning two specific attributes: *n_estimators* and *contamination*.

  The parameter *n_estimators* determines the number of estimators constructed in the ensemble. After conducting thorough experiments, we observed that the default value of 100 for *n_estimators* was highly efficient for our datasets. As a result, we decided to retain the default value for this parameter, ensuring that the algorithm maintains its computational efficiency. The *contamination* parameter, on the other hand, represents the proportion of outliers present in the dataset. To ensure accurate anomaly detection, we meticulously adjusted the *contamination* parameter to match the specific characteristics of each dataset. By carefully setting this parameter, we aimed to strike a balance between detecting true anomalies while minimizing false positives.

  To assess the effectiveness of the Isolation Forest model, we generated outlier scores for instances using the *decision function* method. Subsequently, we defined a threshold of 0 to classify instances as anomalies or normal data. Based on the outlier scores, we converted the predictions into binary labels, where 1 represents anomalies and 0 denotes normal instances.

  In our experiments, we found that isolation forest is highly efficient and adept at handling large, high-dimensional datasets, requiring less memory due to its storage of random partition structures. This makes it a practical option for anomaly detection, especially with extensive datasets and high-dimensional data.

- One Class Support Vector Machine: In our study, we utilized the implementation of One Class SVM provided by the sci-kit learn library in Python. The implementation offers ten parameters that can be configured and optimized. Among these parameters, we kept the default settings for eight, and focused on adjusting two key parameters.

  The first parameter we modified was the *kernel* parameter, which allowed us to choose from four available kernels: *linear*, *poly*, *rbf*, and *sigmoid*. Through experimentation, we found that the Radial Bias Function (RBF) kernel yielded the most promising results for both datasets. Kernels play a crucial role by transforming the feature space, enabling the data to become linearly separable, and consequently improving the SVM's classification performance [14].

  The second parameter we adjusted was *nu* which represents the fraction of training errors or, in other words, the number of anomalous instances within the margin. By tuning this parameter, we could control the trade-off between maximizing the margin and capturing the anomalies effectively.

  One noteworthy observation was that as training instances increased, the One Class SVM's training time grew exponentially due to the extensive calculations required for the distance matrix. This highlights the importance of considering computational cost for efficient training with larger datasets.

- Local Outlier Factor: Local Outlier Factor (LOF) algorithm, evaluates the local density deviation of each data point in relation to its neighboring points to identify the anomalies. The sci-kit learn implementation was employed for this purpose. Among the nine available parameters, we focused on adjusting two key parameters: $n\_neighbors$ and $contamination$.

For our specific implementation, we set the $n\_neighbors$ parameter to the default value of 20, while the $contamination$ parameter was set to 0.1. The $contamination$ rate represents the proportion of outliers in the dataset. Throughout our experimentation, we made significant observations and obtained noteworthy findings.

A reduced $nneighbors$ value, set at 10, improved the True Positive Rate (TPR) and anomaly detection by effectively identifying outliers with LOF scores different from 1, showcasing the algorithm's ability to discern anomalies from the majority of the data.

4.4 Explainability

Interpretability and explainability are essential in understanding classification models, as conventional models lack transparency in their decisions. The genera tion of rules, like if-then cases or hierarchical trees, aids in illuminating decision processes, enabling users to grasp the model's predictions and the underlying rationale.

Explainable Anomaly Detection (XAD) techniques can be classified into three categories: Pre-model, In-model, and Post model [8]. Pre-model techniques involve feature selection and feature representation, and they operate solely on data without the use of any machine learning model. In-model techniques, on the other hand, utilize supervised or unsupervised models with built-in explainability, such as decision trees, which employ hierarchical trees to identify if-then cases and provide explanations for specific decisions. By focusing on the Post-model techniques, we aimed to gain a deeper understanding of the decision logic of our anomaly detection algorithms. Anchors generates simple and concise if-then rules that sufficiently explain the decisions made by the anomaly detection model. These rules are specific to local instances, meaning that even if a feature value changes, the predictions or rules remain mostly unchanged. This localized approach ensures robustness and stability in the explanations.

5 Results

Figure 5 presents a bar chart illustrating the comparison of recall scores among the three algorithms. Remarkably, the Local Outlier Factor exhibits subpar performance with a value of 0.08 on the Heart Disease dataset but surprisingly outperforms the other two algorithms on the Credit Card Fraud dataset with a value of 90%. This intriguing observation leads us to attribute the contrasting outcomes to the datasets varying densities, primarily stemming from class imbalance issues.

Conversely, the Isolation Forest secures the second position concerning recall scores on both datasets, and the obtained scores exhibit notable similarities which equals 45% and 44% on Heart Disease dataset and Credit Card Fraud dataset respectively. Interestingly, the One Class SVM's performance proves to be unsatisfactory on the Credit Card Fraud dataset which equals to 30%. Whereas the algorithm showcases a recall value of 44% on the Heart Disease dataset. We posit that this discrepancy occurs due to the One Class SVM's inability to effectively discern the class distributions of the anomalous instances and normal data points within this particular dataset.

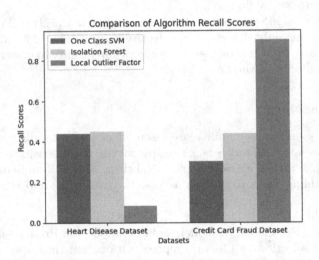

Fig. 5. Recall Score Comparisons

In Fig. 6, the bar chart illustrates the training times of the three algorithms. Notably, Local Outlier Factor demonstrates the shortest or nominal training duration on both datasets with values of 2.07 and 1.48 s for Heart Disease and Credit Card Fraud datasets respectively. This efficiency can be attributed to the fact that most of the computations in Local Outlier Factor occur during the prediction phase.

Isolation Forest ranks second in terms of training times on both datasets. However, it is worth mentioning that the training time for Isolation Forest is not constant and may vary across different runs on the same dataset with one being 46.57 and 35.89 s respectively for Heart Disease and Credit Card Fraud datasets. This variability could be attributed to the randomness involved in the feature selection process.

On the other hand, One Class SVM exhibits the longest training time on both datasets with values of 48.24 s in Heart Disease dataset and 57.42 s in Credit Card Fraud dataset. This is primarily due to the extensive number of calculations involved in defining the distance matrix and support vectors during the training phase. As a result, the computational complexity of One Class SVM contributes to its higher training time compared to the other two algorithms.

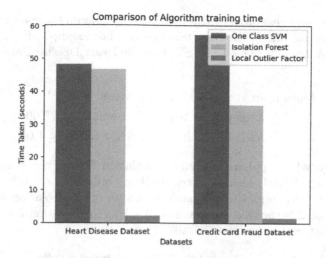

Fig. 6. The Time Taken for Training

5.1   Explainability Results

The generated anchors provide clear if-then rules based on specific features from
the data that were used by the anomaly detection model to arrive at its decision.
For instance, an anchor might state "if gender = Female and age < 20, then
normal patient," or "if gender = male and age > 40, then check fitness." These
rules are informative and human-readable, facilitating better understanding of
the model's behavior.

Additionally, Anchors quantifies the accuracy and coverage of the generated
rules, allowing users to gauge the reliability and scope of each explanation. This
helps in assessing the trustworthiness of the rules and gaining insights into their
impact on the model's predictions.

```
Anchor: city_pop_log <= 9.89 AND long > 0.20 AND category_enc <= 10.00 AND state_enc > 28.00
Precision: 0.94
Coverage: 0.18
```

Fig. 7. Rule generated by Anchors for Isolation Forest

Figure 7 shows the rules generated by Anchors for a trained isolation forest
model. We input one instance to the anchors and it gives one rule, which provides
explanation about why this particular instance is classified as anomaly or normal.
Anchors take into consideration overall dataset and generates this rule. It also
specifies the precision rate and the amount of instances covered. From the above
generated rule, we see that the rule it is generated is 94 percent accurate.

Upon applying anchors to one class SVM, we observed that it effectively identifies the key features and their corresponding values associated with anomalous

instances. The rules generated by this method shed light on the specific attributes that play a crucial role in the detection process. For example, Eq. 1 is the rule produced by Anchors for One Class SVM on the Heart Disease Dataset, achieving a precision of 0.82.

$$\text{Anchor: ap\_lo} > 80.00 \text{AND age\_years} \leq 46.22$$
$$\text{AND gender} > 1.00 \text{AND weight} \leq 79.00 \tag{1}$$
$$\text{AND height\_ft} > 5.58 \text{ AND active} \leq 1.00$$

Conversely, when applying anchors to Isolation Forest, the generated rules appeared to be relatively shallow, often involving only one or two features. This outcome can be attributed to the intrinsic design of the Isolation Forest algorithm. Equation 2 is the rule generated by Anchors for Isolation forest on Heart Disease Dataset with a precision of 0.97

$$\text{Anchor : gluc} > 1.00 \text{ AND cholesterol} > 1.00 \tag{2}$$

6   Related Work

In this section, we present a review of the existing literature related to anomaly detection, aimed at identifying patterns or data points that deviate from the normal pattern within a dataset. Different approaches to anomaly detection, including supervised and unsupervised, have been explored in prior research, and various algorithms have been applied for outlier detection. While considering the refined problem statement of applying data mining techniques for anomaly detection, our goal is to implement and compare different effective methods for identifying anomalies that significantly deviate from the majority of the data.

The overview research done on the machine learning techniques for detecting the anomalies [11] helped us in narrowing to the domain of unsupervised techniques over the supervised ones with proper justifications. The supervised techniques require a significant amount of labelled data, where anomalies are explicitly identified and labelled. During the classification phase, the trained model is used to predict whether new instances are normal or anomalous by comparing them to the learned patterns. In [13], it is explained that the significant limitations of using the supervised techniques in anomaly detection. The important aspect being stated was that the acquiring of labelled data for anomalies can be challenging and time-consuming, especially in real-world scenarios where anomalies may be rare or evolving. Apart from that these techniques rely heavily on labelled training data, which makes them less effective in identifying anomalies that were not present in the training set.

Pointing down to the unsupervised algorithms, we understood that they make use of the unlabelled data where the anomalies are not predefined and align greater with the real-world scenarios of complex feature structures. These techniques focus on detecting instances that deviate significantly from the expected or normal patterns in the data. Therefore Unsupervised algorithms explore the data characteristics and identify anomalies based on their deviation from what is considered normal. The existing studies motivated us to focus reliably on the three algorithms which were specifically used for anomaly detection. They are:

1. Isolation Forest
2. One class support vector machine
3. Local Outlier Factor

In [9], the application of Isolation Forest detection to anomalies is discussed. Being an unsupervised model it builds an ensemble of isolation trees (iTrees) for a given dataset. The property of identifying the instances with short average path lengths on the isolation trees as anomalies becomes more peculiar to the algorithm. The algorithm which works on the principal idea of isolating anomalies rather than profiling normal instances has the ability to handle high dimensional datasets along with being computationally efficient and having interpretable results.

One class Support vector machine [14] gives significant insights into the pros of the algorithm in this specific context of anomalies. It captures the underlying structure of a target class and differentiates it from the rest of the data and also tries to find a function that is positive for regions with high density of points, and negative for small densities. The main advantage of the algorithm is the capability to handle non-linear relationships and imbalanced datasets.

Apart from these two algorithms, the work done by [1] helped us to delve more into the local outlier factor methods which have been specifically designed and contextually relevant in outlier or anomaly detection. It assigns each object in the dataset a degree of being outlier (Local Outlier Factor) and the data points with LOF values above a certain threshold are identified as Outliers We found the most significant advantage of the algorithm to be its usefulness for identifying anomalies in datasets with varying densities and complex geometric structures.

7   Conclusion

Our comprehensive study has yielded valuable insights into the suitability of diverse algorithms for varying dataset characteristics. Isolation Forests demonstrate robustness in handling static datasets with efficient training times and satisfactory performance. One Class Support Vector Machines (SVM) stand as a potent choice for well-separable datasets with substantial computational resources. Local Outlier Factor (LOF) excels in addressing datasets with fluctuating densities, showcasing its strength in scenarios marked by density variations. Each algorithm's distinct attributes render them well-suited to specific dataset traits, underlining the significance of tailored algorithm selection.

For future endeavors, delving deeper into the decision-making mechanisms of Isolation Forest and One Class SVM warrants exploration through the implementation of alternative rule generation systems like Scalable Bayesian Rule Lists, SHAP, hypercubes, among others. Such methods can provide a holistic comparison of these algorithms. Additionally, customizing rule generation with domain-specific insights holds potential for enhancing interpretability and relevance, further enriching our understanding of their performance.

References

1. Breunig, M.M., Kriegel, H.P., Ng, R.T., Sander, J.: LOF: identifying density-based local outliers (2000)
2. Chen, G., Zhang, X., Wang, Z.J., Li, F.: Robust support vector data description for outlier detection with noise or uncertain data. Knowl.-Based Syst. **90**, 129–137 (2015)
3. Degirmenci, A., Karal, O.: Efficient density and cluster based incremental outlier detection in data streams. Inf. Sci. **607**, 901–920 (2022)
4. Dey, A., Totel, E., Costé, B.: Daemon: dynamic auto-encoders for contextualised anomaly detection applied to security monitoring. In: Meng, W., Fischer-Hübner, S., Jensen, C.D. (eds.) ICT Systems Security and Privacy Protection - SEC 2022. IFIP Advances in Information and Communication Technology, vol. 648, pp. 53–69. Springer, Cham (2022). https://doi.org/10.1007/978-3-031-06975-8_4
5. Kartik: Fraud detection dataset (2020). What about this: https://www.kaggle. com/datasets/kartik2112/fraud-detection?select=fraudTrain.csv
6. Kavitha, M., Srinivas, P., Kalyampudi, P.L., Srinivasulu, S., et al.: Machine learning techniques for anomaly detection in smart healthcare. In: 2021 Third International Conference on Inventive Research in Computing Applications (ICIRCA), pp. 1350–1356. IEEE (2021)
7. Li, K.L., Huang, H.-K., Tian, S.F., Xu, W.: Improving one-class SVM for anomaly detection (2003)
8. Li, Z., Zhu, Y., van Leeuwen, M.: A survey on explainable anomaly detection, October 2022. http://arxiv.org/abs/2210.06959
9. Liu, F.T., Ting, K.M., Zhou, Z.H.: Isolation forest, pp. 413–422 (2008). https://doi.org/10.1109/ICDM.2008.17
10. Manevitz, L.M., Yousef, M., Cristianini, N., Shawe-Taylor, J., Williamson, B.: One-class SVMs for document classification (2001)
11. Omar, S., Ngadi, M., Jebur, H., Benqdara, S.: Machine learning techniques for anomaly detection: an overview. Int. J. Comput. Appl. **79** (2013). https://doi.org/10.5120/13715-1478
12. Pang, G., Shen, C., van den Hengel, A.: Deep anomaly detection with deviation networks. In: Proceedings of the 25th ACM SIGKDD International Conference on Knowledge Discovery and Data Mining, pp. 353–362 (2019)
13. Rewicki, F., Denzler, J., Niebling, J.: Is it worth it? Comparing six deep and classical methods for unsupervised anomaly detection in time series. Appl. Sci. (Switzerland) **13** (2023). https://doi.org/10.3390/app13031778
14. Schölkopf, B., Platt, J.C., Shawe-Taylor, J., Smola, A.J., Williamson, R.C.: Estimating the support of a high-dimensional distribution. Neural Comput. **13**(7), 1443–1471 (2001). https://doi.org/10.1162/089976601750264965
15. Telo, J.: Ai for enhanced healthcare security: an investigation of anomaly detection, predictive analytics, access control, threat intelligence, and incident response. J. Adv. Anal. Healthc. Manag. **1**(1), 21–37 (2017)
16. ulianova, S.: Cardiovascular disease dataset (2019). https://www.kaggle.com/datasets/sulianova/cardiovascular-disease-dataset

Towards Fairness and Privacy: A Novel Data Pre-processing Optimization Framework for Non-binary Protected Attributes

Manh Khoi Duong$^{(\boxtimes)}$ and Stefan Conrad

Heinrich Heine University, Universitätsstraße 1, 40225 Düsseldorf, Germany
{manh.khoi.duong,stefan.conrad}@hhu.de

Abstract. The reason behind the unfair outcomes of AI is often rooted in biased datasets. Therefore, this work presents a framework for addressing fairness by debiasing datasets containing a (non-)binary protected attribute. The framework proposes a combinatorial optimization problem where heuristics such as genetic algorithms can be used to solve for the stated fairness objectives. The framework addresses this by finding a data subset that minimizes a certain discrimination measure. Depending on a user-defined setting, the framework enables different use cases, such as data removal, the addition of synthetic data, or exclusive use of synthetic data. The exclusive use of synthetic data in particular enhances the framework's ability to preserve privacy while optimizing for fairness. In a comprehensive evaluation, we demonstrate that under our framework, genetic algorithms can effectively yield fairer datasets compared to the original data. In contrast to prior work, the framework exhibits a high degree of flexibility as it is metric- and task-agnostic, can be applied to both binary or non-binary protected attributes, and demonstrates efficient runtime.

Keywords: Fairness · Data privacy · Non-binary · Fairness-agnostic · Genetic algorithms

1 Introduction

Machine learning has become an increasingly important tool for decision-making in various applications, ranging from income [17] to recidivism prediction [18]. However, the use of these models can perpetuate existing biases in the data and result in unfair treatment of certain demographic groups. One of the key concerns in the development of fair machine learning models is the prevention of discrimination regarding protected attributes such as race, gender, and religion.

This work was supported by the Federal Ministry of Education and Research (BMBF) under Grand No. 16DHB4020.

© The Author(s), under exclusive license to Springer Nature Singapore Pte Ltd. 2024
D. Benavides-Prado et al. (Eds.): AusDM 2023, CCIS 1943, pp. 105–120, 2024.
https://doi.org/10.1007/978-981-99-8696-5_8

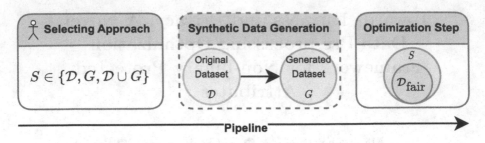

Fig. 1. The pipeline consists of three steps: (1) The user sets the sample set S and other settings, including the objective, discrimination measure, and protected attribute; (2) Synthetic data is generated if needed; (3) A solver optimizes the fairness objective to yield a biased-reduced subset $\mathcal{D}_{\text{fair}}$ from the user-selected set S. If $S = G$ was chosen, the user obtains a bias-reduced synthetic dataset that does not leak privacy-related information.

While most of the existing literature focuses on classification problems where the protected attribute is binary [2,4,6,7,10,20,24,28], the real world presents a more complex scenario where the protected attribute can consist of more than two social groups, making it non-binary. While works that discuss and deal with non-binary protected attributes exist, and we do not neglect their existence [5, 14,29], we view it as a necessity to contribute further to this field by providing a flexible framework that accommodates various fairness notions and applications, including data privacy, to strive for the employment of responsible artificial intelligence in practice.

Since bias is rooted in data, we introduce an optimization framework that pre-processes data to mitigate discrimination. In the context of fairness, pre-processing ensures the generation of a fair, debiased dataset. We address the challenges associated with non-binary protected attributes by deriving appropriate discrimination measures. To prevent discrimination, we formulate a combinatorial optimization problem to identify a subset from a given sample dataset that minimizes a specific discrimination measure, as depicted in Fig. 1. Depending on the provided sample dataset, which may also include synthetically generated data, the framework allows for the removal of such data points or the inclusion of synthetic ones to achieve equitable outcomes. By using generated data, we can utilize our method in applications where data privacy is a concern. Since the discrimination objective is stated as a black box, heuristics, which do not assess the analytical expression of the discrimination measure during optimization, are needed to solve our stated problem. Our formulation makes the framework *fairness-agnostic*, allowing it to be used to pursue any fairness objective.

The experimentation was carried out on the Adult [17], Bank [22], and COMPAS [18] datasets, all known to exhibit discrimination. We compared the discrimination of the datasets before and after pre-processing them with different heuristics on various discrimination measures. The results show that genetic algorithms [12] were most effective in reducing discrimination for non-binary protected attributes. To summarize, the primary contributions of this paper are:

- We present an optimization framework that renders different approaches for yielding fair data. The approaches include removing, adding generated data, or solely using generated data.
- We underscore the framework's ability to handle cases where data privacy is a significant concern.
- Our methodology is designed to handle a protected attribute that can be non-binary, offering broader applicability.
- We carry out an extensive evaluation of the proposed techniques on three biased datasets. The evaluation focuses on their effectiveness in reducing discrimination and their runtimes.
- We publish our implementation at https://github.com/mkduong-ai/fairdo as a documented Python package and distribute it over PyPI.

2 Related Work

Recently, related works have equivalently formulated subset selection problems to achieve fairness goals [7,26]. While in the work of Tang et al. [26], a distribution is generated that represents the selection probability of each feasible set to maximize the global utility on average, our work aims to return a definite subset. To achieve fairness according to any defined criteria, our formulation treats discrimination measures as black boxes. These measures can encompass both group and individual fairness notions, distinguishing our work from that of Tang et al. [26], whose framework is limited to group fairness.

Previous studies have also utilized synthetic data to address fairness and privacy concerns [7,19]. Both of these studies employed heuristics similar to our approach. In particular, Liu et al. [19] specialized on generating synthetic data using a genetic algorithm to satisfy specific privacy definitions [3,8]. While our framework does not generate privacy-preserving data specifically, it utilizes synthetic data, which can be generated with such methods. Similarly to our work, Duong et al. [7] leveraged synthetic data by introducing a sampling-based heuristic for selecting a subset of such data points to minimize discrimination. Our work generalizes the work of Duong et al. [7] as their approach can be viewed as a special case of ours. Additionally, our formulation offers greater flexibility compared to the approach of Duong et al. [7], as it allows for any heuristic to tackle the task and is also not limited to binary protected attributes.

3 Measuring Discrimination

In this section, we introduce the notation used to derive discrimination measures for assessing dataset fairness: A *data point* or *sample* is represented as a triple (x, y, z), where $x \in X$ is the *feature*, $y \in Y$ is the ground truth *label* indicating favorable or unfavorable outcomes, and $z \in Z$ is the *protected attribute*, which is used to differentiate between groups. The sets X, Y, Z typically hold numeric values and are defined as $X = \mathbb{R}^d$, $Y = \{0, 1\}$, and $Z = \{1, 2, \ldots, k\}$ with $k \geq 2$. For instance, in the context of predicting personal attributes, we can use X to represent numeric values that encode particular aspects of a person. Y typically describes the positive or negative outcome that we aim to predict for the person. Z can denote any protected attribute, such as race, which can be used to identify the person as Caucasian, Afro-American, Latin American, or Asian. We assume that z is not included as a feature in x. To be able to differentiate between groups, $k \geq 2$ must hold. If $k > 2$, the protected attribute Z is said to be non-binary. Following the definition, a *dataset*, denoted as $\mathcal{D} = \{d_i\}_{i=1}^n$, consists of data points, where a single sample is defined as $d_i = (x_i, y_i, z_i)$. Machine learning models are trained using these datasets to predict the target variable y based on the input variables x and z. Finally, we denote a discrimination measure with $\psi \colon \mathbb{D} \to [0, 1]$, where $\mathbb{D}$ is the set of all datasets.

In the following, x, y, z are noted as random variables that can take on specific values.

3.1   Absolute Measures

To deal with non-binary groups, Žliobaitė [29] suggested in her work to compare groups pairwise. For this, she presented three possible ways which are comparing each group with another, one against the rest for each group, and all groups against the unprivileged group. The author further discussed options to aggregate the results. Although Žliobaitė [29] stated textually how to measure discrimination for more than two groups, we express them mathematically in this work. To treat groups equally without presuming which group is unprivileged and to get the full picture, we choose to make use of comparing each group with another. We first introduce the common fairness notion *statistical parity* [16, 28], which demands equal positive outcomes for different groups in $Z = \{1, 2, \ldots, k\}$. It is usually defined for binary groups, but we present the non-binary cases [29].

Definition 1 (Statistical parity). *Demanding that each of the k groups have the same probability of receiving the favorable outcome is statistical parity, i.e.,*

$$P(y = 1 \mid z = 1) = \ldots = P(y = 1 \mid z = k)$$
$$\iff \quad P(y = 1 \mid z = i) = P(y = 1 \mid z = j) \quad \forall i, j \in Z.$$

As the group size k grows, the satisfaction of statistical parity becomes less probable. Because of this, the equality constraints are treated softly by deriving differences between the groups. Consequently, smaller differences imply more equality. For binary groups, the difference is often referred to as statistical disparity (SDP) [6].

Definition 2 (Sum of absolute statistical disparities). *Let there be k groups, then the sum of absolute statistical disparities is calculated as follows [29]:*

$$\psi_{SDP\text{-}sum}(\mathcal{D}) = \sum_{\substack{i,j \in Z \\ i \neq j}} |P(y=1 \mid z=i) - P(y=1 \mid z=j)|$$

$$= \sum_{i=1}^{k} \sum_{j=i+1}^{k} |P(y=1 \mid z=i) - P(y=1 \mid z=j)|.$$

Because the total number of comparisons is $\frac{k(k-1)}{2}$ [29], the average discrimination between all groups becomes:

$$\psi_{SDP\text{-}avg}(\mathcal{D}) = \frac{2}{k(k-1)} \cdot \sum_{i=1}^{k} \sum_{j=i+1}^{k} |P(y=1 \mid z=i)$$

$$- P(y=1 \mid z=j)|.$$

Definition 3 (Maximal absolute statistical disparity). *Maximal absolute statistical disparity measures the absolute statistical disparity between all pairs $i, j \in Z$ and returns the maximum value. Specifically, it is given by:*

$$\psi_{SDP\text{-}max}(\mathcal{D}) = \max_{i,j \in Z} |P(y=1 \mid z=i)$$

$$- P(y=1 \mid z=j)|.$$

Žliobaitė [29], after consulting with legal experts, recommends using the maximum function to aggregate disparities, though the choice depends on the ethical context of the specific use case. Discrimination measures can be seen as social welfare functions. Minimizing the sum of absolute statistical disparities is analogous to the utilitarian viewpoint [21], which aims to maximize the general utility of the population. If one decides to care for the least well-off group, then minimizing the maximal absolute statistical disparity corresponds to the Rawlsian social welfare [25].

4   Optimization Framework

Inspired by related works that identify unfair data samples [15,27], we propose a method to remove such samples for fairness. The task is formulated as a combinatorial problem where the aim is to determine a subset $\mathcal{D}_{\text{fair}}$ of a given set S such that the discrimination of the subset $\psi(\mathcal{D}_{\text{fair}})$ is minimal, as shown in Fig. 1. Depending on the application, set S can be the original data $\mathcal{D}$, a synthetic set G with the same distribution as $\mathcal{D}$, or their union $\mathcal{D} \cup G$.

4.1   Problem Formulation

To state the problem mathematically, let note $S = \{s_1, s_2, \ldots, s_{\tilde{n}}\}$ and further introduce a binary vector b with the same length as S, i.e., $b = (b_1, b_2, \ldots, b_{\tilde{n}})$. To define the combinatorial optimization problem, each entry b_i in b is examined whether it is 1 ($b_i = 1$), in which case the corresponding sample s_i in S is included in the subset $\mathcal{D}_{\text{fair}}$. Therefore, the fair set is defined with

$$\mathcal{D}_{\text{fair}} = \{s_i \in S \mid b_i = 1, i = 1 \ldots \tilde{n}\}. \tag{1}$$

The objective $f \colon 0, 1^{\tilde{n}} \to [0, 1]$ can then be expressed by:

$$
\begin{aligned}
f_{S,\psi}(b) &= \psi(\mathcal{D}_{\text{fair}}) \\
\Longleftrightarrow \quad f_{S,\psi}(b) &= \psi(\{s_i \in S \mid b_i = 1, i = 1 \ldots \tilde{n}\}),
\end{aligned}
\tag{2}
$$

where $f_{S,\psi}$ is defined as the discrimination of a subset $\mathcal{D}_{\text{fair}}$ of the given set S and ψ evaluates the level of discrimination on $\mathcal{D}_{\text{fair}}$. Note that the decision variable is b, for which $\mathcal{D}_{\text{fair}}$ can be obtained. The subindices S and ψ of $f_{S,\psi}$ can be seen as settings for the objective. Ignoring the subindices, we write out the combinatorial optimization problem as follows:

$$\min_b \quad f(b) \tag{3}$$

$$\text{subject to} \quad b_i \in \{0, 1\} \quad \forall i = 1, \ldots, \tilde{n}.$$

Because the set of feasible subsets $\mathcal{P}(S)$ grows exponentially regarding the cardinality of S, we employ heuristics to solve our stated problem.

In the following subsections, we discuss different and useful settings of S that serve different purposes with their corresponding advantages and disadvantages.

4.2 Removing Samples ($S = \mathcal{D}$)

By setting $S = \mathcal{D}$, it is intended to determine data points in the training set that can be removed to prevent discrimination. Intuitively, having an overexposure of certain types of samples that fulfill stereotypes can result in a discriminatory dataset. In such situations, the most practical step is to remove the affected samples.

However, this method is not recommended if the given dataset is small. Likewise, some could argue that minorities can be easily removed by this method. Luckily, this can be prevented by choosing the right discrimination measure.

4.3 Employing only Synthetic Data ($S = G$)

To employ synthetic data, this method relies on a statistical model. The statistical model is used to learn the distribution of the original data $P(\mathcal{D})$. By doing so, synthetic samples G can be drawn from the learned distribution $G \sim P(\mathcal{D})$.

Relying solely on synthetic data is particularly important in use cases where data privacy and protection are major concerns and the use of real data is prohibited. Of course, synthetic data is not necessarily disjoint from the original dataset and can therefore be a privacy breach itself. For tabular and smaller datasets, this can be naively mitigated by removing such privacy breaching points from the synthetic data by setting $S = G \setminus \mathcal{D}$. Other ways include populating differential privacy techniques in the data generation process [1,8,13,19].

When generally using synthetic data, one cannot easily ensure that the corresponding label of the features is correct. Training machine learning models on synthetic data can therefore lead to higher error rates when predicting on real data. Despite the distribution of the synthetic data following the distribution of the real dataset, it depends heavily on the method used when it comes to generating qualitative, faithful data.

4.4   Merging Real and Synthetic Data ($S = \mathcal{D} \cup G$)

Another approach to generate a non-discriminatory dataset is to merge the original dataset $\mathcal{D}$ with synthetic data G that has been generated with a statistical model as described in Sect. 4.3. By combining the two sets $S = \mathcal{D} \cup G$, it is possible to increase the size of the resulting dataset while avoiding over-representation of discriminatory samples.

One advantage of this method is that it can improve the quality of the data by utilizing both the real $\mathcal{D}$ and synthetic data G. The resulting dataset can be larger and more diverse, which can lead to greater robustness when training machine learning models. If the dataset is too small to apply removal techniques ($S = \mathcal{D}$) or relying solely on synthetic data ($S = G$) appears unreliable, merging the two sets may be a viable option.

However, this method is not without its limitations and comes with disadvantages when generally using synthetic data, e.g., quality and faithfulness. Different from the method described in Sect. 4.3, this method is not applicable for purposes with privacy concerns as samples from the real data are not omitted.

4.5   Adding Synthetic Data

A different approach that requires a new formulation of the objective is to include synthetic data points without deleting any samples from the real data. As well, a set of generated data points G must be given, and the research question is which of the generated points can lead to a fairer distribution when including them in the original dataset. The possible use case for this problem is to fine-tune machine learning models that have already learned from an unfair dataset. This is mostly useful for large machine learning models where resources are scarce to retrain the whole model. Following the preceding notation, the fair dataset becomes:

$$\mathcal{D}_{\text{fair}}^{\text{add}} = \mathcal{D} \cup \{s_i \in S \mid b_i = 1, i = 1\ldots\tilde{n}\} \qquad (4)$$

and we express the corresponding objective $f_{S,\psi}^{\text{add}}$ by:

$$f_{S,\psi}^{\text{add}}(b) = \psi(\mathcal{D}_{\text{fair}}^{\text{add}})$$
$$\iff \quad f_{S,\psi}^{\text{add}}(b) = \psi(\mathcal{D} \cup \{s_i \in S \mid b_i = 1, i = 1\ldots\tilde{n}\}), \qquad (5)$$

where S is set to G to achieve the described approach. Certainly, S can also be set to $\mathcal{D}$ or any other set operation on $\mathcal{D}$ with G. Although such settings are possible, they do not serve any meaningful purposes. However, one could argue that setting $S = \mathcal{D}$ can act as a reweighing method. Still, we argue against facilitating duplicates in a dataset with intent, as no additional information is provided.

As seen, our framework offers many advantages due to its versatility and therefore potential use in a broad range of applications. By choosing the appropriate objective function, discrimination measure, and sample set, the formulation is tailored to the specific intent and use case. Because the formulation is agnostic to the solver, it can serve multiple purposes without modifying solvers.

Table 1. Overview of Datasets

Dataset	Entries	Cols.	Label	Protected Attribute	Description
Adult [17]	32 561	22	Income	Race: White, Black, Asian-Pacific-Islander, American-Indian-Eskimo, Other	Indicates individuals earning over $50,000 annually
Bank [22]	41 188	53	Term deposit subscription	Job: Admin, Blue-Collar, Technician, Services, Management, Retired, Entrepreneur, Self-Employed, Housemaid, Unemployed, Student, Unknown	Shows whether the client has subscribed to a term deposit.
COMPAS [18]	7 214	8	2-year recidivism	Race: African-American, Caucasian, Hispanic, Other, Asian, Native American	Displays individuals that were rearrested for a new crime within 2 years after initial arrest

5 Heuristics

This section presents heuristics that specifically solve combinatorial optimization problems. These include: a baseline method that returns the original dataset, a simple random heuristic, and genetic algorithms with different operators.

1. **Original**: Uses the original data by returning a vector of ones $b = \mathbf{1}_{\tilde{n}}$.
2. **Random Heuristic**: Generates a user-defined number of random vectors, with each entry having a 50% chance of being zero or one, and then returns the best solution.
3. **Genetic Algorithm (GA)**: The workflow of GAs [9] involves generating an initial population of candidate solutions and then repeatedly performing *selection, crossover,* and *mutation* operations over several generations. In our implementation, the GA terminates earlier if improved solutions were not found within 50 generations. Following operators were used in our experimentation [11]:
    - Selection: *Elitist, Tournament, Roulette Wheel* (see [11] for more details)
    - Crossover: *Uniform* (each entry of the offspring has the same probability of either inheriting the entry from the first or second parent)
    - Mutation: *Bit Flip* (flips a fixed amount of random bits for each vector, that is $\lfloor p_m \cdot \tilde{n} \rfloor$, where $p_m \in [0, 1]$ is the mutation rate)

6 Evaluation

In our evaluation, we conducted multiple experiments to address the following research questions:

- **RQ1** How do the heuristics perform in making the datasets fairer?
- **RQ2** How does runtime vary among the heuristics?
- **RQ3** How stable are the results across the runs?
- **RQ4** Is there a clear winner? If yes, which method is recommended for practical use?

To answer these research questions, we specifically designed experiments for the Adult [17], Bank [22], and COMPAS [18] datasets. Both the Adult and COMPAS datasets include race as a non-binary protected attribute, whereas the Bank dataset utilizes the job as a non-binary protected attribute. All datasets were prepared and cleansed in the same manner: Categorical features were one-hot encoded, with the exception of the protected attribute and the label. Additionally, rows containing missing values were excluded from all datasets. Table 1 shows details about the datasets used in our experiments after the preparation and cleansing steps.

Following the dataset preparation, we executed two distinct experiments. The first experiment (Sect. 6.1) was dedicated to hyperparameter tuning of the GAs, adjusting both population sizes and the number of generations to pinpoint optimal configurations. Armed with these optimal settings, our second experiment (Sect. 6.2) focused on comparing different selection operators within GAs (**RQ1**). Our aim was to determine which operator yielded the best performance. This experiment included comparisons to several baseline methods, one of which simply returned the original data. By expanding our evaluation to multiple discrimination measures in this phase, we can comprehensively assess the effectiveness of GAs in reducing discrimination in datasets.

The experimental methodology involves the application of heuristics to produce a binary mask, which yields fair data. We then measure the discrimination of the resulting dataset. To ensure stability in our findings (**RQ3**), each experiment was repeated 15 times. We additionally recorded the runtime of each trial to tackle **RQ2**. Depending on the experiment, we employed suitable heuristics that aim to solve each objective with the associated discrimination measure, as listed in Table 2. For instance, each heuristic either optimizes $f_{S,\psi}$ or $f_{S,\psi}^{add}$ with varying settings of S and ψ as given in the table. In order to perform experiments with synthetic data, we generated data that has the same size as the original dataset, i.e., $|G| = |\mathcal{D}|$. The statistical model used to generate synthetic data is Gaussian copula [23] which is fast and performs well on tabular data. For privacy-sensitive use cases, we advise utilizing privacy-preserving techniques [1,8,13,19]. All experiments were conducted on an Intel(R) Xeon(R) Gold 5120 processor clocking at 2.20 GHz.

Table 2. Configuration details of heuristics, objectives, and discrimination measures for each experiment.

Experiment	Heuristics	Objectives (f, S)	Disc. Measures (ψ)
Hyperparam	GA	Remove, Merge, Add	Sum SDP
Comparison	Original, Random, GA (Elitist, Tournament, Roulette Wheel)	Remove, Merge, Add	Sum SDP, Max SDP

6.1   Hyperparameter Tuning

For the genetic algorithm, we performed hyperparameter tuning, exploring various population sizes [20, 50, 100, 200] and generations [50, 100, 200, 500], all using tournament selection, uniform crossover, and bit flip mutation at a rate of 5%. These configurations are described in Sect. 5. We evaluated the algorithm on three distinct objectives and set $\psi_{SDP\text{-}sum}$ as the discrimination measure.

Discrimination. As seen in Fig. 2, the heatmaps display the average discrimination (including the standard deviation) of GAs solving various objectives on different datasets. Each heatmap shows hyperparameters that were set for the experimentation. Across the different objectives and datasets, there is a consistent trend indicating that utilizing larger populations combined with a higher number of generations typically results in less discrimination. This is particularly evident when contrasting scenarios with a population size of 20 and 50 generations, which, on average, have discrimination scores higher by 0.1. However, the improvements in discrimination plateau beyond certain thresholds. Specifically, once the number of generations surpasses 200 or when the population size exceeds 100, there is no significant further decrease in discrimination observable.

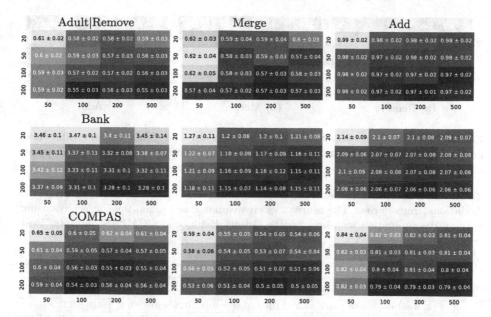

Fig. 2. Heatmaps showing discrimination scores ($\psi_{\text{SDP-sum}}$) after pre-processing with genetic algorithms using different population sizes (y-axis) and generations (x-axis). Rows depict the results of Adult, Bank, and COMPAS datasets, while columns represent the objectives.

Runtime. For brevity reasons, we display the runtimes solely for the Bank dataset in Fig. 3, given its larger size and the similarity of the results across other datasets. The outcome of this analysis pointed towards an optimal setting of a population size of 100 combined with 500 generations. Under our specifications, executing the GA with these settings takes, on average, between 1.5 and 4.5 min. While increasing the population size further did not show significant improvements in reducing the bias in the datasets, it proved to be more efficient in terms of the runtime.

6.2   Comparing Heuristics

After determining that a population size of 100 with 500 generations offered optimal results w.r.t. discrimination and time, this configuration was maintained for all subsequent experiments. Here, three GAs were compared, each differing by their selection operator: elitist, tournament, and roulette wheel selection. All GAs were set with uniform crossover and bit flip mutation at a rate of 5% to perform the experiments. Additionally, we established both the original dataset and the random heuristic as baselines.

Discrimination. Table 3 presents the discrimination results of our experiments. It is evident that all tested algorithms are stable, as reflected by the low standard

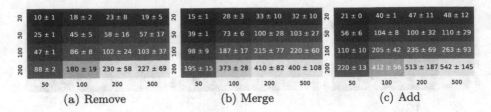

	10 ± 1	18 ± 2	23 ± 8	19 ± 5		15 ± 1	28 ± 3	33 ± 10	32 ± 10		21 ± 0	40 ± 1	47 ± 11	48 ± 12
20	25 ± 1	45 ± 5	58 ± 16	57 ± 17	20	39 ± 1	73 ± 6	100 ± 28	103 ± 27	20	56 ± 6	104 ± 8	100 ± 32	110 ± 29
	47 ± 1	86 ± 8	102 ± 24	103 ± 37		98 ± 9	187 ± 17	215 ± 77	220 ± 60		110 ± 10	205 ± 42	235 ± 69	263 ± 93
	88 ± 2	180 ± 19	230 ± 58	227 ± 69		195 ± 15	373 ± 28	410 ± 82	400 ± 108		220 ± 13	412 ± 56	513 ± 187	542 ± 145

(a) Remove   (b) Merge   (c) Add

Fig. 3. Heatmaps showing runtimes in seconds for the Bank dataset after pre-processing with genetic algorithms using different population sizes (y-axis) and generations (x-axis).

Table 3. Displayed are the mean discrimination scores, accompanied by standard deviations, from 15 runs. The heuristics were evaluated across multiple objectives using varying discrimination measures on the Adult, Bank, and COMPAS datasets. Best results are marked bold.

Objective	Method	Sum SDP			Max SDP		
		Adult	Bank	COMPAS	Adult	Bank	COMPAS
Add	1. Original	1.07 ± 0.02	1.83 ± 0.09	1.17 ± 0.06	0.23 ± 0.00	0.09 ± 0.00	0.17 ± 0.01
	2. Random	1.03 ± 0.02	2.27 ± 0.07	0.94 ± 0.03	0.21 ± 0.00	0.11 ± 0.00	0.15 ± 0.01
	3. Elitist	**0.82 ± 0.02**	**1.54 ± 0.06**	**0.59 ± 0.03**	**0.16 ± 0.00**	**0.07 ± 0.00**	**0.10 ± 0.00**
	4. Tournament	0.97 ± 0.02	2.06 ± 0.06	0.80 ± 0.03	0.20 ± 0.00	0.10 ± 0.00	0.13 ± 0.00
	5. Roulette	1.03 ± 0.02	2.31 ± 0.08	0.94 ± 0.05	0.21 ± 0.00	0.11 ± 0.00	0.15 ± 0.01
Merge	1. Original	1.07 ± 0.02	1.83 ± 0.09	1.17 ± 0.06	0.23 ± 0.00	0.09 ± 0.00	0.17 ± 0.01
	2. Random	0.80 ± 0.03	1.46 ± 0.09	0.76 ± 0.08	0.16 ± 0.01	0.07 ± 0.00	0.12 ± 0.01
	3. Elitist	**0.21 ± 0.04**	**0.42 ± 0.07**	**0.11 ± 0.05**	**0.04 ± 0.01**	**0.02 ± 0.00**	**0.01 ± 0.00**
	4. Tournament	0.58 ± 0.04	1.17 ± 0.09	0.51 ± 0.04	0.11 ± 0.01	0.05 ± 0.00	0.09 ± 0.01
	5. Roulette	0.85 ± 0.05	1.49 ± 0.09	0.79 ± 0.09	0.16 ± 0.01	0.07 ± 0.00	0.12 ± 0.01
Remove	1. Original	0.97 ± 0.00	4.81 ± 0.00	1.89 ± 0.00	0.17 ± 0.00	0.25 ± 0.00	0.27 ± 0.00
	2. Random	0.71 ± 0.02	4.07 ± 0.07	0.72 ± 0.03	0.12 ± 0.00	0.19 ± 0.00	0.12 ± 0.01
	3. Elitist	**0.25 ± 0.02**	**1.41 ± 0.12**	**0.20 ± 0.07**	**0.05 ± 0.00**	**0.07 ± 0.01**	**0.01 ± 0.00**
	4. Tournament	0.57 ± 0.02	3.29 ± 0.08	0.56 ± 0.04	0.11 ± 0.00	0.15 ± 0.01	0.09 ± 0.01
	5. Roulette	0.75 ± 0.03	4.15 ± 0.10	0.75 ± 0.08	0.13 ± 0.00	0.20 ± 0.01	0.12 ± 0.01

deviations (**RQ3**). All heuristics were able to reduce the discrimination available in the datasets in most cases. Elitist selection consistently outperformed other methods, offering notable improvements in fairness compared to the original datasets (**RQ1**). We emphasize that the measures handle non-binary attributes, providing flexibility in targeting various fairness goals. Further, by the range of discrimination measures utilized, our methodology can aim for diverse fairness goals, be it the enhancement of the utilitarian social welfare ($\psi_{\text{SDP-sum}}$) or Rawlsian social welfare ($\psi_{\text{SDP-max}}$), as evidenced. An interesting observation from our study is the varied discrimination levels based on the specific measure used, as seen in the Bank dataset, where its discrimination is either highest or lowest when compared with other datasets. This is due to the higher number of groups, leading to more group comparisons that affect the overall discrimination score. When examining the objectives, removing both the synthetic and original data tends to outperform others. This observation is particularly evident in the

Merge objective. Given the consistent performance of the elitist selection in our tests, we strongly recommend its use for those aiming to achieve the best fairness outcomes (**RQ4**).

Table 4. Mean runtimes in seconds of different methods solving different objectives with varying discrimination measures on the Adult, Bank, and COMPAS datasets.

Objective	Method	Sum SDP			Max SDP		
		Adult	Bank	COMPAS	Adult	Bank	COMPAS
Add	1. Original	0 ± 0	0 ± 0	0 ± 0	0 ± 0	0 ± 0	0 ± 0
	2. Random	50 ± 1	107 ± 12	14 ± 0	51 ± 6	103 ± 7	13 ± 0
	3. Elitist	320 ± 105	605 ± 224	53 ± 21	334 ± 80	636 ± 179	79 ± 23
	4. Tournament	122 ± 38	209 ± 50	39 ± 17	119 ± 37	216 ± 74	34 ± 12
	5. Roulette	82 ± 26	131 ± 46	26 ± 9	82 ± 40	132 ± 48	26 ± 12
Merge	1. Original	0 ± 0	0 ± 0	0 ± 0	0 ± 0	0 ± 0	0 ± 0
	2. Random	46 ± 3	67 ± 1	15 ± 2	44 ± 4	66 ± 1	15 ± 3
	3. Elitist	283 ± 103	359 ± 143	79 ± 25	286 ± 111	397 ± 161	75 ± 28
	4. Tournament	127 ± 39	185 ± 69	36 ± 11	131 ± 61	169 ± 51	44 ± 19
	5. Roulette	69 ± 21	127 ± 53	28 ± 9	83 ± 33	118 ± 31	29 ± 14
Remove	1. Original	0 ± 0	0 ± 0	0 ± 0	0 ± 0	0 ± 0	0 ± 0
	2. Random	23 ± 1	44 ± 1	11 ± 0	22 ± 1	47 ± 11	11 ± 0
	3. Elitist	138 ± 66	281 ± 119	52 ± 18	176 ± 68	290 ± 80	50 ± 15
	4. Tournament	78 ± 13	119 ± 25	24 ± 9	73 ± 27	132 ± 46	25 ± 7
	5. Roulette	58 ± 27	72 ± 10	22 ± 8	52 ± 24	71 ± 24	18 ± 7

Runtime. An analysis of the runtimes is presented in Table 4. The original method consistently took 0 s (rounded) to finish. At second comes the random method and lastly GAs. The elitist operator took the longest, with runtimes approximately three times slower than the quickest operator, the roulette wheel. Tournament selection comes in between. Most experiments were finished in 5 min or less, which is still very efficient. Regarding the measures, the runtimes when optimizing $\psi_{\text{SDP-max}}$ appeared negligibly higher compared to $\psi_{\text{SDP-sum}}$, so it can be disregarded. Generally, larger datasets yielded longer runtimes, revealing a linear relationship between dataset size and runtime. In addressing the research question posed in **RQ4**, it becomes evident that the elitist operator is superior among the tested methods. Despite being the slowest method, it is still very efficient at reducing discrimination on datasets consisting of up to 41 188 samples, as seen in our experimentation.

7   Conclusion

We introduced a novel and flexible optimization framework to reduce discrimination and preserve privacy in datasets. The framework accommodates various intents such as data removal, synthetic data addition, and exclusive use of synthetic data for privacy reasons. Notably, the objectives in our framework are designed to be independent of specific discrimination measures, allowing users and stakeholders to address any form of discrimination without modifying the solvers.

Due to the relatively sparse work existing on dealing with non-binary attributes, particularly regarding established methods, we tackled non-binary protected attributes in our experiments by deriving discrimination measures based on the work of Žliobaitė [29] and showed that our framework allowed the effective and fast reduction of discrimination by employing heuristics.

8 Future Work and Discussion

Future work could include extending the usability of this framework by deriving different discrimination measurements. Thus, handling multiple protected attributes as well as regression tasks can be done without modifying the general methodology. Additionally, formulating and integrating constraints into the objective function can also be done, which further enhances the responsibility of our approach. For instance, we could consider constraints such as group sizes and add penalties if samples of minorities get removed.

Although we aim for fairness and data privacy with our framework, it is still important to engage with diverse stakeholders to identify unintended consequences and address possible ethical implications. Particularly, an extensive discussion and analysis of the used objective and discrimination measure for a specific application should be done to ensure that the data aligns with the desired fairness goals.

References

1. Abay, N.C., Zhou, Y., Kantarcioglu, M., Thuraisingham, B., Sweeney, L.: Privacy preserving synthetic data release using deep learning. In: Berlingerio, M., Bonchi, F., Gärtner, T., Hurley, N., Ifrim, G. (eds.) ECML PKDD 2018. LNCS (LNAI), vol. 11051, pp. 510–526. Springer, Cham (2019). https://doi.org/10.1007/978-3-030-10925-7_31
2. Barocas, S., Hardt, M., Narayanan, A.: Fairness and machine learning. fairmlbook.org (2019). http://www.fairmlbook.org
3. Bun, M., Steinke, T.: Concentrated differential privacy: simplifications, extensions, and lower bounds. In: Hirt, M., Smith, A. (eds.) TCC 2016. LNCS, vol. 9985, pp. 635–658. Springer, Heidelberg (2016). https://doi.org/10.1007/978-3-662-53641-4_24
4. Caton, S., Haas, C.: Fairness in machine learning: a survey. arXiv preprint arXiv:2010.04053 (2020)
5. Celis, L.E., Huang, L., Keswani, V., Vishnoi, N.K.: Fair classification with noisy protected attributes: a framework with provable guarantees. In: Meila, M., Zhang, T. (eds.) Proceedings of the 38th International Conference on Machine Learning. Proceedings of Machine Learning Research, 18–24 July 2021, vol. 139, pp. 1349–1361. PMLR (2021). https://proceedings.mlr.press/v139/celis21a.html

6. Dunkelau, J., Leuschel, M.: Fairness-aware machine learning (2019)
7. Duong, M.K., Conrad, S.: Dealing with data bias in classification: can generated data ensure representation and fairness? In: Wrembel, R., Gamper, J., Kotsis, G., Tjoa, A.M., Khalil, I. (eds.) Big Data Analytics and Knowledge Discovery, DaWaK 2023. LNCS, vol. 14148, pp. 176–190. Springer, Cham (2023). https://doi.org/10.1007/978-3-031-39831-5_17
8. Dwork, C.: Differential privacy. In: Bugliesi, M., Preneel, B., Sassone, V., Wegener, I. (eds.) ICALP 2006. LNCS, vol. 4052, pp. 1–12. Springer, Heidelberg (2006). https://doi.org/10.1007/11787006_1
9. Eiben, A.E., Smith, J.E.: Introduction to Evolutionary Computing. NCS, Springer, Heidelberg (2015). https://doi.org/10.1007/978-3-662-44874-8
10. Friedrich, F., et al.: Fair diffusion: instructing text-to-image generation models on fairness. arXiv preprint at arXiv:2302.10893 (2023)
11. Goldberg, D.E.: Genetic Algorithms in Search, Optimization and Machine Learning, 1st edn. Addison-Wesley Longman Publishing Co., Inc, USA (1989)
12. Holland, J.: Adaptation in Natural and Artificial Systems (1975)
13. Jordon, J., Yoon, J., Van Der Schaar, M.: PATE-GAN: generating synthetic data with differential privacy guarantees. In: International Conference on Learning Representations (2019)
14. Kamani, M.M., Haddadpour, F., Forsati, R., Mahdavi, M.: Efficient fair principal component analysis. Mach. Learn. 111, 3671–3702 (2022). https://doi.org/10.1007/s10994-021-06100-9
15. Kamiran, F., Calders, T.: Data preprocessing techniques for classification without discrimination. Knowl. Inf. Syst. 33(1), 1–33 (2012)
16. Kamishima, T., Akaho, S., Asoh, H., Sakuma, J.: Fairness-aware classifier with prejudice remover regularizer. In: Flach, P.A., De Bie, T., Cristianini, N. (eds.) ECML PKDD 2012. LNCS (LNAI), vol. 7524, pp. 35–50. Springer, Heidelberg (2012). https://doi.org/10.1007/978-3-642-33486-3_3
17. Kohavi, R.: Scaling up the accuracy of Naive-Bayes classifiers: a decision-tree hybrid. In: KDD 1996, pp. 202–207. AAAI Press (1996)
18. Larson, J., Angwin, J., Mattu, S., Kirchner, L.: Machine bias, May 2016. https://www.propublica.org/article/machine-bias-risk-assessments-in-criminal-sentencing
19. Liu, T., Tang, J., Vietri, G., Wu, S.: Generating private synthetic data with genetic algorithms. In: International Conference on Machine Learning, pp. 22009–22027. PMLR (2023)
20. Mehrabi, N., Morstatter, F., Saxena, N., Lerman, K., Galstyan, A.: A survey on bias and fairness in machine learning. ACM Comput. Surv. (CSUR) 54(6), 1–35 (2021)
21. Mill, J.S.: Utilitarianism. Parker, Son, and Bourn (1863)
22. Moro, S., Cortez, P., Rita, P.: A data-driven approach to predict the success of bank telemarketing. Decis. Support Syst. 62, 22–31 (2014)
23. Patki, N., Wedge, R., Veeramachaneni, K.: The synthetic data vault. In: 2016 IEEE International Conference on Data Science and Advanced Analytics (DSAA), October 2016, pp. 399–410 (2016). https://doi.org/10.1109/DSAA.2016.49
24. Prost, F., Qian, H., Chen, Q., Chi, E.H., Chen, J., Beutel, A.: Toward a better trade-off between performance and fairness with kernel-based distribution matching. CoRR abs/1910.11779 (2019). http://arxiv.org/abs/1910.11779
25. Rawls, J.: A Theory of Justice. Belknap Press (1971)
26. Tang, S., Yuan, J.: Beyond submodularity: a unified framework of randomized set selection with group fairness constraints. J. Comb. Optim. 45(4), 102 (2023)

27. Verma, S., Ernst, M.D., Just, R.: Removing biased data to improve fairness and accuracy. CoRR abs/2102.03054 (2021). https://arxiv.org/abs/2102.03054
28. Zemel, R., Wu, Y., Swersky, K., Pitassi, T., Dwork, C.: Learning fair representations. In: International Conference on Machine Learning, pp. 325–333. PMLR (2013)
29. Žliobaitė, I.: Measuring discrimination in algorithmic decision making. Data Min. Knowl. Disc. **31**(4), 1060–1089 (2017). https://doi.org/10.1007/s10618-017-0506-1

MStoCast: Multimodal Deep Network for Stock Market Forecast

Kamaladdin Fataliyev[1]([✉])(ID) and Wei Liu[2](ID)

[1] University of Technology Sydney, Sydney, Australia
kamaladdin.fataliyev@student.uts.edu.au
[2] University of Technology Sydney, Sydney, Australia
wei.liu@uts.edu.au

Abstract. Stock market analysis is a complex task that involves various types of data, such as web news, historical prices, and technical market indicators. Recent research in this area focuses on analyzing these modalities either separately or all together, but the underlying correlation patterns in the multimodal data were not captured. To address this issue, we propose MStoCast, a Multimodal Stock Market Forecast model that uses innovatively designed deep networks. First, we propose a common network that captures cross-modality and joint information. Then, we construct a unique network that discovers bi-modal information from the inputs. These pieces of information are then integrated and processed through a fully connected layer to predict the direction of the closing price movement. Experiments using real world datasets show that our MStoCast model significantly outperforms other state-of-the-art models.

Keywords: Stock market forecast · Multimodal deep learning · Word embeddings

1 Introduction

The price movements in stock markets are influenced by various data sources, including historical price data and technical indicators [29], financial news [25], social media [5], and official announcements [9]. It has been demonstrated that analyzing these multiple data modalities collectively can aid in capturing the underlying patterns of stock movements [17,27], thereby making stock market prediction a multimodal learning task [28]. Employing effective multimodal representation and learning techniques to capture the influence of these diverse data modalities is crucial for model performance. Therefore, it has become essential to appropriately integrate these various types of data (e.g., financial news, stock market data, and technical indicators) to predict price movements [17,18].

While the common approach involves concatenating raw features from input modalities [2], this vector-based method might struggle to extract the interconnectedness of the data sources [35]. Although matrices have been utilized to

© The Author(s), under exclusive license to Springer Nature Singapore Pte Ltd. 2024
D. Benavides-Prado et al. (Eds.): AusDM 2023, CCIS 1943, pp. 121–136, 2024.
https://doi.org/10.1007/978-981-99-8696-5_9

handle two modalities, they pose challenges when dealing with three or more input modalities, and they may also fail to capture intra-modal information. However, it's imperative to capture both intra-modal and inter-modal features [30].

Lastly, capturing bi-modal relationships among the input data is a crucial aspect that should also be taken into consideration, especially when dealing with three or more modalities. However, existing models often might fall short when dealing with three or more modalities.

Moreover, capturing bi-modal relationships among the input data is also significant. However, existing models might fall short when dealing with three or more modalities. To address these challenges, we propose MStoCast, a stock movement prediction model that employs both CNNs (such as ResNet [11]) and RNNs (such as BERT [6]) to analyze multiple input modalities (i.e., financial news, stock market data, and technical indicators). It's important to note that, although we have chosen ResNet and BERT (two of the most advanced models at present) for this research, they can be substituted with other CNN and RNN models to accommodate the requirements of different problems. Our objective is to capture inter-modal and intra-modal information, as well as bi-modal relationships among the input data modalities, by introducing common and unique sub-networks.

MStoCast utilizes two types of information extracted from multimodal input sequences. The first type of information involves the bi-modal interactions among pairs of input modalities, representing distinct and unique insights. The second type of information encompasses common patterns, capturing both inter-modal relationships and intra-modal details derived from all input modalities. To achieve this, our approach involves designing both a unique network and a common network, each aimed at extracting the respective kinds of information crucial for accurate stock market prediction.

The unique network begins by implementing early information fusion where we combine the input features in pairs to create compound matrices. These matrices are then processed through three separate ResNets, enabling the extraction of bi-modal relations from the modalities. Conversely, the common network commences by employing Long Short-Term Memory networks (LSTMs) to extract intra-modal and modality-specific information from each input modality. These extracted intra-modal details are amalgamated into a multi-modal tensor through their outer product. Subsequently, Convolutional Neural Networks (CNNs), specifically ResNets, are employed to extract cross-modal information from this tensor. The resulting information from both the common and unique networks is then concatenated, forming a compound vector. Through the utilization of global average pooling, a feature map is obtained. The final stages encompass the creation of two fully connected layers, responsible for analyzing the feature vector. Additionally, another fully connected network is employed to facilitate market movement prediction.

In our market prediction model, we incorporate historical market data, a selection of seven technical market indicators, and financial news. MStoCast effectively analyzes these three input modalities to anticipate the directional movement of closing prices. Textual data is encoded using BERT sentence embeddings, while a combination of CNNs and RNNs is leveraged to construct a robust multi-modal representation model tailored for market prediction. It's worth emphasizing that, although we opt for ResNet and BERT-two of the most advanced models currently-in our approach, these models can be substituted with alternative CNN and RNN models that align with the unique requirements of other problems. Our primary focus remains on capturing the crucial inter-modal and intra-modal information, as well as comprehending the intricate bi-modal relationships within the input data, all achieved through the integration of the proposed common and unique sub-networks.

The rest of the paper is organized as follows. In Sect. 2, we show the related work in stock market prediction domain, and then introduce the specifics of the proposed model in Sect. 3. Sections 4 and 5 cover experiments that includes market prediction tests. In Sect. 6, we give conclusion and our recommendation for future work.

2 Related Work

Quantitative indicators, such as historical market data and technical indicators have been widely explored and have shown to be effective for stock market prediction [3,22]. A novel State Frequency Memory recurrent network is proposed by [34] to make long and short-term predictions using the historical market data. The study by [10] employ LSTM to predict the movement direction of S&P500 index prices where they show that LSTM outperforms Random Forest, DNN and Logistic Regression Classifier based models. Another study [31] uses raw financial trading data in their novel hybrid model called CLVSA: A Convolutional LSTM Based Variational Sequence-to-Sequence Model with Attention.

The fusion of market data and technical indicators has also garnered attention in the stock market prediction research. [21] delve into the application of LSTM networks to forecast price trends by leveraging historical price data and technical indicators. Employing LSTM networks with an attention mechanism, [4] focuses on predicting Hong Kong stock movements through market data and technical indicators, highlighting the effectiveness of the attention mechanism in LSTM-based prediction models. In a similar vein, [15] propose an RNN-based strategy for predicting three prominent Chinese stock market indexes, using a multi-task RNN for feature extraction from raw market data.

Taking a different approach, [26] introduce a novel market prediction model, AZFinText, which harnesses proper nouns for new data representation. Leveraging the success of deep learning in other domains, researchers have begun exploring their utility in Natural Language Processing (NLP) tasks. A study by [29] utilize Word2Vec word embeddings to encode textual news data for CNN and LSTM-based prediction model.

The integration of financial news data, using NLP techniques, alongside quantitative indicators has become a prevalent theme in stock market prediction. The research by [5] incorporate sentiment dictionaries to extract features from social media news, while [19] demonstrate the superiority of using article summaries over complete article bodies for prediction. Analyzing events extracted from news articles in conjunction with technical indicators, [23] predict FTSE 350 index prices. The utilization of CNN-based event embeddings for market prediction via financial news articles is explored by [7]. In a noteworthy multimodal study, [35] employ tensors to jointly model news articles and social media sentiments, providing predictive insights for market prices in China A-share and Hong Kong Stock Market.

3   Model Design of MStoCast

In the design of our MStoCast method, we aim to utilize both common and unique types of information. This entails capturing modality-specific as well as joint information, while also delving into the modeling of bi-modal relationships across various data modalities. We initially take historical market data and financial news as our main data sources. Then we derive a list of seven technical indicators from the market data and utilize three data modalities: market, technical indicators and financial news, where sentence embeddings are employed to encode the textual financial news data. The three raw input features are denoted as Z_T, Z_M, Z_N, representing technical indicators, market price data, and news data, respectively. These feature vectors are analysed with three separate LSTMs in the common network and concatenated as raw features to form bi-modal matrices in the unique network. By performing early information fusion in the unique network and feature fusion in the common network, MStoCast aims to capture joint and cross-modal information from the input modalities.

The information gathered from both the common and unique sub-networks is then amalgamated through a fusion layer to perform multimodal market movement prediction. The design of our method is presented in Fig. 1.

3.1   Unique Network

In the unique network, raw concatenated input matrices are analyzed through convolutional networks, such as ResNets, to capture the bi-modal relationships among the input data, which we detail as follows.

We begin by pairing up the raw input features, thereby generating three distinct input matrices. These matrices are as follows:

$$Z_{TM} = Z_T \oplus Z_M$$
$$Z_{NM} = Z_N \oplus Z_M \tag{1}$$
$$Z_{NT} = Z_N \oplus Z_T$$

where Z_{NM} is formed by the outer product (denoted by $\oplus$) of raw features from the market data Z_M and news embeddings Z_N; Z_{TM} by using raw features from

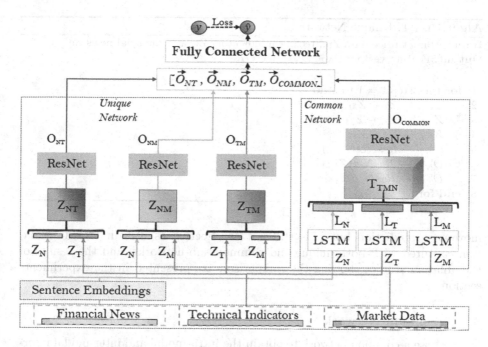

Fig. 1. The framework of our MStoCast model

the market data Z_M and technical indicators Z_T; and Z_{NT} is formed using news embeddings Z_N and technical indicators Z_T.

We then proceed to construct three distinct ResNets, each responsible for extracting underlying cross-modal information from the input matrices. Each network in the unique sub-network includes two residual blocks. In each residual block, the input is first processed using convolutional layers, and then we apply batch normalization and analyze the output with an activation layer using Rectified Linear Unit (ReLU) [1] function to add non-linearity. Batch normalization lets the network to train fast by keeping the mean output close to 0 and the output standard deviation close to 1 and the standard ReLU function returns the element-wise maximum of 0 and the input. The output goes through another round of convolution and batch normalization processes. We then add the initial input to the output of second batch normalization layer and feed it into the last activation layer.

We then produce three feature matrices O_{NT}, O_{NM}, O_{TM} by processing the three bi-modal matrices through three ResNets:

$$
\begin{aligned}
O_{TM} &= ResNet(Z_{TM}) \\
O_{NM} &= ResNet(Z_{NM}) \\
O_{NT} &= ResNet(O_{NT})
\end{aligned}
\tag{2}
$$

The overall algorithm of our unique network is give in Algorithm 1. The unique network can be further extended to analyze these outputs for market

---

Algorithm 1. Unique Network

Input: Market price data Zm, technical indicators Zt, and financial news Zn
Output: Feature vectors O_{NT}, O_{NM} and O_{TM}

1: **for** time step t = 1 to n **do**
2:     $Z_{TM} \leftarrow Z_T \oplus Z_M$
3:     $Z_{NM} \leftarrow Z_N \oplus Z_M$
4:     $Z_{NT} \leftarrow Z_N \oplus Z_T$
5:     $O_{NT}^t \leftarrow ResNet(Z_{NT})$
6:     $O_{NM}^t \leftarrow ResNet(Z_{NM})$
7:     $O_{TM}^t \leftarrow ResNet(Z_{TM})$
8: **end for**

---

movement prediction but it is still not as effective as utilizing the information captured with both unique and common sub-networks, and the results of the ablation studies showing the comparison are presented in the experiments section.

3.2 Common Network

We propose a common network to obtain the intra-modal and inter-modal information from the inputs. The common network includes three key elements: LSTM layers, feature tensors, and ResNet blocks. First, the inputs are separately processed using LSTMs to capture modality specific features.

$$L_M = LSTM(Z_M)$$
$$L_T = LSTM(Z_T) \tag{3}$$
$$L_N = LSTM(Z_N)$$

where L_M is the output of LSTM layer using market input features Z_M, L_T is the output of LSTM layer using the technical indicators, and L_N is the output of LSTM layer using news embeddings Z_N.

The latent features obtained from with the three LSTMs are then brought together via an outer product to form a multi-mode tensor. These tensors represent the inter-modal and intra-modal information within the multiple modalities.

$$T_{TM} = \sum_j L_{T_{(ijk)}} \times L_{M_{(ijk)}}^T$$
$$T_{TMN} = \sum_k T_{TM_{(ijk)}} \times L_{N_{(ikl)}} \tag{4}$$

Lastly, the joint information from the tensor is retrieved through using a ResNet. In this common network, the ResNet takes T_{TMN} as its input and produces the output O_{COMMON}:

$$O_{COMMON} = ResNet(T_{TMN}) \tag{5}$$

The overall algorithm of our common network is given in Algorithm 2.

Algorithm 2. Common Network

Input: Market price data Zm, technical indicators Zt, and financial news Zn
Output: Feature vector O_{COMMON}

1: **for** time step t = 1 to n **do**
2:    $L_M \leftarrow LSTM(Z_T)$
3:    $L_N \leftarrow LSTM(Z_N)$
4:    $L_M \leftarrow LSTM(Z_M)$
5:    $T_{TM} \leftarrow \sum_j L_{T_{(ijk)}} \times L^T_{M_{(ijk)}}$
6:    $T_{TMN} \leftarrow \sum_k T_{TM_{(ijk)}} \times L_{N_{(ikl)}}$
7:    $O^t_{COMMON} \leftarrow ResNet(T_{TMN})$
8: **end for**

3.3 Fusion Layer

The feature vectors from the common and unique sub-networks are integrated through vector concatenation to form O_{merged}. We first flatten this feature vector via global average pooling operation and then process it with two fully connected layers using ReLU as the activation function. At the last step another fully connected layer is used to make a prediction. The overall network is a binary classification model that predicts the movement direction of the stock prices and the weights are optimized by minimizing the binary crossentropy loss. The direction of the movement is defined as the difference between Close prices on day t+1 and day t.

4 Experimental Setup

4.1 Dataset

In our experiments, we utilized real-world datasets comprising financial news, market data, and technical indicators spanning from January 1, 2013, to December 31, 2019, encompassing 1761 trading days. The financial news was sourced from Reuters[1], where each article included a title, body, and the publishing date. The date served to synchronize the articles with the daily market data. Following preprocessing, including the removal of duplicates and unrelated articles, the dataset consisted of 527,047 news headlines. The minimum number of articles per day was 38, while the maximum was 998.

We employed the headlines from the financial news, as research has indicated that utilizing news titles can yield superior prediction results compared to using the entire article body [27]. Our approach involved utilizing 5-day windows as input, meaning that data from five consecutive days were used at each step to predict the movement of the closing price for the subsequent day. As the number of news titles per trading day varied, we amalgamated all the titles for a given day into a single coherent sentence. Subsequently, we harnessed BERT to encode

---

[1] https://www.reuters.com/business/finance/.

the textual data into feature vectors, ultimately generating a singular sentence embedding vector per trading day.

Market data pertaining to the S&P 500 index and individual stock data for five companies within the index were extracted for the corresponding dates from Yahoo Finance[2]. These market data were represented using five attributes: Open, High, Low, Close prices, and Volume. The data were normalized to fall within the [0, 1] range.

In addition, we computed seven technical indicators for each trading day, derived from the prices over the preceding five days. These indicators encompassed Stochastic %K, Stochastic %D, Momentum, Rate of Change, William's %R, A/D Oscillator, and Disparity 5. These particular indicators have demonstrated effectiveness in market prediction [14]. Refer to Table 1 for the list of selected technical indicators and their descriptions.

Our experiments were centered on predicting the directional movement of the S&P index price and the individual stock prices of five companies (AAPL, MSFT, AMZN, TSLA, GOOGL). For index prediction, we initially employed an 80–20% split for training and testing. Additionally, we evaluated yearly performance by utilizing the first 10 months of each year for training and the last 2 months for testing. In the case of individual stock prediction, the 80–20% split was again used for training and testing purposes.

Table 1. Technical indicators and their descriptions

Indicator	Description
Stochastic %K	The %K is the percentage of the difference between its highest and lowest values over a certain time period
Stochastic %D	Moving average of Stochastic %K
Momentum	The change in a security's price over a given time period
Rate of Change	The percentage difference between the current price and the price n days ago
William's %R	A momentum indicator that measures overbought/oversold levels
A/D Oscillator	A momentum indicator that associates changes in price
Disparity 5	The distance of current price and the moving average of 5 days

4.2 Settings

In this study, the standard measure of accuracy (Acc) and Matthews Correlation Coefficient are selected to evaluate the performance of the models for

---

[2] https://finance.yahoo.com/.

S&P500 index and individual stock prediction [7,16,17,27,35]. MCC is generally employed when the sizes of classes y = 1 and y = 0 differ. These two metrcis are defined as follows:

$$ACC = \frac{n}{N} \tag{6}$$

where n is the number of correct predictions and N is the number of total predictions, and

$$MCC = \frac{TP \times TN - FP \times FN}{\sqrt{(TP + FP)(TP + FN)(TN + FP)(TN + FN)}} \tag{7}$$

where TP, TN, FP, and FN are the number of true positives, true negatives, false positives, and false negatives, respectively.

In the common network, we employ 64 LSTM layers to analyze the sentence embeddings from the news titles, while the technical indicator and market data inputs are processed using 32 LSTM layers.

The design of the ResNet is consistent across both the unique and shared networks. In the common network, we utilize 64 filters and set the kernel size to 3 in the convolutional layers. For the unique network, we employ convolutional layers with 32 filters and a kernel size of 2. In both sub-networks, a stride of 1 is applied, and we perform padding convolution.

The initial two fully connected layers in the fusion layer contain 32 output neurons each and employ the ReLU activation function to process the inputs. The Adam optimizer is utilized to optimize the network parameters, and we set the epoch size to 500 epochs with a batch size of 64.

4.3 Baseline Methods

We compare our approach with the following baselines on predicting individual stocks and S&P500 index.

- **Recurrent Convolutional Neural Network (RCNN)** [29]: a CNN and LSTM based market forecast model that utilizes technical indicators and financial news.
- **Event Embeddings-RCN (EB-RCN)** [24]: a similar model to RCNN that uses market data alongside with event embeddings [7] from news data.
- **Bidirectional Gated Recurrent Unit (BGRU)** [13]: a market forecast model with financial news and historical market data.
- **LSTM-based Recurrent State Transition (ANRES)** [20]: a market movement prediction model utilizing events from news data.
- **Hybrid Attention Network (HAN)** [12]: a state-of-the-art stock market forecast model with hierarchical attention utilizing news data.
- **Adversarial Attentive LSTM (Adv-LSTM)** [8]: a market forecast model utilizing attentive LSTMS with adversarial training strategy.

These machine learning models are built for stock market prediction. But in order to test the effectiveness of MStoCast as a multimodal learning system, we compare it against the following two state of the art multimodal learning models that have been successful in other domains. Both of these models have been proposed for sentiment analysis but we adopt their architectures and customize them for stock market prediction.

- **Tensor Fusion Network (TFN)** [32] utilizes the bi-modal and unimodal information from the input modalities.
- **Early Fusion LSTM (EF-LSTM)** [33] concatenates the inputs from different modalities and employs a single LSTM to analyse the combined input.

We also perform ablation studies by creating two models using our common (MStoCast-Common) and unique (MStoCast-Unique) sub-networks separately and evaluate their performances.

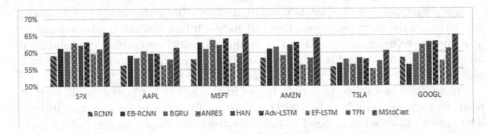

Fig. 2. Accuracy on index and individual stock prediction (the higher, the better).

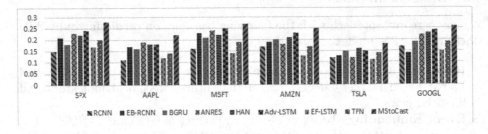

Fig. 3. MCC results on index and individual stock prediction (the higher, the better).

5   Results and Analaysis

The primary contribution of this research lies in its innovative multimodal approach, which leverages both common and unique information from various data modalities to predict price movements in stock markets. To demonstrate the efficacy of the proposed MStoCast and its components, we conducted experiments using real-world datasets. Initially, we performed ablation tests to ascertain the significance of the novel multimodal design. Then, we compared the performance of MStoCast against several state-of-the-art models from stock market prediction domain literature. The results of the experiments are given in Fig. 2 and Fig. 3.

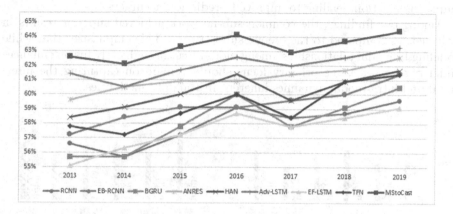

Fig. 4. Accuracy results on S&P500 index prediction (the higher, the better).

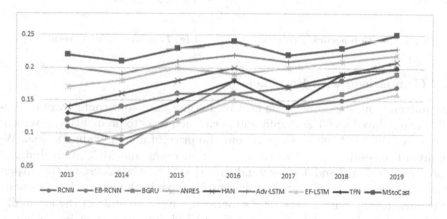

Fig. 5. MCC results on S&P500 index prediction (the higher, the better).

5.1 Ablation Study

We conducted ablation studies to assess the effectiveness of incorporating both unique and common information within MStoCast for predicting stock trends. To this end, we carried out experiments to compare the performance of the unique sub-network (MStoCast-Unique) and the common sub-network (MStoCast-Common) against the complete MStoCast architecture. We focused on predicting the movement direction of both the S&P index and five individual stocks. The outcomes of these experiments are presented in Fig. 6 and Fig. 7.

A notable observation is that MStoCast consistently outperforms both the unique and common sub-networks in both stock index prediction and individual stock trend prediction. These results underscore the effectiveness of our proposed design, which capitalizes on inter-modal and intra-modal information as common knowledge, while exploiting bi-modal relationships among input modalities as unique information, leading to improved prediction accuracy.

Among our findings, the common sub-network achieved superior results in both metrics compared to the unique sub-network. This emphasizes that while modeling bi-modal relationships among modalities through early fusion enhances model performance, the primary focus should remain on capturing the inter-modal and intra-modal dynamics inherent in the data modalities.

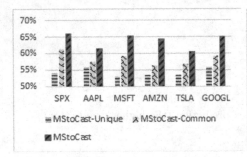
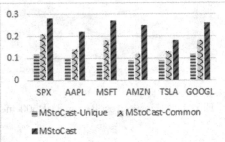

Fig. 6. Ablation accuracy results.   **Fig. 7.** Ablation MCC results.

5.2   Comparison with Baselines

We conducted our experiments under two distinct data split configurations. Initially, we employed an 80–20% split and performed tests to predict the movement direction of the S&P 500 index price and the price of five individual stocks. We evaluated the results using two key metrics: accuracy and MCC. Our findings, as presented in Fig. 2 and 3, clearly indicate that MStoCast consistently outperforms baseline stock market prediction models across both metrics. Furthermore, the results demonstrate MStoCast's superiority over two state-of-the-art multimodal deep learning models, TFN and EF-LSTM. The consistent improvement in performance against baseline models underscores the efficacy of the proposed design for both index and individual stock prediction, leveraging multimodal data.

Among the baseline models, attention-based HAN and Adv-LSTM stand out, demonstrating better performance in most tickers in terms of both accuracy and MCC. These findings highlight the significance of the attention mechanism in focusing on crucial data aspects when capturing latent features. However, even against these models, MStoCast exhibits substantial enhancements in both accuracy and MCC metrics. We attribute this to the innovative multimodal learning design that models intra-modal and inter-modal relationships within the input data.

We expanded the comparative analysis to include two successful multimodal deep learning models from other domains: TFN and EF-LSTM. MStoCast outperforms these models significantly in predicting both index and individual stock price movements. While TFN outperforms EF-LSTM across all tickers, both models are often outperformed by other baseline methods.

In our experiments, models like EB-RCN and ANRES aimed to extract event-based information from financial news and integrate it into prediction models. The results highlighted that this approach yields superior performance compared to employing basic word embeddings for textual data encoding. This underscores the crucial role of effective textual representation techniques in multimodal stock market prediction.

Another observation is that MStoCast-Common consistently outperforms the RCNN model in most tickers, although it is often surpassed by EB-RCN. This suggests that while tensors and sentence embeddings contribute to prediction accuracy, employing more sophisticated textual representation methods, such as event embeddings, can yield even better results.

Among the baseline models, EF-LSTM and MStoCast-Unique networks employed early information fusion, where raw input features were concatenated at the input stage prior to analysis. Conversely, other baseline methods adopted a late fusion technique, wherein models independently analyzed input modalities before fusing data at a later stage. The outcomes indicated that late fusion techniques produce better results by capturing more crucial latent features from input modalities.

We also evaluated models' yearly prediction performance for S&P 500 index prediction, using the first 10 months of each year for training and the last two months for testing. The results, illustrated in Fig. 4 and 5, showcase MStoCast's consistent superiority over baseline models for each year, in both directional accuracy and MCC. The same trend from the above tests also continue here where Adv-LSTM performs the best among the baselines. Although the yearly results are slightly lower than the initial test results, this can be attributed to the smaller test sample size inherent in the yearly setup.

6   Conclusion and Future Work

In this research, we proposed a novel multimodal stock market prediction model named MStoCast which effectively utilizes stock market data, financial news data, and technical indicator data. The model design focuses on extracting inter-modal and intra-modal information, and explores the bi-modal relations among the inputs. In the experiments and trading simulation demonstrations with real-world datasets, MStoCast model considerably outperformed the baseline models. The results show the strong potential of MStoCast in multimodal learning.

References

1. Agarap, A.F.: Deep learning using rectified linear units (ReLU). arXiv arXiv:1803.08375 (2018)
2. Akita, R., Yoshihara, A., Matsubara, T., Uehara, K.: Deep learning for stock prediction using numerical and textual information. 2016 IEEE/ACIS 15th International Conference on Computer and Information Science (ICIS), pp. 1–6 (2016)
3. Bustos, O., Quimbaya, A.P.: Stock market movement forecast: a systematic review. Exp. Syst. Appl. **156**, 113464 (2020). https://api.semanticscholar.org/CorpusID: 219097702
4. Chen, S., Ge, L.: Exploring the attention mechanism in LSTM-based Hong Kong stock price movement prediction. Quant. Finan. **19**, 1507–1515 (2019), https:// api.semanticscholar.org/CorpusID:199364412
5. Chen, W., Yeo, C., Lau, C., Lee, B.S.: Leveraging social media news to predict stock index movement using RNN-boost. Data Knowl. Eng. **118**, 14–24 (2018)
6. Devlin, J., Chang, M.W., Lee, K., Toutanova, K.: BERT: pre-training of deep bidirectional transformers for language understanding. arXiv arXiv:1810.04805 (2019)
7. Ding, X., Zhang, Y., Liu, T., Duan, J.: Deep learning for event-driven stock prediction. In: Proceedings of the 24th International Joint Conference on Artificial Intelligence (IJCAI) (2015)
8. Feng, F., Chen, H., He, X., Ding, J., Sun, M., Chua, T.S.: Enhancing stock movement prediction with adversarial training. In: International Joint Conference on Artificial Intelligence (2018)
9. Feuerriegel, S., Gordon, J.: Long-term stock index forecasting based on text mining of regulatory disclosures. Decis. Support Syst. **112**, 88–97 (2018)
10. Fischer, T.G., Krauss, C.: Deep learning with long short-term memory networks for financial market predictions. Eur. J. Oper. Res. **270**, 654–669 (2017). https:// api.semanticscholar.org/CorpusID:207640363
11. He, K., Zhang, X., Ren, S., Sun, J.: Deep residual learning for image recognition. In: Proceedings of the IEEE Conference on Computer Vision and Pattern Recognition, pp. 770–778 (2016)
12. Hu, Z., Liu, W., Bian, J., Liu, X., Liu, T.Y.: Listening to chaotic whispers: a deep learning framework for news-oriented stock trend prediction. In: Proceedings of the Eleventh ACM International Conference on Web Search and Data Mining (2017)
13. Huynh, H.D., Dang, L.M., Duong, D.: A new model for stock price movements prediction using deep neural network. In: Proceedings of the Eighth International Symposium on Information and Communication Technology (2017)
14. Kim, K.: Financial time series forecasting using support vector machines. Neurocomputing **55**, 307–319 (2003)
15. Li, C., Song, D., Tao, D.: Multi-task recurrent neural networks and higher-order Markov random fields for stock price movement prediction: Multi-task RNN and higer-order MRFs for stock price classification. In: Proceedings of the 25th ACM SIGKDD International Conference on Knowledge Discovery & Data Mining (2019). https://api.semanticscholar.org/CorpusID:198952502
16. Li, Q., Chen, Y., Jiang, L., Li, P., Chen, H.: A tensor-based information framework for predicting the stock market. ACM Trans. Inf. Syst. **34**, 11:1–11:30 (2016)
17. Li, Q., Tan, J., Wang, J., Chen, H.: A multimodal event-driven LSTM model for stock prediction using online news. IEEE Trans. Knowl. Data Eng. **33**, 3323–3337 (2020)

18. Li, Q., Wang, T., Li, P., Liu, L., Gong, Q., Chen, Y.: The effect of news and public mood on stock movements. Inf. Sci. **278**, 826–840 (2014)
19. Li, X., Xie, H., Song, Y., Zhu, S., Li, Q., Wang, F.L.: Does summarization help stock prediction? A news impact analysis. IEEE Intell. Syst. **30**, 26–34 (2015)
20. Liu, X., Huang, H., Zhang, Y., Yuan, C.: News-driven stock prediction with attention-based noisy recurrent state transition. arXiv arXiv:2004.01878 (2020)
21. Nelson, D.M.Q., Pereira, A.M., de Oliveira, R.A.: Stock market's price movement prediction with LSTM neural networks. In: 2017 International Joint Conference on Neural Networks (IJCNN), pp. 1419–1426 (2017). https://api.semanticscholar.org/CorpusID:206919491
22. Nti, I.K., Adekoya, A.F., Weyori, B.A.: A systematic review of fundamental and technical analysis of stock market predictions. Artif. Intell. Rev. **53**(4), 3007–3057 (2019). https://doi.org/10.1007/s10462-019-09754-z
23. Nuij, W., Milea, V., Hogenboom, F., Frasincar, F., Kaymak, U.: An automated framework for incorporating news into stock trading strategies. IEEE Trans. Knowl. Data Eng. **26**, 823–835 (2014)
24. Oncharoen, P., Vateekul, P.: Deep learning for stock market prediction using event embedding and technical indicators. In: 2018 5th International Conference on Advanced Informatics: Concept Theory and Applications (ICAICTA), pp. 19–24 (2018)
25. Schumaker, R.P., Chen, H.: Textual analysis of stock market prediction using breaking financial news: the AZFin text system. ACM Trans. Inf. Syst. **27**, 12:1–12:19 (2009)
26. Schumaker, R.P., Zhang, Y., Huang, C., Chen, H.: Evaluating sentiment in financial news articles. Decis. Support Syst. **53**, 458–464 (2012)
27. Shi, L., Teng, Z., Wang, L., Zhang, Y., Binder, A.: DeepClue: visual interpretation of text-based deep stock prediction. IEEE Trans. Knowl. Data Eng. **31**, 1094–1108 (2019)
28. Thakkar, A., Chaudhari, K.: Fusion in stock market prediction: a decade survey on the necessity, recent developments, and potential future directions. Int. J. Inf. Fus. **65**, 95–107 (2020). https://api.semanticscholar.org/CorpusID:221323781
29. Vargas, M.R., Lima, B.S.L.P.D., Evsukoff, A.: Deep learning for stock market prediction from financial news articles. In: 2017 IEEE International Conference on Computational Intelligence and Virtual Environments for Measurement Systems and Applications (CIVEMSA), pp. 60–65 (2017)
30. Verma, S., Wang, C., Zhu, L., Liu, W.: DeepCU: integrating both common and unique latent information for multimodal sentiment analysis. In: International Joint Conference on Artificial Intelligence (2019)
31. Wang, J., Sun, T., Liu, B., Cao, Y., Zhu, H.: CLVSA: a convolutional LSTM based variational sequence-to-sequence model with attention for predicting trends of financial markets. arXiv arXiv:2104.04041 (2019). https://api.semanticscholar.org/CorpusID:199465738
32. Zadeh, A., Chen, M., Poria, S., Cambria, E., Morency, L.P.: Tensor fusion network for multimodal sentiment analysis. arXiv arXiv:1707.07250 (2017)
33. Zadeh, A., Liang, P.P., Poria, S., Cambria, E., Morency, L.P.: Multimodal language analysis in the wild: CMU-MOSEI dataset and interpretable dynamic fusion graph. In: Annual Meeting of the Association for Computational Linguistics (2018)

34. Zhang, L., Aggarwal, C.C., Qi, G.J.: Stock price prediction via discovering multi-frequency trading patterns. In: Proceedings of the 23rd ACM SIGKDD International Conference on Knowledge Discovery and Data Mining (2017). https://api. semanticscholar.org/CorpusID:28307599

35. Zhang, X., Zhang, Y., Wang, S., Yao, Y., Fang, B.X., Yu, P.S.: Improving stock market prediction via heterogeneous information fusion. Knowl. Based Syst. **143**, 236–247 (2017). https://api.semanticscholar.org/CorpusID:3708860

Few-Shot and Transfer Learning with Manifold Distributed Datasets

Sayed Waleed Qayyumi[✉][iD], Laurence A. F. Park[iD], and Oliver Obst[iD]

Centre for Research in Mathematics and Data Science, School of Computer,
Data and Mathematical Sciences, Western Sydney University, Locked Bag 1797,
Penrith, NSW 2751, Australia
{s.qayyumi,l.park,o.obst}@westernsydney.edu.au
https://www.westernsydney.edu.au/crmds/

Abstract. A manifold distributed dataset with limited labels makes it difficult to train a high-mean accuracy classifier. Transfer learning is beneficial in such circumstances. For transfer learning to succeed, the target and base datasets should have a similar manifold structure. A novel method is presented in this paper for determining the similarity between two manifold structures. To determine whether target and base datasets have similar manifolds and are suitable for transfer learning, this method can be used. A novel few-shot algorithm is then presented that uses transfer learning to classify manifold distributed datasets with a limited number of labels. Using the base and target datasets, the manifold structure and its relevant label distribution are learned. Using this information in combination with the few labels and unlabeled data from the target dataset, we can develop a classifier with high mean accuracy.

Keywords: Few-shot learning · Transfer learning · manifold distributed datasets · measuring similarity

1 Introduction

A few-shot learning method is used in machine learning and artificial intelligence to train models with a limited number of examples. In contrast to traditional machine learning, few-shot learning utilizes prior knowledge from related tasks in order to gain insight from scarcely labeled data. Traditional machine learning requires a large amount of labeled data in order to be accurate; however, in real-world scenarios, it is often impossible or prohibitively expensive to acquire extensive labeled datasets for each new task or category. The goal of few-shot learning is to develop algorithms and techniques that allow models to learn from a limited number of examples per class in an efficient manner. In contrast to traditional machine learning, few-shot learning enables machines to learn as humans do, by generalizing information from past experiences to new situations even when only a few examples are provided. It is therefore fair to say that few-shot learning bridges the gap between traditional machine learning and human-like learning abilities.

© The Author(s), under exclusive license to Springer Nature Singapore Pte Ltd. 2024
D. Benavides-Prado et al. (Eds.): AusDM 2023, CCIS 1943, pp. 137–149, 2024.
https://doi.org/10.1007/978-981-99-8696-5_10

The process of transfer learning in machine learning and deep learning refers to the use of knowledge gained from one task to improve performance on another. Instead of starting from scratch, transfer learning uses pre-trained models or representations to accelerate training, enhance generalization, and improve performance on target tasks. Traditionally, machine learning models are trained on specific datasets for specific tasks, which requires a considerable amount of labeled data. Annotating large datasets can, however, be time-consuming, expensive, or impractical in certain circumstances. A transfer learning method can overcome this limitation by transferring knowledge from an abundantly labeled source domain to a scarcely labeled target domain. Transfer learning is based on the principle that knowledge gained from solving one task can be applied to solving another related task. Utilizing this knowledge, models can extract meaningful features, capture important patterns, and generalize well to new data even when limited labeled data is available.

Manifold distributed datasets are those whose samples are located on or near a low-dimensional manifold within a high-dimensional space. In traditional datasets, data points are often assumed to be independent and identically distributed (i.i.d.). However, real-life data often demonstrate complex structures and correlations that can be better understood by considering the underlying manifold. A manifold is a curved or folded surface embedded in a higher-dimensional space. The manifold represents the underlying structure of data, where each point represents a separate sample. By mapping high-dimensional data points to a lower-dimensional manifold representation, manifold learning uncovers the intrinsic structure and geometry of the data.

A novel method is presented in this paper for determining the similarity between two manifold structures. To determine whether target and base datasets have similar manifolds and are suitable for transfer learning, this method can be used. Using the mentioned method, the paper then presents a novel algorithm for few-shot learning of manifold distributed datasets. Using a similar base labeled manifold distributed dataset, the algorithm learns the manifold structure of the data. Then, using the manifold structure learned from the base dataset, along with the labeled and unlabeled data from the target manifold distributed dataset, a few-shot learning classifier is trained.

This article makes the following contributions:

- A novel method for calculating the similarity between two manifold structures is proposed. The purpose of this method is to determine whether a base dataset is suitable for transfer learning. (Sect. 3)
- A novel algorithm is presented for the transfer learning when data is manifold distributed and labels are limited in few-shot learning scenarios. Transfer learning is achieved using graphs and random walks. (Sect. 4).

This article will proceed in the following manner: Sect. 2 examines current research on few-shot learning, transfer learning, and classification of manifold-distributed data with few labeled samples. Section 3 presents a method to compare the similarity of manifold structures between two manifold distributed datasets. Section 4 outlines our proposed algorithm. Section 5 describes the experiments conducted on synthetic and real-life datasets. Our results are also compared with other state-of-the-art methods in this section.

2   Background and Related Work

Recent advances in deep learning, meta-learning, and transfer learning have led to significant progress in few-shot learning. In order to address the few-shot learning problem, numerous approaches and techniques have been developed, each with its own strengths and characteristics.

A common strategy used in few-shot learning is meta-learning, also known as learning to learn. Through the use of meta-learning algorithms, models can rapidly adapt to new tasks with limited labeled data by accumulating knowledge from multiple related tasks during a pre-training phase. Models become more adept at generalizing to new classes as they learn how to learn.

Metric learning is another popular approach to few-shot learning. A metric-based method seeks to infer a similarity metric or embedding space in which samples from the same class are more similar than samples from different classes. Using these methods, effective classification and recognition can be achieved with a minimum amount of training data.

Memory-augmented architectures have also been shown to be promising for few-shot learning. Both training and testing are conducted using models that use external memory structures as a dynamic storage medium. Learning through few-shots is made possible by the ability to retain important knowledge and adapt it to new situations.

Further, few-shot learning has been applied to improve model performance, adaptability, and generalization through the use of generative modeling, attention mechanisms, and knowledge distillation.

In recent years, several influential papers have significantly advanced the few-shot learning technique. Several innovative algorithms and methodologies are presented that have pushed the boundaries of what is possible even with limited training examples. Here are some interesting ideas presented in a number of popular papers. In [13] the authors introduced matching networks for one-shot learning. The authors propose a trainable model that compares examples from a support set with examples from a target set. Similarities between samples are computed using a differentiable nearest-neighbor algorithm. As demonstrated in this study, memory-augmented architectures can be useful for few-shot learning.

In [10] the authors presented a few-shot learning approach utilizing prototypical networks. The algorithm proposed is based on the learning of a metric space where samples from the same class are closer together than samples from other classes. By representing classes with learned prototypes, this method enables efficient inference and generalization of new classes. [8] introduced the concept of meta-learning using few-shot learning as a context. In order to facilitate rapid adaptation to new tasks with limited labeled data, the authors propose a meta-learning algorithm that learns the initialization of the model's parameters as the task progresses. As a result of training on multiple related tasks, generalization and adaptation to new classes are enhanced.

In [3] the authors introduced the model-agnostic meta-learning (MAML) algorithm, which provides a general framework for few-shot learning. The use of MAML allows models to be quickly adapted to new tasks with limited data by optimizing their parameters. MAML learns from few examples efficiently by iteratively updating its initialization based on task-specific gradients.

The transfer learning technique is a powerful tool in machine learning and computer vision for improving the performance of related tasks based on the knowledge gained from the first task. The transfer of representations, features, or knowledge from one domain to another reduces the requirement for expensive training processes and large labeled datasets. Models are believed to be capable of transferring knowledge gained from one task to another, improving performance, generalization, and convergence.

A number of influential papers have contributed to the development and advancement of transfer learning techniques. In these papers, several approaches and architectures have been proposed which have had a significant impact on the field. In [7] the authors used deep convolutional neural networks (CNNs) to introduce the concept of transfer learning. The study demonstrated the effectiveness of pre-training CNNs on a large-scale dataset (ImageNet) and fine-tuning them for specific target tasks. A large-scale dataset could be generalized to different tasks, resulting in significant improvements in accuracy.

In [9] the authors introduced VGGNet, which achieved excellent performance on the ImageNet challenge. The study demonstrated that transfer learning is effective when it is applied to very deep networks. According to the authors, VGGNet learns rich representations on ImageNet that are generalizable to other tasks after pre-training.

In [6] the authors introduced the ResNet architecture, which addresses the problem of training deep neural networks by introducing residual connections. Using ResNet, the authors demonstrated superior performance on ImageNet and highlighted the importance of deep networks in transfer learning. The network could learn residual mappings by using residual connections, allowing features from pre-trained models to be reused more effectively.

Goodfellow et al. [5], despite not exclusively addressing transfer learning, introduced the concept of generative adversarial networks (GANs), which are widely used for transfer learning. An adversarial GAN consists of a generator and a discriminator network. Generators learn to generate synthetic samples that cannot be distinguished from real samples, while discriminators learn to distinguish real samples from fake ones. GANs have been used for a variety of transfer learning tasks, including domain adaptation and style transfer.

In few-shot learning using manifold-distributed data, it is difficult to determine the manifold structure of the data accurately due to the limited number of observations available for training. There is no mathematical possibility of estimating the manifold structure in zero-shot and one-shot learning, for instance. Consequently, it is critical to determine the minimum number of samples required to accurately determine the manifold structure. It may not be possible to capture the nonlinear, complex relationships between manifold distributed data using a small number of examples. Consequently, the model may be overfit or underfit, resulting in poor performance when new, unseen data is introduced. Transfer learning success is heavily dependent on selecting a suitable source dataset. However, the algorithm used also contributes to this success [4].

Another challenge is that few-shot learning assumes that the training examples are representative of the entire manifold distribution, which is not necessarily true for manifold distributed data. When the training data is biased towards particular parts of the manifold or does not cover the full range of variations in the data, a model may fail to generalize to new data points that are not included in the training set.

Furthermore, when using few-shot learning with manifold distributed data, it is essential to select the appropriate distance metric or similarity metric. The selection of an appropriate distance metric can be challenging due to the fact that different manifolds require different distance metrics.

Finally, few-shot learning may require complex and computationally expensive models if there is a large amount of distributed data. The training data for these models must be collected in large quantities in order to avoid overfitting, which is in conflict with the few-shot approach.

Manifold-distributed data can be represented intuitively and naturally with graphs. Edges represent pairwise relationships between data points, while nodes represent data points. Graphs help capture the complex, nonlinear relationships between data points of manifold distributed data. Graphs can also be used to incorporate additional information about the data into the classification process, such as pairwise similarities, distances, or labels. Pairwise relationships between data points can be used to assign edge weights. In summary, graphs offer an effective framework for addressing many of the challenges posed by manifold distributed data classification [1]. Using the constructed graph, we can determine the likelihood that unlabeled observations belong to a particular class based on their proximity to labelled observations. Graph-based classification has proven to be a state-of-the-art approach to classifying manifold distributed data [11, 12].

3 Measuring Similarity of Manifolds

Measurement of the similarity between two datasets and manifolds is crucial to determining the efficiency and feasibility of knowledge transfer. Using knowledge from one domain, typically called the source domain, transfer learning is aimed at improving the performance of a model in another, but related, domain, called the target domain. For a deeper understanding of the underlying structures and distributions in both domains, it is necessary to assess the similarity between datasets. Transfer learning is more effective when datasets have similar statistical characteristics, such as the distribution of data, the representation of features, and the semantic relationships among them. It is possible to effectively transfer knowledge from the source domain to the target domain, enhancing generalization and performance.

It is also crucial to measure the similarity between manifolds, which represent the intrinsic low-dimensional structures of data. In high-dimensional data, manifolds can be used to capture geometric patterns and relationships, which are necessary for understanding the underlying structures of target domains and source domains. By analyzing the geometric similarities between manifolds, we can determine whether data in both domains have similar geometric structures, allowing for the transfer of meaningful and relevant knowledge.

Several works have stressed the importance of assessing datasets and the many similarities between them when it comes to transfer learning. In [2], the authors give a comprehensive overview of transfer learning techniques and emphasize the importance of selecting the appropriate source and target domains on the basis of similarity measures. In addition, [15] present a variety of transfer learning paradigms and emphasize the importance of measuring dataset similarity in order to achieve successful knowledge transfer.

In order to achieve effective transfer learning, it is essential to measure the similarity between datasets and manifolds. Understanding the statistical and geometric relationships between domains can help us improve model performance in the target domain. Taking into account the above, we examined various similarity measures in order to select datasets from the source domain that would be most suitable for transfer learning. Various measures can be used to measure the similarity between datasets from a source and a target domain. However, these measures do not work when data is manifold distributed. Therefore, we examined a measure that can be effective when data from the source domain and the target domain are manifold distributed.

A graph is a natural mechanism for unraveling the structure of a manifold [12, 14]. Therefore, we examined a measure based on graphs. Distance can be measured by random walks over graphs, and that is what we are doing in our distance measurement. We have presented our approach to measuring distance between two manifold distributed datasets in Algorithm 1. To determine the effectiveness of our measure, we compared it to all potential measures that can be used in transfer learning. We used both synthetic and real-world datasets. A comparison of our method of measuring similarity with all other methods using synthetic data of a Swiss roll is presented in Fig. 1. We began by comparing identical Swiss rolls. We continued to measure similarity by keeping one Swiss roll the same and adding normal distributed noise to the other Swiss roll. Figure 1 illustrates the results of the experiments.

We also evaluated the performance of our approach to similarity using images from the miniimagenet dataset. In Table 1, a single image of each species is compared with 50 images of other species and its own. Using the set of 50 images, similarity is determined by their average distance (using our method). Based on our results, we are able to identify similar manifold structures and distinguish between different manifolds based on similarity of their structures.

In Table 2, we compare our method with other similarity measures. In the results, we can see that similarity decreases as we compare an image of a bird with other bird-like species and with non-bird-like species. In contrast, other measures of similarity are not able to accomplish this. When we compare similar manifolds between two objects and the number of manifolds is the same, our approach to comparing similarity between images works well. Our method, for example, works when comparing a duck image to another duck image, but fails when comparing a duck image with an image containing more than one duck. The reason for this is that the manifold structure of an image of a single duck differs from the manifold structure of an image with many ducks (Fig. 2).

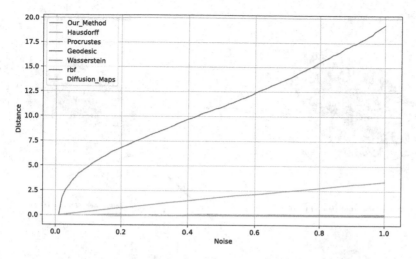

Fig. 1. Distance calculated using various methods as more noise is added to one of the manifolds

Table 1. MiniImageNet Dataset- A single image of each species is compared with 50 images of other species and its own. Similarity is based on the average distance (using our method) from the set of 50 images per class.

	Birds(50)	Ducks(50)	Toco(50)	Snakes(50)	Dogs(50)	Lions(50)
	384.42	411.45	531.44	1474.24	1967.65	956.75
	627.5	187.62	477.95	1597.2	1222.55	1739.42
	576.21	596.57	385.16	801.92	863.38	752.33
	1392.87	1455.15	1660.04	609.33	1461.78	1683.36
	1323.62	870.06	2908.67	1674.11	332.47	669.92
	2156.54	2423.52	2189.73	2569.52	384.42	310.96

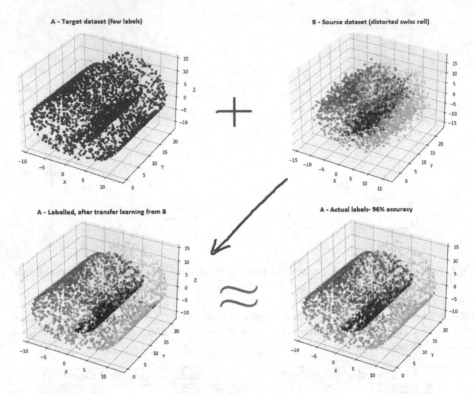

Fig. 2. Transfer Learning - Swiss roll manifold

4   Transfer Learning

We describe our approach to transfer learning between two manifold distributed datasets $X1$ and $X2$ with labels $y1$ and $y2$ in Algorithm 3. Target dataset $X1$ is a manifold distributed dataset with a few labels $y1$ whereas source dataset $X2$ is another manifold distributed dataset with many labels $y2$. The classifier is trained using the source dataset and a portion of the target dataset that is labeled. There are three parts to the entire method.

Algorithm 1 calculates distances between two manifold structures. The method serves as a distance measure for the classifier we use, which is the k neighbor classifier. Furthermore, the method determines whether the source dataset has a similar manifold structure to the target dataset and can be used for transfer learning. For transfer learning to be effective, it is recommended that the source dataset has a level of manifold similarity.

Algorithm 2 constructs graphs of manifold distributed data. A random walk distance measure is used in training classifiers using these graphs. A random walker walks over these graphs, distance is calculated which is used by the k neighbor classifier to determine which neighbor is closest.

Algorithm 3 performs transfer learning. Using Algorithm 1 and 2, this algorithm trains a k neighbor classifier using random walk as the distance metric. In this algorithm, the knowledge of manifold structure (source dataset) and the relationship between manifolds and labels (source dataset) is transferred to the target dataset. In conjunction with the limited labels available in the target dataset, these information are used to train a classifier that is then used to classify the unlabeled data in the target dataset.

Algorithm 1: Measuring Similarity

Data: adj1, adj2, t
Input : adj1, adj2, t
Output: distance

1 Get and assign to $n1, n2$ number of rows in *adj1, adj2*;
2 Create identity matrices of size $n1 \times n1$ and $n2 \times n2$;
3 Create weight matrices W1, W2 : $W1 =$ Inverse of $(I1 - t \times adj1)$ and $W2 =$ Inverse of $(I2 - t \times adj2)$;
4 Calculate distance $d =$ norm of $(W1 - W2)$;
5 **return** d;

Algorithm 2: Construction of Graphs

Data: dataset d, k_neighbors
Input : dataset d, k_neighbors
Output: graph g

1 Create a new empty graph;
2 Calculate the Euclidean distance matrix for d;
3 **for** *i in range(length(d))* **do**
4   *neighbors* ← Indices of *k_neighbors* nearest neighbors of data point *i*;
5   **for** *j in neighbors* **do**
6   Add an edge between data point *i* and data point *j* in *g*;

7 **return** g;

We have provided a detailed explanation of the transfer learning process in 2. There are two Swiss roll manifolds A and B. A is a Swiss roll dataset with very few labels (color class). B is a distorted Swiss roll dataset, but all labels are available. There is a similarity in the manifold structure of both datasets. Using the 1 we are able to measure the similarity between these datasets and determine if we should use them as a learning resource. Considering that the manifold in this case is very similar (70%), we have employed it for transfer learning. It is evident from the output (A-labelled) that transfer learning has added a great deal of value in this instance, achieving an accuracy of 96% when compared to the actual labels.

Algorithm 3: Transfer Learning

Data: X1, y1, X2, y2, k,DT

Input : X1, y1, X2, y2, k,DT

Output: $\mu\alpha$ (mean accuracy of classification)

1   Assign random walk distance between $X1, X2$ to d;

2   **if** $d \leq DT$ **then**

3 | Scale the features of $X1, X2$ to the range [0, 1];

4 | Construct graphs $g1, g2$ from scaled $X1, X2$ with k nearest neighbors;

5 | Convert $g1, g2$ to adjacency matrices;

6 | Create a new k Neighbors classifier with k neighbors and Algorithm 1 as distance
    | metric;

7 | Train classifier on $X2$ and $X1\_labeled$ with labels $y2, y1$;

8 | Classify $X1\_unlabeled$ using the trained classifier;

9 | **return** Prediction ($y1$);

---

5   Experimental Results

We conducted experiments on both synthetic and real-world manifold distributed datasets. These experiments were conducted to evaluate the effectiveness of our method and determine when we should stop using transfer learning in few-shot learning with manifold distributed datasets. Few-shot learning with manifold distributed datasets makes it extremely difficult to construct a classifier with a high mean accuracy. Therefore, it is logical to learn from a source dataset and transfer knowledge to a target dataset.

There is a breakeven point after which transfer learning does not add value, and may even not be necessary. There may or may not be a need for transfer learning depending on the number of labels per class available in the target dataset and how different the source dataset is in terms of its manifold structure. In order to simulate this scenario, we experimented with $10, 20, 30$ and 40 labels per class for synthetic and real-world datasets. We add noise to the source dataset as we increase the number of labelled data points per class. We begin with a noise level of 1 and increase the noise level by 1 every time. It is worth noting that an increase in noise represents an increase in standard deviation. As noise increases, standard deviation increases as follows: $0 \rightarrow 1 = 0.078$, $0 \rightarrow 2 = 0.29$, $0 \rightarrow 3 = 0.64$, and $0 \rightarrow 4 = 1$. A target dataset is created by removing all labels except the number required per class. An identical copy of the same dataset is used as the base dataset. Noise is added to all features except the class.

5.1   Synthetic Datasets

Several experiments were conducted using the data generated for Swiss roll, moon shape, and S curve shape manifolds. We generated two identical datasets: one as a target and the other as a source. The target dataset is modified by removing all labels except the number of labels required for each class. We modify the base dataset by adding noise to all features except the class feature. In the Table 3, you can see the structure

Table 2. MiniImageNet Dataset: The image of a bird is compared to that of other birds and other species based on our approach to similarity and other distance measures

A	B	ours	cos	rbf	pd	wd	hausdorff
		0.00	0.00	20.00	0.00	0.00	0
		253.61	0.10	0.00	0.45	77.37	57066
		582.62	0.15	0.00	0.42	31.00	77164
		2187.30	0.11	0.00	0.45	73.44	72403
		4491.55	0.15	0.00	0.61	23.94	63515

Table 3. % Mean accuracy ($\mu\alpha$) with and without transfer learning (TL) on different manifold structures (1000 data points, 20 iterations) - σ represents noise added. We also compare our approach with neural networks - domain adoption

Shape	Manifold	Sample(L)	σ - Noise	$\mu\alpha$%	$\mu\alpha$ (TL-Our App)%	$\mu\alpha$ NN-DA%
	Moon	10	1	2.00	86.1	79.9
	Moon	20	2	7.00	73.2	68.3
	Moon	30	3	9.10	60.9	60.0
	Moon	40	4	22.2	56.7	63.5
	S Curve	10	1	14.1	39.4	33.9
	S Curve	20	2	19.5	23.2	25.7
	S Curve	30	3	28.8	18.5	16.9
	S Curve	40	4	38.2	15.5	18.3
	Swiss roll	10	1	2.24	88.1	22.5
	Swiss roll	20	2	7.00	74.8	15.3
	Swiss roll	30	3	17.8	62.4	12.2
	Swiss roll	40	4	32.2	58.2	9.8

of the manifold, the number of labeled samples in the target dataset, the noise added to the base dataset, as well as the mean accuracy with and without transfer learning. Based on the difference in mean accuracy with and without transfer learning, it can be seen how beneficial transfer learning can be for few-learning with manifold distributed data. Furthermore, the table illustrates how the manifold difference between the source and target datasets (noise) impacts the efficacy of transfer learning.

5.2   Real World Datasets

We conducted experiments with banknotes, Pendigits, and Satlog datasets. You can access the datasets at (http://archive.ics.uci.edu). We generated two identical datasets: one as a target and the other as a source. The target dataset is modified by removing all labels except the number of labels required for each class. We modify the base dataset by adding noise to all features except the class feature. In the Table 4, you can see the structure of the manifold, the number of labeled samples in the target dataset, the noise added to the base dataset, as well as the mean accuracy with and without transfer learning. Based on the difference in mean accuracy with and without transfer learning, it can be seen how beneficial transfer learning can be for few-learning with manifold distributed data. Furthermore, the table illustrates how the manifold difference between the source and target datasets (noise) impacts the efficacy of transfer learning.

Table 4. % Mean accuracy ($\mu\alpha$) with and without transfer learning (TL) on different manifold structures (1000 data points, 20 iterations) - σ represents noise added. We also compare our approach with neural networks - domain adoption

Shape	Dataset	Sample(L)	σ- Noise	$\mu\alpha$%	$\mu\alpha$ (TL - Our App)%	$\mu\alpha$ NN-DA%
	Banknotes	10	1	86.7	97.1	87.3
	Banknotes	20	2	87.1	94.0	88.5
	Banknotes	30	3	87.7	90.8	86.8
	Banknotes	40	4	88.8	87.2	82.4
	Pendigits	10	1	87.4	99.1	2.4
	Pendigits	20	2	90.9	99.2	5.0
	Pendigits	30	3	91.8	99.1	9.7
	Pendigits	40	4	93.8	98.7	17.3
	Satlog	10	1	55.6	90.0	2.9
	Satlog	20	2	59.8	88.7	10.7
	Satlog	30	3	67.6	88.5	18.0
	Satlog	40	4	70.1	85.2	19.6

6   Conclusions

The development of a classifier with high mean accuracy from a manifold distributed dataset with limited labels is not an easy task. In order for transfer learning to be effective, the manifold structure of the base dataset must be similar to the target dataset. As a means of solving these problems, this paper proposes a novel method for calculating similarity between two manifold distributed datasets. It also proposes a novel transfer learning algorithm in few-shot learning scenarios with manifold distributed datasets. Graphs and random walks are used in our method for calculating similarity between manifold structures. Using the manifold information along with the label distribution from the base manifold distributed dataset and the limited labels from the target dataset, we can construct a high mean accuracy classifier.

In this paper, we discuss the transfer learning in few-shot learning scenarios with manifold distributed datasets. Future research will focus on zero-shot, one-shot, and two-shot learning scenarios that are more complex with respect to estimating manifold structure.

References

1. Cai, D., He, X., Han, J., Huang, T.S.: Graph regularized nonnegative matrix factorization for data representation. IEEE Trans. Pattern Anal. Mach. Intell. **33**(8), 1548–1560 (2010)
2. Cao, B., Pan, S.J., Zhang, Y., Yeung, D.Y., Yang, Q.: Adaptive transfer learning. In: Proceedings of the AAAI Conference on Artificial Intelligence, vol. 24, pp. 407–412 (2010)
3. Finn, C., Abbeel, P., Levine, S.: Model-agnostic meta-learning for fast adaptation of deep networks. In: International Conference on Machine Learning, pp. 1126–1135. PMLR (2017)
4. Ganin, Y., et al.: Domain-adversarial training of neural networks. J. Mach. Learn. Res. **17**(1), 2030–2096 (2016)
5. Goodfellow, I., et al.: Generative adversarial networks. Commun. ACM **63**(11), 139–144 (2020)
6. He, K., Zhang, X., Ren, S., Sun, J.: Deep residual learning for image recognition. In: Proceedings of the IEEE Conference on Computer Vision and Pattern Recognition, pp. 770–778 (2016)
7. Krizhevsky, A., Sutskever, I., Hinton, G.E.: ImageNet classification with deep convolutional neural networks. In: Advances in Neural Information Processing Systems, vol. 25 (2012)
8. Ren, M., et al.: Meta-learning for semi-supervised few-shot classification. arXiv preprint arXiv:1803.00676 (2018)
9. Simonyan, K., Zisserman, A.: Very deep convolutional networks for large-scale image recognition. arXiv preprint arXiv:1409.1556 (2014)
10. Snell, J., Swersky, K., Zemel, R.: Prototypical networks for few-shot learning. In: Advances in Neural Information Processing Systems, pp. 4077–4087 (2017)
11. Tu, E., Cao, L., Yang, J., Kasabov, N.: A novel graph-based k-means for nonlinear manifold clustering and representative selection. Neurocomputing **143**, 109–122 (2014)
12. Tu, E., Zhang, Y., Zhu, L., Yang, J., Kasabov, N.: A graph-based semi-supervised k nearest-neighbor method for nonlinear manifold distributed data classification. Inf. Sci. **367–368**, 673–688 (2016)
13. Vinyals, O., Blundell, C., Lillicrap, T., Wierstra, D., et al.: Matching networks for one shot learning. In: Advances in Neural Information Processing Systems, pp. 3630–3638 (2016)
14. Vishwanathan, S.V.N., Schraudolph, N.N., Kondor, R., Borgwardt, K.M.: Graph kernels. J. Mach. Learn. Res. **11**, 1201–1242 (2010)
15. Weiss, K., Khoshgoftaar, T.M., Wang, D.: A survey of transfer learning. J. Big data **3**(1), 1–40 (2016)

Mitigating the Adverse Effects of Long-Tailed Data on Deep Learning Models

Din Muhammad Sangrasi[1], Lei Wang[1(✉)], Markus Hagenbuchner[1], and Peng Wang[2]

[1] University of Wollongong, Wollongong, Australia
haseebsangrasi@gmail.com, {leiw,markus}@uow.edu.au
[2] University of Electronic Science and Technology China, Chengdu, China

Abstract. When the data distribution in a dataset is highly imbalanced or long-tailed, it can severely affect the effectiveness of a deep network model. This drop in performance is caused due to the biased classifier, which favours the head-class samples because these samples have more dominant features compared to the tail-class samples. Addressing this challenge requires not only capturing subtle inter-class differences and intra-class similarities but also effectively utilising limited data for the minority classes. Supervised contrastive learning (SCL) and transfer of angle information from head classes to tail classes have recently been proposed to address the problem of long-tail classification. For a well-balanced dataset, SCL demonstrates effectiveness by pulling together samples from the same classes while pushing away samples from different classes. However, when applied to long-tailed datasets, SCL could become biased towards the head-class samples. On the other hand, the method of transfer of angle information aims to address the challenges posed by long-tailed image classification; however, it lacks in achieving both intra-class compactness and inter-class separability. To address the shortcomings and exploit the strengths of both of these approaches, we propose a unique hybrid method that seamlessly integrates supervised contrastive learning and angular variance to mitigate the adverse effects of long-tailed data on deep learning models for image classification. We name our method as Supervised Angular Contrastive Learning (SACL). In our experiments on long-tailed datasets with different class imbalance ratios, we demonstrate that our method outperforms most of the existing baseline approaches.

Keywords: Long-Tail Image Classification · Supervised Contrastive Learning · Imbalanced Data · Feature Representation

1 Introduction

The recent advancement in computer vision technologies has enabled significant breakthroughs across many domains, including areas like autonomous driving,

© The Author(s), under exclusive license to Springer Nature Singapore Pte Ltd. 2024
D. Benavides-Prado et al. (Eds.): AusDM 2023, CCIS 1943, pp. 150–162, 2024.
https://doi.org/10.1007/978-981-99-8696-5_11

medical diagnosis, industrial inspection, etc. One of the fundamental tasks in this domain is image classification, where the objective is to assign labels to images based on their visual characteristics. Traditional image classification methods heavily rely on manually engineered features. However, with the advancement of deep learning methods, end-to-end learning approaches have revolutionised the field of computer vision, enabling automatic feature extraction directly from source data. Even though deep learning models have achieved exceptional performance in image classification, they still face challenges when dealing with long-tail class distributions.

In many real-world situations, the distribution of classes is highly long-tailed [21], with a few dominant classes and a long tail of minority classes. This imbalance distribution can severely affect the performance of deep network models due to the biased classifiers, as they tend to perform better on the dominant classes, i.e. head classes, but show degraded performance for tail classes, i.e. majority classes with limited samples. The common strategies used to address the challenges of long-tail distribution in the existing literature are resampling methods [2,4,10,11,24] and cost-sensitive methods [7,13,28,32]. Although the existing methods have addressed the long-tail distribution problem to some extent, however, there is still vast opportunity for enhancing the performance as resampling techniques address the long-tail problem by either under-sampling the head-class samples or over-sampling the tail-class samples. Under-sampling method could result in the loss of crucial information and, consequently, causes the model to underfit, while the over-sampling method increases the samples of tail classes with the aim to create a more balanced dataset between head and tail classes. However, it could cause overfitting to the model. Due to the limitations of the above methods, many other methods have been proposed in the literature, such as information augmentation, which has two types, i.e. transfer learning [6,30,32] and data augmentation [27,29].

One of the promising strategies recently introduced consists of two-stage methods (representation learning and classifier learning) [15,34]. Supervised contrastive learning (SCL) [16] method also falls in this category, which learns an enhanced feature representation in the first stage using the supervised contrastive loss. It is aimed to decrease intra-class distances and increase the inter-class distances, which is one of the essential aspects for learning the better feature representation and enables the classifier to classify better in the second stage. However, with severe class imbalance, the samples from the tail classes are overlapped with the samples of head classes. Another problem with the severe class imbalance lies at that tail class samples occupy less spatial span [20] due to the absence of intra-class variation.

In our research, we delve into effective contrastive learning and angular variance strategies that are recently proposed in the literature [16,20], customizing them to improve the quality and distinctiveness of image representations, especially when dealing with long-tailed data. The aim is to exploit the strengths of both approaches, i.e. the intra-class compactness and inter-class separability of supervised contrastive learning and from the angular variance strategy, we transfer the average of head class angular variance to tail classes to enrich the feature space. The overall objective is to enhance the effectiveness and capability of our model to address the problem of long-tailed image classification. Our approach leads to a novel hybrid network structure. This structure combines a contrastive loss and angular variance to facilitate the learning of distinct image representations. It is achieved by adding the average of head class angular variance with the features of anchor and augmented images in supervised contrastive learning. Additionally, we employ a cross-entropy loss to train the classifier. Our integration of angular variance with supervised contrastive learning has produced superior results when compared to the performance of supervised contrastive learning alone. Our guiding principle is the understanding that superior features lead to superior classifiers. To put this into practice, we adopt a curriculum-based approach. This means we gradually shift the learning process from initially focusing on feature representation learning to later concentrating on classifier learning.

Inspired by the works of [16, 20], our contribution are summarised as below:

- We propose a novel hybrid method that seamlessly integrates supervised contrastive learning and angular variance to address the long-tailed data problem on deep learning models for image classification.
- Experiments on the imbalanced datasets show that our method performs better than baseline methods for most of the imbalance ratios.

2   Related Work

Our work is related to addressing the image classification problem for long-tailed datasets. Recently, the long-tail image classification area has drawn the attention of the computer vision community. Different methods for addressing long-tailed classification are categorised as below:

2.1   Data Re-balancing

Data re-balancing techniques deal with artificially balancing the long-tailed data. They are classified into two categories, i.e. over-sampling [1, 23, 25] and under-sampling [9, 10, 19]. Over-sampling increases the tail class samples by replicating them, which could result in overfitting the model [24]. Under-sampling techniques balance the data by reducing head class samples, which could cause the loss of valuable information in the presence of severe data imbalance [24].

2.2   Information Augmentation

Information Augmentation [36] supplements the model training with additional information for enhancing the model's performance. It is divided into two categories: data augmentation and transfer learning.

Data Augmentation techniques apply different operations or transformations to the training data to create new, synthetic samples. These augmented samples retain the same label as the original samples but introduce variations in the existing data. The primary goal of data augmentation is to diversify the dataset with a limited number of samples, thereby improving the model's capability to generalize to unfamiliar data. Examples of data augmentation are random scaling, image flipping, cropping, rotating, etc. The approach proposed in [22] augments the long-tail image classification pipeline using retrieval augmented classification. In [25], new tail class samples are generated using a convex combination of existing samples. Various methods of noise injections have also been considered as means of optimizing the decision boundaries by moving them away from minority classes [39].

Transfer Learning. Transfer learning techniques [26,30,32] transfer the knowledge from a source domain (e.g., datasets or head classes) to improve the training of the model in a target domain. In the context of deep long-tailed learning, these methods are typically grouped into different categories, i.e., head-to-tail knowledge transfer, model pre-training, knowledge distillation, and self-training. The work in [33] addresses the minor intra-class variance of features of tail-class samples by exploiting the knowledge of head-class intra-class variance so that features of tail-class samples have enough intra-class variance, resulting in better model performance. In [17], perturbation-based optimisation technique for enhancing the representation of the minority class samples by transforming samples from the majority classes into ones that resemble the minority classes. In [8], the authors initially train the model using all the long-tailed samples to develop a foundation for representation, followed by fine-tuning the model, resulting in transferring the feature from head class samples to tail class samples.

Metric Learning. techniques aim at designing distance metrics to determine the similarity and dissimilarity among the samples. In the context of long-tailed classification, the metric learning-based method designs distance-based loss functions for learning more discriminative feature space. The work in [35] enhances representation learning by considering the distances between all pairs of samples within a single mini-batch. It is aimed to increase the distance between different classes by increasing the distances between the centres of any two classes within the mini-batch. Simultaneously, it minimises the larger distances between samples within the same class, thereby reducing intra-class variations resulting in a discriminative set of features. To improve the generalisation of the model and to address the severe class imbalance problem, the authors in [14] proposed the variation of contrastive loss called k-positive contrastive loss. Authors in [31] proposed a method based on contrastive learning for addressing long-tail classification problems.

Decoupled Learning. Decoupled learning techniques are one of the recent techniques addressing long-tail classification challenges. In these techniques deep long-tailed model is trained in two stages, i.e. representation learning followed by classifier learning. In [15], the authors proposed a model that addresses the long-tail problem using two separate stages for representation and classification learning. They used the cross-entropy loss function in both stages; the findings suggest that random data sampling is beneficial for feature learning, whereas class-balanced sampling proves more effective when it comes to training the classifier.

3   Proposed Framework

Our work has been inspired by [16,20]. Deep learning algorithms require enormous, diverse and balanced data for training. In the case of the long-tailed dataset, majority classes (tail) have limited samples while minority (head) classes have numerous samples, so this unequal distribution between head and tail classes and lack of intra-class diversity in tail classes severely affect the feature space representation in which tail classes usually occupy a small spatial span while head classes usually occupy a large spatial span, resulting in a biased classifier [20].

The angles between the features and the corresponding class centres determine spatial span and separate the features inside the spatial spans, so a larger spatial span has a larger angular variance, and a smaller spatial span has a smaller angular variance. Since head classes often have a wide range of diversity, this property is also apparent in the angles formed between the features of head-class samples and their corresponding class centres. However, tail classes need more intra-class diversity due to fewer samples in the tail classes. In order to mitigate the problem of intra-class diversity in the tail classes, the angular variance of head classes can be applied to tail classes [20], along with data augmentation techniques in order to enhance the diversity of tail classes resulting in enriched feature space. Figure 1 shows the framework of our proposed method for addressing the long-tail image classification problem.

3.1   Supervised Contrastive Learning

Supervised contrastive (SC) loss [16] is an extended version of unsupervised contrastive (UC) loss [5]. The primary difference between the SC loss and the UC loss lies in how they assemble the positive and negative samples related to a particular anchor image. In the UC loss, the positive image is typically an alternative augmented version of the anchor image and all other images, even from the same class, are considered negative images. However, the SC loss takes a different approach. In addition to the augmented image, it includes other images belonging to the same class as positive images.

Following SCL [16], we use a backbone network, i.e. ResNet [12] to extract features and generate an image representation $i \in \mathbb{R}^D$ for each image x. Then, we employ a projection head encoder to convert this image representation i into a more appropriate vector representation $z \in \mathbb{R}^{D_E}$. This transformation is

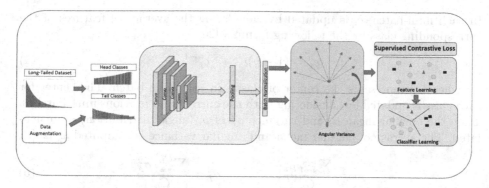

Fig. 1. Overview of the proposed framework. The first part of the network applies different augmentation techniques, followed by calculating the average head-class angular variance and then transferring it to the tail classes. The second part of the network (SCL) learns and classifies the features respectively.

carried out through a multi-layer perceptron (MLP) with one hidden layer, as it has been shown to enhance the quality of the preceding layer's [16]. Subsequently, we perform normalization on z to prepare it for measuring distances using inner product calculations. Classification is achieved through a single linear layer applied to the image representation to generate predictions for class logits. Cross entropy loss $\mathcal{L}_{CE}$ is applied on these logits for the prediction of samples. For a mini-batch size of N, SCL is represented as [16]:

$$\mathcal{L}_{SCL} = \sum_{i=1}^{2N} \mathcal{L}_{SCL}(z_i) \tag{1}$$

$$\mathcal{L}_{SCL}(z_i) = -\frac{1}{2N_{y_i} - 1} \sum_{j=1}^{2N} \log\left(\frac{\exp(z_i \cdot z_j/\tau)}{\sum_{k=1, k \neq i}^{2N} \exp(z_i \cdot z_k/\tau)}\right) \tag{2}$$

where z_i and z_j represent the positive pair of images and z_k is the negative image and $\tau \geq 0$ is temperature parameter. The integration of angle variance with supervised contrastive loss is discussed in Sect. 3.3. This integrated approach facilitates feature learning and classification.

3.2 Angular Variance

Following the literature [20], to enhance the intra-class variance of tail class samples, we transfer the average of head-class angular variance to tail class samples. Let f_i^m represent the m-th feature of class i, and c_i is the center of class i. Then the distribution of angles formed between the class center of i-th class and features of m-th sample of class i is given as:

$$\rho_{i,m} = arccos\left(\frac{f_i^m \cdot c_i}{\| f_i^m \| \| c_i \|}\right) \tag{3}$$

In each mini-batch, c_i is updated by calculating the average of features of the corresponding class by the following formula [20].

$$c_i^n = (1 - \gamma)c_i^n + \gamma c_i^{n-1} \tag{4}$$

where c_i^n represents the center of class i in n-th mini-batch. The center for each class is updated by considering both the current and previous mini-batches. Then the angular mean μ_i and variance σ_i^2 of ρ_i (angle distribution) are calculated. The average of angular mean and angular variance is computed as [20]:

$$\mu_h = \frac{\sum_{s=1}^{C_h} \mu_s}{C_h}, \qquad \sigma_h^2 = \frac{\sum_{s=1}^{C_h} \sigma_s^2}{C_h} \tag{5}$$

where C_h is the total number of head classes, μ_s and σ_s^2 are the angular mean and angular variance of s-th head class, respectively. The n-th tail class angular distribution is defined as a normal distribution $N\left(\mu_t^n, \sigma_t^{n^2}\right)$. Due to the availability of many samples in head classes, they have more intra-class angular diversity than the tail classes. To enrich the tail classes with angular diversity, the average of head class variance is transferred to the tail class samples. This is achieved by creating a feature cloud around each sample feature of the t-th tail class. A feature cloud is a set of virtual feature vectors around the tail class samples and its generation process adheres to the probability distribution that has been learned from the more prevalent head class. It helps to make the enlarged space for the tail class during training and ensures that the actual tail class samples are kept at a distance from the other class samples.

The angle between the feature of n-th tail class and a feature sampled from its corresponding feature cloud is defined as θ_Δ, and its distribution is modelled as $\theta_\Delta \sim \left(0, \sigma_h^2 - \sigma_t^{n^2}\right)$ and $\theta_\Delta \in \mathbb{R}^{1 \times C}$.

3.3   Integrating Angular Variance into Supervised Contrastive Loss

In the context of angular variance [20], once the features of all samples in a mini-batch have been extracted, we will compute the class centre for each class present within that mini-batch. By utilizing a threshold value on the number of samples in each class, our model effectively distinguishes between tail and head class samples. Once this distinction is established, we measure the angles formed between the features and their corresponding class centres. This process results in the formation of an angle distribution [20]. From this angle distribution, we calculate the angular mean and variance. The average of head class angular variance is transferred to tail class samples. In the final step, we incorporate the average of head class angular variance alongside the features into the supervised contrastive loss [16]. For a tail sample $z_j \in c_t^j$, sample the perturb angle θ_Δ from $N\left(0, \sigma_h^2 - \sigma_t^2\right)$ and our final loss (SACL) is given as:

$$\mathcal{L}_{SACL} = -\frac{1}{2N_{yi} - 1} \sum_{j=1}^{2N} \log\left(\frac{\exp\left((z_i \cdot z_j + \theta_\Delta)/\tau\right)}{\sum_{k=1, k \neq i}^{2N} \exp(z_i \cdot z_k/\tau)}\right) \tag{6}$$

All the steps of our proposed method are listed in the following algorithm.

Algorithm 1: Angular Variance Based Augmentation for Supervised Contrastive Learning

---

    Input : All training images, number of epoch N_e, threshold T used to split head classes and tail classes, and temperature τ

    Output: Trained Model

1 **Prerequisite:** Assign a threshold value T to divide the dataset into head and tail classes

2 Initialisation

3 **for** $ep \leftarrow 0$ **to** N_e **do**

4   **foreach** $mini-batch$ **do**

5   Apply data augmentation twice to obtain two copies of the batch and extract the features

6   Tranform the features to normalised embedding through encoder network

7   Calculate the angular variance for each class, and obtain the angular variance for head classes σ_h^2 based on T

8   For each tail class c_t, calculate the angular variance gap to the head classes as $\sigma_h^2 - \sigma_t^2$

9   For z_i, let z_p denotes a positive sample with $p \in P(i)$, where $P(i) \equiv \{p \in A(i) : y_i = y_p\}, A(i) \equiv I/\{i\}$

10   For a tail sample $z_i \in c_t^j$, sample the perturb angle θ_Δ from $N(0,\ \sigma_h^2\text{-}\sigma_t^2)$ and calculate the loss ($\mathcal{L}_{SACL}$) via equation 6

11   Update the weights

12   **end**

13 **end**

14 Train the classifier

Our proposed loss is listed in step 10 of our algorithm. For head class samples, θ_Δ is 0, however, for tail class samples, θ_Δ is the difference between average head class variance and the variance corresponding to a tail class. Our final loss incorporates the strengths of supervised contrastive learning [16] and angular variance [20]. Combining the two techniques helps the deep learning model to address the long-tail image classification problem with better performance, as demonstrated by our experiments.

4 Experiments

In this section, we provide an overview of the long-tailed datasets used in both our model and the baseline methods. We then provide the details about the implementation of our model. Lastly, we present a comparative analysis of the results obtained from our methods and the baseline methods.

4.1 Datasets

The original CIFAR-10 and CIFAR-100 datasets each contain a total of 60,000 images. Among these, 50,000 are designated as training images, while the remaining 10,000 serve as test images. These images are 32×32 pixels in size, and the datasets consist of 10 and 100 classes, respectively. To ensure a fair comparison, we have generated long-tailed versions of both datasets using the method described in the literature [3] with the same class imbalance ratios, i.e. 10, 50, and 100 for each dataset. Figure 2 illustrates the long-tailed versions of both datasets with an imbalance ratio of 100. Class imbalance ratio (β) is the ratio between the maximum number of samples in the most frequent class to the minimum number of samples in the least frequent class. In other words, β is calculated as $\beta = N_{max}/N_{min}$.

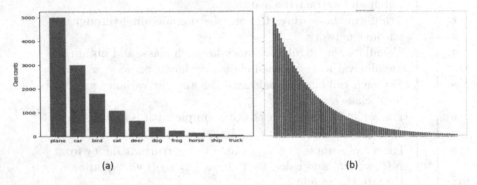

(a)   (b)

Fig. 2. (a) Long-tail CIFAR-10 with imbalance factor of 100 (b) Long-tail CIFAR-100 with imbalance factor of 100

4.2 Implementation Details

For feature extraction, we have used ResNet-32 [12] as a backbone network. Different data augmentation techniques, such as random cropping with a resolution of 32×32, horizontal flip, colour jittering and random grayscale with a probability of 0.2, are used on each dataset. We use Stochastic gradient descent (SGD) optimizer with a momentum of 0.9. The learning rate and temperature scaling values are 0.5 and 0.1, respectively. The training has been performed on two NVIDIA 1080Ti GPUs for 200 epochs with a batch size of 128, whereas testing has been performed with a balanced version of the test dataset using a batch size of 64. For a fair comparison, we adopted the hyper-parameter settings and backbone architecture from the baseline methods.

4.3 Comparison to the Baseline Methods

In this part, we assess our model by comparing it with existing baseline methods for addressing long-tail image classification. We conduct these comparisons on the long-tailed CIFAR-10 and CIFAR-100 datasets, which exhibit varying degrees of class imbalance. The results of this comparison are presented in Table 1. Various techniques have been suggested in the literature for addressing the long-tail distribution problem in deep learning models. The traditional cross-entropy [37] loss, and its various forms lack the constraint of intra-class and inter-class distances, so the head-class samples dominate the tail-class samples, resulting in a biased classifier. Focal loss [18] uses a reweighting strategy and assigns higher weights to tail classes while lower weights to head classes to address the extreme class imbalance encountered during the training. The other set of methods for comparison consists of Cross Entropy-based resampling (CE-DRS) and Cross Entropy-based reweighting (CE-DRW), both of which are proposed in previous state-of-the-art [3]. These methods consist of two stages. In the first stage, the network is trained on imbalanced data using cross-entropy loss, while in the second stage, class resampling and reweighting are applied during training in CE-DRS and CE-DRW, respectively. The closest baseline method to our approach is supervised contrastive (SC) loss [16]. Our proposed method is also based on the supervised contrastive loss; it also incorporates the angular variance to enhance the diversity of tail samples. Methods compared with our proposed method are listed in Table 1. For a fair comparison, we have adopted the hyper-parameter settings and the choice of backbone architecture from the baseline method [38] as it has the best performance compared to other baseline methods. In Table 1, the top-performing results are highlighted in bold, while the second-best results are indicated in italics. It can be seen our method consistently outperforms the supervised contrastive loss across all imbalance ratios in both long-tailed datasets. Additionally, it surpasses the other baseline methods in the case of an imbalance ratio of 10 for both datasets. However, for the remaining imbalance ratios, our method's results are comparable to those of the baseline methods.

Table 1. Top-1 accuracy (%) on different imbalance ratios of CIFAR datasets.

Dataset	Long-Tailed Cifar100			Long-Tailed Cifar10		
Imbalance Ratio	10	50	100	10	50	100
CE	55.71	43.85	38.32	86.39	74.81	70.36
Focal [18]	44.38	44.32	38.41	86.66	76.72	70.38
Mixup	58.02	44.99	39.54	87.91	77.82	73.06
Manifold Mixup	56.55	43.09	38.25	87.03	77.95	73.1
CE-DRW [3]	58.12	45.29	41.51	87.56	79.21	76.34
CE-DRS [3]	58.11	45.48	41.61	87.38	79.97	75.61
CB-Focal	57.99	45.17	39.6	87.1	79.81	74.57
LDAM-DRW [3]	58.71	46.62	*42.04*	88.16	*81.03*	*77.03*
BBN [38]	59.12	**47.02**	**42.56**	88.32	**82.18**	**79.82**
SupContrasive	*60.13*	46.59	40.11	*89.36*	80.14	74.09
SACL (Ours)	**61.24**	*46.81*	*41.13*	**89.91**	80.24	74.55

5   Conclusion

In this work, we developed a novel method for mitigating the adverse effects of long-tailed data for image classification on a deep learning model. Our method integrates the average of head class angular variance with the supervised contrastive loss. Experiments on long-tailed versions of datasets demonstrated that our method has outperformed most of the baseline methods for some imbalance ratios, and its results are comparable for the remaining imbalance ratios. To the best of our knowledge, this is the first-ever work that explores the direction of combining supervised contrastive learning with angular variance for addressing the long-tail classification problem. A limitation of our work is that it requires a predefined threshold value to divide the dataset into head and tail classes. However, we plan to tackle this issue in our future research.

References

1. Barua, S., Islam, M.M., Yao, X., Murase, K.: MWMOTE-majority weighted minority oversampling technique for imbalanced data set learning. IEEE Trans. Knowl. Data Eng. **26**(2), 405–425 (2012)
2. Byrd, J., Lipton, Z.: What is the effect of importance weighting in deep learning? In: International Conference on Machine Learning, pp. 872–881. PMLR (2019)
3. Cao, K., Wei, C., Gaidon, A., Arechiga, N., Ma, T.: Learning imbalanced datasets with label-distribution-aware margin loss. Adv. Neural Inf. Process. Syst. **32** (2019)
4. Chawla, N.V., Bowyer, K.W., Hall, L.O., Kegelmeyer, W.P.: SMOTE: synthetic minority over-sampling technique. J. Artif. Intell. Res. **16**, 321–357 (2002)
5. Chen, T., Kornblith, S., Norouzi, M., Hinton, G.: A simple framework for contrastive learning of visual representations. In: International Conference on Machine Learning, pp. 1597–1607. PMLR (2020)
6. Chu, P., Bian, X., Liu, S., Ling, H.: Feature space augmentation for long-tailed data. In: Vedaldi, A., Bischof, H., Brox, T., Frahm, J.-M. (eds.) ECCV 2020, Part XXIX. LNCS, vol. 12374, pp. 694–710. Springer, Cham (2020). https://doi.org/10.1007/978-3-030-58526-6_41
7. Cui, Y., Jia, M., Lin, T.Y., Song, Y., Belongie, S.: Class-balanced loss based on effective number of samples. In: Proceedings of the IEEE/CVF Conference on Computer Vision and Pattern Recognition, pp. 9268–9277 (2019)
8. Cui, Y., Song, Y., Sun, C., Howard, A., Belongie, S.: Large scale fine-grained categorization and domain-specific transfer learning. In: Proceedings of the IEEE Conference on Computer Vision and Pattern Recognition, pp. 4109–4118 (2018)
9. Dal Pozzolo, A., Caelen, O., Johnson, R.A., Bontempi, G.: Calibrating probability with undersampling for unbalanced classification. In: 2015 IEEE Symposium Series on Computational Intelligence, pp. 159–166. IEEE (2015)
10. Drummond, C., Holte, R.C., et al.: C4. 5, class imbalance, and cost sensitivity: why under-sampling beats over-sampling. In: Workshop on Learning from Imbalanced Datasets II, vol. 11, pp. 1–8 (2003)
11. He, H., Garcia, E.A.: Learning from imbalanced data. IEEE Trans. Knowl. Data Eng. **21**(9), 1263–1284 (2009). https://doi.org/10.1109/TKDE.2008.239
12. He, K., Zhang, X., Ren, S., Sun, J.: Deep residual learning for image recognition. In: Proceedings of the IEEE Conference on Computer Vision and Pattern Recognition, pp. 770–778 (2016)

13. Huang, C., Li, Y., Loy, C.C., Tang, X.: Learning deep representation for imbalanced classification. In: Proceedings of the IEEE Conference on Computer Vision and Pattern Recognition, pp. 5375–5384 (2016)
14. Kang, B., Li, Y., Xie, S., Yuan, Z., Feng, J.: Exploring balanced feature spaces for representation learning. In: International Conference on Learning Representations (2020)
15. Kang, B., et al.: Decoupling representation and classifier for long-tailed recognition. arXiv preprint arXiv:1910.09217 (2019)
16. Khosla, P., et al.: Supervised contrastive learning. Adv. Neural. Inf. Process. Syst. **33**, 18661–18673 (2020)
17. Kim, J., Jeong, J., Shin, J.: M2M: imbalanced classification via major-to-minor translation. In: Proceedings of the IEEE/CVF Conference on Computer Vision and Pattern Recognition, pp. 13896–13905 (2020)
18. Lin, T.Y., Goyal, P., Girshick, R., He, K., Dollár, P.: Focal loss for dense object detection. In: Proceedings of the IEEE International Conference on Computer Vision, pp. 2980–2988 (2017)
19. Lin, W.C., Tsai, C.F., Hu, Y.H., Jhang, J.S.: Clustering-based undersampling in class-imbalanced data. Inf. Sci. **409**, 17–26 (2017)
20. Liu, J., Sun, Y., Han, C., Dou, Z., Li, W.: Deep representation learning on long-tailed data: a learnable embedding augmentation perspective. In: Proceedings of the IEEE/CVF Conference on Computer Vision and Pattern Recognition, pp. 2970–2979 (2020)
21. Liu, Z., Miao, Z., Zhan, X., Wang, J., Gong, B., Yu, S.X.: Large-scale long-tailed recognition in an open world. In: Proceedings of the IEEE/CVF Conference on Computer Vision and Pattern Recognition, pp. 2537–2546 (2019)
22. Long, A., et al.: Retrieval augmented classification for long-tail visual recognition. In: Proceedings of the IEEE/CVF Conference on Computer Vision and Pattern Recognition, pp. 6959–6969 (2022)
23. Mohammed, R., Rawashdeh, J., Abdullah, M.: Machine learning with oversampling and undersampling techniques: overview study and experimental results. In: 2020 11th International Conference on Information and Communication Systems (ICICS), pp. 243–248. IEEE (2020)
24. More, A.: Survey of resampling techniques for improving classification performance in unbalanced datasets. arXiv preprint arXiv:1608.06048 (2016)
25. Mullick, S.S., Datta, S., Das, S.: Generative adversarial minority oversampling. In: Proceedings of the IEEE/CVF International Conference on Computer Vision, pp. 1695–1704 (2019)
26. Pan, S.J., Yang, Q.: A survey on transfer learning. IEEE Trans. Knowl. Data Eng. **22**(10), 1345–1359 (2009)
27. Perez, L., Wang, J.: The effectiveness of data augmentation in image classification using deep learning. arXiv preprint arXiv:1712.04621 (2017)
28. Ren, M., Zeng, W., Yang, B., Urtasun, R.: Learning to reweight examples for robust deep learning. In: International Conference on Machine Learning, pp. 4334–4343. PMLR (2018)
29. Shorten, C., Khoshgoftaar, T.M.: A survey on image data augmentation for deep learning. J. Big Data **6**(1), 1–48 (2019)
30. Tan, C., Sun, F., Kong, T., Zhang, W., Yang, C., Liu, C.: A survey on deep transfer learning. In: Kůrková, V., Manolopoulos, Y., Hammer, B., Iliadis, L., Maglogiannis, I. (eds.) ICANN 2018, Part III. LNCS, vol. 11141, pp. 270–279. Springer, Cham (2018). https://doi.org/10.1007/978-3-030-01424-7_27

31. Wang, P., Han, K., Wei, X.S., Zhang, L., Wang, L.: Contrastive learning based hybrid networks for long-tailed image classification. In: Proceedings of the IEEE/CVF Conference on Computer Vision and Pattern Recognition, pp. 943–952 (2021)
32. Wang, Y.X., Ramanan, D., Hebert, M.: Learning to model the tail. Adv. Neural Inf. Process. Syst. **30** (2017)
33. Yin, X., Yu, X., Sohn, K., Liu, X., Chandraker, M.: Feature transfer learning for face recognition with under-represented data. In: Proceedings of the IEEE/CVF Conference on Computer Vision and Pattern Recognition, pp. 5704–5713 (2019)
34. Zhang, J., Liu, L., Wang, P., Shen, C.: To balance or not to balance: a simple-yet-effective approach for learning with long-tailed distributions. arXiv preprint arXiv:1912.04486 (2019)
35. Zhang, X., Fang, Z., Wen, Y., Li, Z., Qiao, Y.: Range loss for deep face recognition with long-tailed training data. In: Proceedings of the IEEE International Conference on Computer Vision, pp. 5409–5418 (2017)
36. Zhang, Y., Kang, B., Hooi, B., Yan, S., Feng, J.: Deep long-tailed learning: a survey. IEEE Trans. Pattern Anal. Mach. Intell. **45**, 10795–10816 (2023)
37. Zhang, Z., Sabuncu, M.: Generalized cross entropy loss for training deep neural networks with noisy labels. Adv. Neural Inf. Process. Syst. **31** (2018)
38. Zhou, B., Cui, Q., Wei, X.S., Chen, Z.M.: BBN: bilateral-branch network with cumulative learning for long-tailed visual recognition. In: Proceedings of the IEEE/CVF Conference on Computer Vision and Pattern Recognition, pp. 9719–9728 (2020)
39. Zur, R.M., Jiang, Y., Pesce, L.L., Drukker, K.: Noise injection for training artificial neural networks: a comparison with weight decay and early stopping. Med. Phys. **36**(10), 4810–4818 (2009)

Shapley Value Based Feature Selection to Improve Generalization of Genetic Programming for High-Dimensional Symbolic Regression

Chunyu Wang, Qi Chen[(✉)], Bing Xue, and Mengjie Zhang

Centre for Data Science and Artificial Intelligence and School of Engineering and
Computer Science, Victoria University of Wellington, PO Box 600, Wellington 6400,
New Zealand
{chunyu.wang,qi.chen,bing.xue,mengjie.zhang}@ecs.vuw.ac.nz

Abstract. Symbolic regression (SR) on high-dimensional data is a challenging problem, often leading to poor generalization performance. While feature selection can improve the generalization ability and efficiency of learning methods, it is still a hard problem for genetic programming (GP) for high-dimensional SR. Shapley value has been used in an additive feature attribution method to attribute the difference between the output of the model and an average baseline to the input features. Owing to its solid game-theoretic principles, Shapley value has the ability to fairly compute each feature importance. In this paper, we propose a novel feature selection algorithm based on the Shapley value to select informative features in GP for high-dimensional SR. A set of experiments on ten high-dimensional regression datasets show that, compared with standard GP, the proposed algorithm has better learning and generalization performance on most of the datasets. A further analysis shows that the proposed method evolves more compact models containing highly informative features.

Keywords: Feature selection · generalization · genetic programming · symbolic regression

1 Introduction

Nowadays, with the development of data collection, the high dimensionality of data becomes more common [1]. However, learning from high-dimensional data presents several inherent challenges, including the issues of overfitting and high computational cost [1]. Feature selection is a process to identify relevant features related to the output variable(s). There have been many works in the feature selection field [2,3]. However, most of them focus on classification. Symbolic regression (SR) is a type of regression analysis and has the task of simultaneously identifying the model structure/type and the associated coefficients. Genetic programming (GP) [4] as an evolutionary computation technique is capable to

© The Author(s), under exclusive license to Springer Nature Singapore Pte Ltd. 2024
D. Benavides-Prado et al. (Eds.): AusDM 2023, CCIS 1943, pp. 163–176, 2024.
https://doi.org/10.1007/978-981-99-8696-5_12

generate an optimal solution without the need for prior information on the form and size of solutions, which makes GP particularly well-suited for SR. In GP for SR (GPSR), only a limited number of existing works explicitly consider feature selection [5–7]. The main reason is that GP possesses an implicit feature selection ability, which automatically selects features during the GP tree model building process. But this natural ability is not strong enough when the dimensionality of the problem/data is high.

Generalization refers to the performance of the learnt models on the unseen data. Models with a good generalization ability are expected to obtain similar performance on the unseen data to that on the training data. Although generalization is considered an important performance metric in many machine learning tasks, compared with the rapid development of GP, how to effectively improve the generalization in GP is still an open issue [8–10]. Moreover, as the increase of dimensionality, it is harder for GP to learn a good model.

Feature selection is able to reduce noise, irrelevant and redundant features, which is likely to reduce the risk of overfitting and improve the generalization ability. Despite its importance, there are only a limited number of existing works using GP with explicit feature selection that can be found in the literature, e.g., a GP with permutation importance feature selection method for high-dimensional regression [8]. However, the permutation importance cannot identify the correlated features, as it measures the impact of individual features on the model performance. Thus, when there are anomalous data instances along with the presence of correlated features, the permutation importance might not accurately reflect the feature's true importance [11]. To address the problem, we adopt the Shapley value, which comes from the game theory and attributes the payoff of a cooperative game to the players of the game in a mathematically fair and unique way, to compute the feature importance. Shapley value attributes the features based on changes in predictions for individual points, which is suitable for identifying correlated features and obtains stable results of feature importance [12].

This paper aims to propose a novel Shapley value based feature selection method in GP for high-dimensional symbolic regression problems. To be specific, this paper investigates the following research objectives:

(1) whether feature selection based on Shapley value can improve the generalization ability of GP on the unseen data.
(2) whether feature selection based on Shapley value can improve the learning ability of GP on the training data.
(3) whether GP with the novel feature selection algorithm can reduce the dimensionality of data and evolve more compact models.

2   Background

2.1   Genetic Programming for Symbolic Regression

Genetic programming (GP) is an evolutionary computation algorithm that aims to create computer programs for problem solving [11]. Initially, GP generates a population of programs. This population then goes through an evolutionary process consisting of evaluation, fitness-based selection, and breeding until reaches some terminal criteria. At the end, GP gets the optimal program which has the best fitness as the solution for the problem.

Symbolic regression (SR) is a type of regression analysis and has the task of function identification and coefficient fitting [6]. Compared with classical regression, SR is able to generate a mathematical model without prior information on the form and size of the model. Thus, instead of finding a set of model parameters for a predefined regression model structure, GP has the ability to adaptively find an optimal solution without the prior information of the form and size of solutions, which makes GP flexible and suits to SR. Thus, many researchers propose different kinds of GP algorithms to enhance the efficiency and effectiveness for SR [13,14]. However, how to enhance the generalization ability of GPSR is still an open issue.

2.2 Genetic Programming for Feature Selection

GP has an 'embedded' ability to explore the feature space and select important features, which makes GP become a good choice for feature selection [3,5,15]. However, this natural ability is not strong enough for high-dimensional data [11]. Thus, some existing works explore many methods to improve the efficiency of GP for feature selection. For example, Neshatian and Zhang [3] proposed a multi-objective GP method for classification which searches for feature subsets rather than single features where the terminals in GP are regarded as the set of selected features. The two objectives of their multi-objective GP methods are to find out the maximal relevance of feature subsets and to minimize the size of feature subsets. Sandin et al. [15] proposed a novel GP-based method for unbalanced high-dimensional data for classification. The method searches through the space of possible combinations of feature selection metrics, i.e., information gain, χ^2 and odds ratio, to decide an unbiased estimator of the discriminative power of the feature. Chen et al. [6] proposed a novel permutation importance based feature selection method for high-dimensional GPSR. The permutation importance evaluates feature importance by shuffling feature values in the learnt model. Specifically, when the values of one feature are shuffled, the increase of the model error indicates that the feature is important. This algorithm enhances the generalization of GPSR. However, the permutation of features may generate unrealistic data instances which are impossible in the real world when two or more features are correlated. For example, if there are positively two correlated features, i.e., height and weight of a person and one of the features is shuffled, a new instance with two meter person weighing 30 kg may be created that is unlikely or even physically impossible. If the unrealistic data instances are used to measure the importance, the result is unreliable, which could lead to select irrelevant features [11].

2.3 Generalization in Genetic Programming

Generalization is one of the most important performance measurements for learning models in supervised learning problems. A limited number of researches are

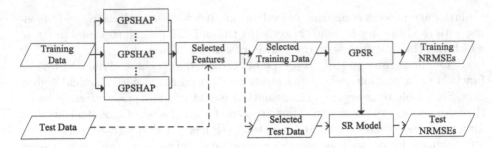

Fig. 1. The data flowchart of the proposed GP system.

centered on generalization in GP for symbolic regression [13,14,16]. Chen et al. [16] proposed a multi-objective method to improve the generalization in GP. The training error and the Rademacher complexity of the models are regarded as the two objectives. The Rademacher complexity of a model is computed by the maximum correlation between the model and the Rademacher variables on the training instances. Haeri et al. [13] proposed a layered GP based on the variance for improving generalization. The evolutionary process is divided into hierarchical layers where the first layer adopts the smoothest training set and the next layers are more complex than the previous layers. Furthermore, the method uses the variance of the output value to measure and control the functional complexity, thus reducing overfitting. The results show that the method not only reduces the complexity but also improves the generalization of GP. Mousavi et al. [14] proposed a multi-objective GP algorithm to improve the generalization. In addition to optimizing the commonly used error measure, the method obtains better generalization with the first-order derivative of a candidate model as another objective.

3   The Proposed Method

This paper proposes a new feature selection algorithm named GP with Shapley value (GPSHAP) which belongs to the embedded method. It is based on the fact that GP is able to explore the search space in order to automatically select important features [5]. This paper assumes that individuals with high fitness are more likely to have relevant features but not all the features in these individuals are relevant and some of them may be redundant. Thus, these features can be regarded as the candidate important feature subset. By further measuring the importance of these candidate features, we can identify truly relevant features.

Figure 1 shows the flowchart for the new GPSR system, which is divided into two sequential stages. In the first stage, the new feature selection method, GPSHAP, is used to select a subset of important features. After that, standard GPSR performs on the training data with only the selected features.

In GPSHAP, the Shapley value is adopted to evaluate the importance of features. The Shapley value originates from the cooperative game theory which denotes a conception of fairly allocating a total game gain to every game player. Shapley value is an additive feature attribution algorithm satisfying mathematical axioms, i.e., efficiency, symmetry, dummy and additivity, which together can

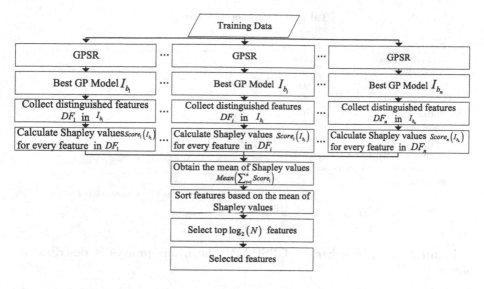

Fig. 2. The flowchart of GPSHAP.

be considered a definition of a fair payout [11,17]. Thus Shapley value has the potential to fairly attribute the feature contribution. It attributes the features to compute changes in predictions by adding the feature into the feature subset, which is suitable for computing each feature importance and obtaining stable results of feature importance [12]. The prediction of SR is regarded as the game gain and different features in the best individual are regarded as players. Let $F = \{v_1, ..., v_k\}$ denotes the k distinguished features in the best individual I_b. S is a subset of F, i.e., $S \subseteq F$. Besides, $f_{S \cup \{i\}}$ refers to a GP model trained with the ith feature and f_S is a model trained without the ith feature. Thus, predictions are represented as $f_{S \cup \{i\}}(x_{S \cup \{i\}}) - f_S(x_S)$, where x_S is the values of the features in S. The Shapley value is defined in Eq. (1), which considers all possible combinations.

$$\phi_i = \sum_{S \subseteq F \backslash \{i\}} \frac{|S|!(|F| - |S| - 1)!}{|F|!} [f_{S \cup \{i\}}(x_{S \cup \{i\}}) - f_S(x_S)] \qquad (1)$$

However, computing the Shapley value is time-consuming for high-dimensional data as it enumerates all possible combinations. To improve the efficiency, in this work, the approximation of ϕ_i is computed based on sampling [18]. The approximation of ϕ_i is the average marginal contribution for multiple sampling steps defined in Eq.(2).

$$\phi_i = \frac{1}{T} \sum_{t=1}^{T} (f_{S \cup \{i\}}(x_{S \cup \{i\}}) - f_S(x_S)) \qquad (2)$$

where T is the total number of sampling steps.

Table 1. Parameter Settings

Parameter	Values
Population Size	512
Generations	50
Crossover Rate	0.9
Mutation Rate	0.1
Elitism Rate	0.01
Maximum Tree Depth	10
Initialization	Ramped-Half&Half
Minimum Initialization Depth	2
Maximum Initialization Depth	6
Function Set	$+, -, *, Inv(\frac{1}{x})$, sqrt
Terminal Set	Features (Selected Features), Random Constant$\in [-1.0, 1.0]$
Fitness Function	NRMSE

Figure 2 is the flowchart of GPSHAP. The whole process is described as follows.

(1) Perform n GPSR runs parallelly, each generates one best-of-run individual I_{b_i}.
(2) The terminal nodes are collected from each I_{b_i} as the distinguished feature subset DF_i.
(3) Shapley values of every feature $(Score_i(I_{b_i}))$ in DF_i are computed based on Eq. (2).
(4) All parallel Shapley values are collected. The mean of these Shapley values $(Mean(\sum_{i=1}^{n} Score_i(I_{b_i})))$ is computed as the feature importance on the training data, where if a feature does not appear in any best-of-run individuals, its Shapley value is set to 0.
(5) Features are ranked based on the mean of Shapley values.
(6) Following previous work on finding a good threshold [19], we choose to select the top $log_2(N)$ features as the final features where N means the total number of available features on the dataset.

4   Experiment Design

To investigate and confirm both the feature selection ability and the generalization ability, GPSHAP is compared with standard GP and a GP with permutation importance method, named as GPPI [11] on ten datasets including two synthetic and eight real-world datasets.

4.1   Evaluation Measure — Fitness Function

GPSHAP, standard GP and GPPI all adopt *the Normalized Root Mean Square* (NRMSE) defined in Eq. (3) as the fitness function on every dataset.

$$NRMSE = \frac{\sqrt{\frac{1}{N}\sum_{i=1}^{N}(f(X_i) - Y_i)^2}}{Y_{max} - Y_{min}} \tag{3}$$

where N is the number of instances, $f(X_i)$ is the outputs of the model and Y_i is the target output. $Y_{max} - Y_{min}$ denotes the range of the target output on the training dataset.

Table 2. Two Synthetic Functions

Functions	Training Samples	Test Sample	Noise
$F_1 = -g\frac{X_1 X_2}{X_3^2}$	70 points	30 points	50 input variables
	$X_1, X_2 = rnd[0,1]$	$X_1, X_2 = rnd[0,1]$	$= rnd[0,1]$
	$X_3 = rnd[1,2]$	$X_3 = rnd[1,2]$	
$F_2 = \frac{30X_1 X_3}{(X-10)X_2^2}$	1000 points	10000 points	50 input variables
	$X_1, X_3 = rnd(-1,1)$	$X_1, X_3 = rnd(-1,1)$	$= rnd[0,1]$
	$X_2 = rnd(1,2)$	$X_2 = rnd(1,2)$	

Table 3. Benchmark Problems

Name	#Features	#Total Instances	#Training Instances	#Test Instances
F_1	53	100	70	30
F_2	53	11000	1000	10000
CCN	122	1994	1395	599
CCUN	124	1994	1395	599
4544_GOM	117	1059	767	328
505_tecator	124	240	168	72
QT-1	1024	427	299	128
QT-2	1024	395	277	118
QT-3	1024	692	484	208
QT-4	1024	389	272	117

4.2 Parameters and Datasets

To guarantee a fair comparison, parameter settings in GPSHAP, standard GP and GPPI are the same as shown in Table 1. The top 1% population are reserved as the elites and the rest of populations are generated by genetic operators.

For the two synthetic datasets as shown in Table 2, F_1 is Newton's law of gravitation and g is the gravitational constant with the value of $6.67408E - 11$. F_2 originates from [20]. In addition, to examine the ability of the methods in selecting relevant features, noisy features are added to each dataset.

The Communities and Crime unnormalized dataset (CCUN) and the Communities and Crime normalized dataset (CCN) are taken from UCI [21], two high-dimensional datasets, 4544_GeographicalOriginalofMusic (4544_GOM) and 505_tecator are from Penn Machine Learning Benchmarks (PMLB) [22] and the rest of four real-world datasets are taken from OpenML [23].

Table 3 shows a summary of the ten datasets. In this paper, a popular splitting method in machine learning is used where 70% of instances are selected randomly as the training sets and the rest of 30% of instances are regarded as the test sets [6]. During the feature selection progress, all methods only use the training sets. The experiments are run independently 30 times for each method on each dataset.

5   Results and Discussions

This section discusses the SR results of GPSHAP, standard GP and GPPI with comparisons of NRMSEs on the training sets and test sets. In addition, further analyses on the model size, the number of features and distinguished features, as well as the computational time are also provided.

Table 4. Statistical Significant Test Results

Datasets	Method	Training NRMSE (Median ± MAD)	Test NRMSE (Median ± MAD)	p_value (Training)	p_value (Test)	Significant Test (with GPSHAP) (Training, Test)
F_1	GP	0.086 ± 0.03	0.179 ± 0.069	1.7E-08	8.35E-08	$(-, -)$
	GPPI	0.09 ± 0.02	0.2 ± 0.07	1.84E-08	1.69E-08	$(-, -)$
	GPSHAP	5.06E-17 ± 2.9E-17	9.12E-17 ± 6.48E-17			
F_2	GP	0.068 ± 0.03	0.064 ± 0.03	8.29E-06	4.12E-06	$(-, -)$
	GPPI	0.048 ± 0.003	0.041 ± 0.027	4.22E-03	2.75E-03	$(-, -)$
	GPSHAP	0.041 ± 0.003	0.038 ± 0.002			
CCN	GP	0.094 ± 0.001	0.078 ± 0.002	0.06	0.16	$(=, =)$
	GPPI	0.093 ± 0.001	0.08 ± 0.003	0.34	7.62E-03	$(=, -)$
	GPSHAP	0.093 ± 9.45E-04	0.077 ± 0.001			
CCUN	GP	0.143 ± 0.002	0.138 ± 0.002	0.01	0.05	$(-, =)$
	GPPI	0.142 ± 0.001	0.136 ± 0.002	0.49	0.55	$(=, =)$
	GPSHAP	0.142 ± 6.45E-04	0.137 ± 5.23E-04			
4544_GOM	GP	0.092 ± 0.004	0.093 ± 0.006	7.89E-04	8.91E-05	$(-, -)$
	GPPI	0.091 ± 0.005	0.09 ± 0.004	3.89E-03	2.65E-03	$(-, -)$
	GPSHAP	0.086 ± 0.001	0.084 ± 0.003			
505_tecator	GP	0.115 ± 0.017	0.111 ± 0.02	4.46E-04	3.32E-06	$(-, -)$
	GPPI	0.1 ± 0.02	0.111 ± 0.019	3.01E-04	8.29E-06	$(-, -)$
	GPSHAP	0.087 ± 0.004	0.084 ± 0.006			
QT-1	GP	0.211 ± 0.004	0.224 ± 0.006	1.09E-05	0.11	$(-, =)$
	GPPI	0.2 ± 0.004	0.219 ± 0.004	0.06	0.75	$(=, =)$
	GPSHAP	0.202 ± 0.004	0.22 ± 0.006			
QT-2	GP	0.113 ± 0.006	0.105 ± 0.008	8.48E-09	1.25E-07	$(-, -)$
	GPPI	0.104 ± 0.003	0.095 ± 0.004	0.36	0.01	$(=, -)$
	GPSHAP	0.102 ± 0.002	0.091 ± 0.003			
QT-3	GP	0.168 ± 0.004	0.163 ± 0.005	0.84	1.11E-03	$(=, -)$
	GPPI	0.162 ± 0.003	0.158 ± 0.004	4.44E-07	0.36	$(+, =)$
	GPSHAP	0.169 ± 0.002	0.158 ± 0.003			
QT-4	GP	0.172 ± 0.006	0.192 ± 0.005	0.49	0.11	$(=, =)$
	GPPI	0.165 ± 0.004	0.19 ± 0.008	2.15E-06	0.59	$(+, =)$
	GPSHAP	0.174 ± 0.003	0.192 ± 0.005			

Figure 3 and Fig. 5 show the distribution of NRMSEs of the best-of-run individuals from the 30 GP runs on the training sets and the test sets, respectively. Besides, Fig. 4 and Fig. 6 show the evolution plots for the training errors and the corresponding test errors from 30 GP runs, respectively. The lowest NRMSEs on the training set are collected at every generation and the corresponding test NRMSEs are recorded. Note that the test NRMSEs are recorded to analyse the generalization during the evolutionary process but not used during this process. In this paper, the Mann-Whitney U Test is used to compare the training and test NRMSEs of the 30 best-of-run models for GPSHAP, standard GP and GPPI. The significance level is 0.05. "−" means that GPSHAP is significantly better than comparison methods, "=" means that there is no significant difference between them and "+" denotes that GPSHAP is significantly worse than comparison methods.

5.1 Results on the Training Sets - Learning Ability

As shown in Fig. 3, GPSHAP has better training performance than GP and GPPI on six training sets. Based on the statistical significance tests, GPSHAP has significantly better training performance in seven of the ten datasets than GP and in four of ten datasets than GPPI. On QT-3 and QT-4, GPSHAP has slightly worse training performance than GP but not significant while it has significantly higher training error than GPPI on these two datasets.

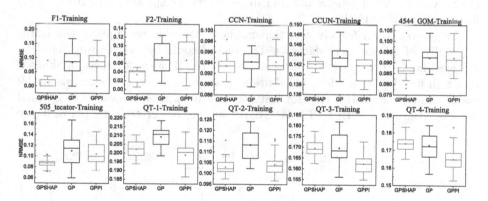

Fig. 3. The distributions of **training** NRMSEs of the 30 best-of-run individuals.

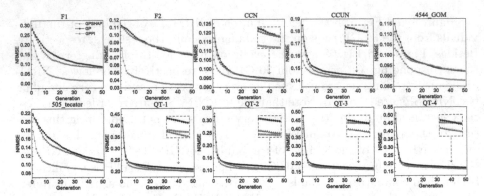

Fig. 4. The evolution plots on the mean NRMSE of 30 best-of-run individuals on the **Training Sets**.

Figure 4 denotes the evolutionary process of GPSHAP, GP and GPPI on the training sets where GPSHAP converges faster than GP and GPPI on most cases. GPSHAP has a much better starting point because the feature selection selects the top $log_2(N)$ important features. On F_1, F_2, 4544_GOM and 505_tecator, GPSHAP has obviously better performance than GP and GPPI from stem to stern. On QT-3 and QT-4, although the final results of GPSHAP are worse than GPPI, the GPSHAP generally gets better performance than GPPI from the very first several generations and the convergence speed of GPSHAP is also faster than that of GPPI.

In general, both feature selection methods are able to improve the learning ability of GP on most datasets. The reason could be that feature selection reduces the feature space, so the search space of GP also decreases, making it easier to find better models. In addition, feature selection also discards irrelevant features and retains relevant features, which helps GP converge to optimal models. Compared with GPPI, GPSHAP obtains better training performance

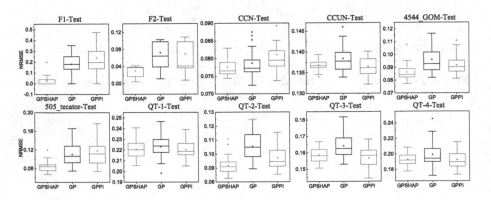

Fig. 5. The distributions of *test* NRMSEs of the 30 best-of-run individuals.

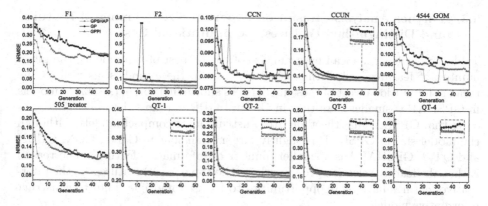

Fig. 6. The evolution plots of the mean NRMSE of 30 best-of-run individuals during every generation on the *Test Sets*.

on six training sets. The reason could be that the permutation importance can be biased due to unrealistic data instances, which causes unstable results.

5.2   Results on the Test Sets - Generalization Ability

The overall patterns on the test sets are similar to those on the training sets. Specifically, Fig. 5 shows that GPSHAP has a smaller test error than GP on all the ten sets, which means GPSHAP has a better generalization ability on these datasets. Although GPSHAP has worse training performance than GP on QT-3 and QT-4, it has a better generalization ability on the two datasets. Compared with GPPI, GPSHAP also performs significantly better on six test sets.

On the two synthetic datasets, the generalization performance of GPSHAP is obviously better than that of GP and GPPI, which shows that the Shapley value can identify important features and keep relevant features during the evolutionary process. The results suggest that GPSHAP is good at exploring relationships between input variables and target variables on the synthetic datasets.

On the eight real-world datasets, the generalization performance of GPSHAP is also better than those of GP and GPPI on four datasets. As can be seen from the evolution plots in Fig. 6, although both GPSHAP and GP have the issue of overfitting on CCN, which is indicated by the increasing test error at the end of the evolutionary process, the overfitting extent of GPSHAP is much smaller than that of GP. Although GPSHAP performs slightly worse than GPPI on CCUN, QT-1 and QT-4, there is no significant difference between them which is confirmed by the significance tests in Table 4. On the 4544_GOM dataset, the fluctuation of GPSHAP shows more obvious than GP and GPPI, but the GPSHAP performs better learning ability. GPSHAP, GP and GPPI have similar performance at final several generations on QT-1, QT-3 and QT-4 while GPSHAP reduces the error faster at initial several generations. In general, GPSHAP can improve the generalization of GP as it discards noisy or irrelevant features, which makes GP have a higher probability to use relevant features to construct models.

5.3 Further Analysis - Result on Program Size, Number of Features and Distinguished Features, Computational Cost

Table 5 describes the mean program size of the 30 best-of-run models, the mean number of features and distinguished features in these models, as well as the mean computational cost of these models. As shown in Table 5, the size of models in GPSHAP is smaller than that in GP and GPPI on most datasets. This indicates that GPSHAP has the ability to construct more compact models. Although the model size in GPSHAP is slightly larger than that in GP on CCN, CCUN and QT-2, GPSHAP has a smaller number of informative features. The reason is that the feature selection process selects the top $log_2(N)$ features, so the informative features have more opportunities to be used multiple times to construct informative models.

In terms of the number of features, GPSHAP, GP and GPPI all only use a small number of features compared with the total number of features on the

Table 5. Program Size, Number of Features and Distinguished Features, and Computational Cost

Dataset	Method	#Nodes (Mean ± Std)	#Features	#Distinguished Features	#Time(Second) (Mean ± Std)
F_1	GPSHAP	**37.86 ± 18.65**	**12.03 ± 7.12**	**3.7 ± 0.9**	305.32 ± 178.45
	GPPI	42.56 ± 17.17	16.96 ± 7.35	7.13 ± 2.67	123.46 ± 58.7
	GP	43.4 ± 15.01	16.76 ± 6.2	6.7 ± 2.26	**55.81 ± 41.3**
F_2	GPSHAP	**42.8 ± 11.51**	**14.36 ± 5.51**	**4 ± 0.81**	3752.7 ± 2647.25
	GPPI	53.26 ± 23.84	21.6 ± 11.43	8 ± 4.2	1640.21 ± 348.51
	GP	53.36 ± 19.93	23.03 ± 9.62	8.3 ± 3.33	**474.04 ± 192.03**
CCN	GPSHAP	41.5 ± 13.69	**15.26 ± 5.29**	**5.93 ± 0.85**	4106.19 ± 2455.07
	GPPI	46.83 ± 15.6	19.3 ± 7.25	7.5 ± 1.68	1575.65 ± 569.18
	GP	**39.96 ± 11.30**	16.26 ± 4.83	8.06 ± 2.23	**756.45 ± 403.63**
CCUN	GPSHAP	36.83 ± 12.21	14.4 ± 5.29	**5.36 ± 0.87**	2219.26 ± 1314.37
	GPPI	37.16 ± 11.85	15.7 ± 5.13	8.03 ± 2.27	1270.6 ± 390.27
	GP	**34.1 ± 10.9**	**14.3 ± 5.22**	7.0 ± 1.84	**635.32 ± 273.01**
4544_GOM	GPSHAP	33.23 ± 13.69	10.9 ± 4.88	4.6 ± 1.56	1764.88 ± 1961.72
	GPPI	**24.56 ± 13.81**	**8.33 ± 4.77**	**4.26 ± 2.42**	1223.34 ± 387.6
	GP	33.23 ± 15.27	11.43 ± 6.02	6.36 ± 4.08	**691.75 ± 267.91**
505_tecator	GPSHAP	**65.26 ± 26.44**	**25.9 ± 10.28**	**5.83 ± 0.93**	805.23 ± 820.78
	GPPI	73.86 ± 24.68	31.9 ± 11.2	10.86 ± 3.52	548.21 ± 148.44
	GP	80.53 ± 34.01	34.23 ± 14.26	11.93 ± 5.85	**204.25 ± 121.61**
QT-1	GPSHAP	**92.1 ± 23.84**	**31.56 ± 9.1**	**8.46 ± 1.23**	6101.61 ± 2551.59
	GPPI	98.76 ± 13.88	37.86 ± 6.1	19.66 ± 3.96	3171.85 ± 402.72
	GP	102.86 ± 25.83	39.1 ± 10	20.1 ± 7.66	**1293.18 ± 249.66**
QT-2	GPSHAP	91.83 ± 21.39	**31.13 ± 8.75**	**7.96 ± 1.32**	4865.35 ± 1796.86
	GPPI	94.93 ± 25.97	35.13 ± 10.88	15.6 ± 5.65	2674.88 ± 379.5
	GP	**85.4 ± 22.52**	31.46 ± 9.76	14.46 ± 5.88	**939.64 ± 250.81**
QT-3	GPSHAP	**86.76 ± 20.68**	**29.13 ± 8.08**	**7.76 ± 1.43**	10468.4 ± 5516.17
	GPPI	102.06 ± 22.69	37.76 ± 8.35	18.76 ± 5.98	5308.61 ± 844.83
	GP	108.86 ± 40.12	41.46 ± 16.7	21.53 ± 11.27	**2133.83 ± 656.44**
QT-4	GPSHAP	**79.53 ± 29.22**	**26.4 ± 12.74**	**7.33 ± 1.22**	5327.27 ± 2026.03
	GPPI	92.96 ± 25.9	33.83 ± 9.77	16.83 ± 5.48	2915.54 ± 425.31
	GP	92.66 ± 22.1	34.93 ± 9.78	20.06 ± 7.85	**1133.04 ± 307.77**

datasets. However, GPSHAP uses a smaller number of features to construct models than GP and GPPI, which means the Shapely value can better identify important features and the feature selection process removes more irrelevant or more redundant features and retains relevant features.

As for the computational cost, GPSHAP obviously spends a higher cost than GP, because GPSHAP needs the additional computational cost for feature selection including computing the Shapley value, collecting features from best-of-run individual and ranking features. GPSHAP is also more time-consuming than GPPI because computing Shapley value is more expensive than computing permutation importance. Although GPSHAP needs more computational cost, it substantially improves the learning and generalization abilities and reduces the number of features for constructing models, which evolves more compact models. We will consider shortening the training time in the future.

6   Conclusions and Future Work

This paper proposed a novel Shapley value based feature selection method for GPSR, which is named GPSHAP, to improve the generalization of high-dimensional GPSR. Experimental results show that compared with standard GP and GP with permutation importance, GPSHAP not only selects a smaller number of good features to construct models, but also has better learning and generalization performance. The results confirm that the feature selection process based on the Shapley value is able to remove irrelevant features and selects informative features, which helps GP evolve more compact models with better generalisation ability.

For future work, to improve the efficiency of the proposed Shapley value based feature selection method, we will consider combining it into the evolutionary process in the future instead of using an independent stage. In order to make the method more effective on high-dimensional datasets, we will adopt the Shapley value to compute subtrees in order to control the genetic operators.

Acknowledgement. This work is supported in part by the Marsden Fund of New Zealand Government under Contract MFP-VUW2016, MFP-VUW1914 and MFP-VUW1913.

References

1. Ray, P., Reddy, S., Banerjee, T.: Various dimension reduction techniques for high dimensional data analysis: A review. Artif. Intell. Review. **54**, 3473–3515 (2021)
2. Zhang, H., Zhou, A., Chen, Q., Xue, B., Zhang, M.: SR-Forest: a genetic programming based heterogeneous ensemble learning method. IEEE Trans. Evol. Comput. (2023). https://doi.org/10.1109/TEVC.2023.3243172
3. Neshatian, K., Zhang, M.: Pareto front feature selection: Using genetic programming to explore feature space. In: Proceedings of the Genetic and Evolutionary Computation Conference, pp. 1027–1034 (2009)

4. Koza, J.: Genetic Programming: On the Programming of Computers by Means of Natural Selection. MIT Press, Cambridge, MA, USA (1992)
5. Chen, Q., Xue, B., Niu, B., Zhang, M.: Improving generalisation of genetic programming for high-dimensional symbolic regression with feature selection. In: Proceedings of the IEEE International Conference on Evolutionary Computation, pp. 3793–3800 (2016)
6. Chen, Q., Zhang, M., Xue, B.: Feature selection to improve generalization of genetic programming for high-dimensional symbolic regression. IEEE Trans. Evol. Comput. 21(5), 792–806 (2017)
7. Helali, B., Chen, Q., Xue, B., Zhang, M.: Genetic programming-based selection of imputation methods in symbolic regression with missing values. In: AI 2020: Advances in Artificial Intelligence, pp. 12576 (2020)
8. Zhang, H., Zhou, A., Zhang, H.: An evolutionary forest for regression. IEEE Trans. Evol. Comput. 26(4), 735–749 (2022)
9. Zhang, H., Zhou, A., Qian, H., Zhang, H.: PS-tree: a piecewise symbolic regression tree. Swarm Evol. Comput. 71, 101061 (2022)
10. O'Neill, M., Vanneschi, L., Gustafson, S., Banzhaf, W.: Open issues in genetic programming. Genet. Program. Evol. Mach. 11(3), 339–363 (2010)
11. Molnar, C.: Interpretable machine learning: a guide for making black box models explainable (2nd ed.). https://christophm.github.io/interpretable-ml-book (2022)
12. Heskes, T., Sijben, E., Bucur, I., Claassen, T.: Causal shapley values: exploiting causal knowledge to explain individual predictions of complex models. Adv. Neural Info. Proc. Syst. 33, 4778–4789 (2020)
13. Haeri, M., Ebadzadeh, M., Folino, G.: Improving GP generalization: a variance-based layered learning approach. Genet. Program. Evol. Mach. 16(1), 27–55 (2015)
14. Astarabadi, S., Ebadzadeh, M.: Avoiding overfitting in symbolic regression using the first order derivative of GP trees. In: Proceedings of the Genetic and Evolutionary Computation Conference, pp. 1441–1442 (2015)
15. Sandinetal, I.: Aggressive and effective feature selection using genetic programming. In: Proceedings of the IEEE International Conference on Evolutionary Computation, pp. 1–8 (2012)
16. Chen, Q., Xue, B., Zhang, M.: Rademacher complexity for enhancing the generalization of genetic programming for symbolic regression. IEEE Trans. Cybern. 52(4), 2382–2395 (2022)
17. Lundberg, S., Lee, S.: A unified approach to interpreting model predictions. Adv. Neural Inf. Process. Syst. 30 (2017)
18. Strumbelj, E., Kononenko, I.: Explaining prediction models and individual predictions with feature contributions. Know. Inf. Syst. 41(3), 647–665 (2014)
19. Seijo-Pardo, B., Porto-Díaz, I., Bolón-Canedo, V., Alonso-Betanzos, A.: Ensemble feature selection: homogeneous and heterogeneous approaches. Knowl.-Based Syst. 118, 124–139 (2017)
20. Keijzer, M.: Improving symbolic regression with interval arithmetic and linear scaling. In: Proceedings of the European Conference on Genetic Programming, pp. 70–82 (2003)
21. Lichman, M.: UCI Machine Learning Repository. http://archive.ics.uci.edu/ (2013)
22. Olson, R., Cava, W., Orzechowski, P., Urbanowicz, R., Moore, J.: PMLB: a large benchmark suite for machine learning evaluation and comparison. BioData Mining. 10, 1–13 (2017)
23. Vanschoren, J., Rijn, J., Bischl, B., Torgo, L.: OpenML: networked science in machine learning. ACM SIGKDD Explo. Newsletter. 15(2), 49–60 (2014)

Hybrid Models for Predicting Cryptocurrency Price Using Financial and Non-Financial Indicators

Tulika Shrivastava[1], Basem Suleiman[1,2]([✉]),
and Muhammad Johan Alibasa[1,3]

[1] The University of Sydney, Camperdown, NSW 2050, Australia
tshr3215@uni.sydney.edu.au
[2] University of New South Wales, Sydney, NSW 2052, Australia
b.suleiman@unsw.edu.au
[3] Telkom University, Jawa Barat 40257, Indonesia
alibasa@telkomuniversity.ac.id

Abstract. Cryptocurrency has become very popular and widely used by major businesses as digital currency for online investments and services. However, the price prediction of such digital currencies as Bitcoin and Ethereum is challenging. It involves financial indicators and nonfinancial indicators, such as historical data and social media data, respectively. In this paper, we propose deep learning and hybrid models that effectively incorporate both types of indicators and introduce the optimal algorithms for long-term price prediction of Bitcoin and Ethereum. We conduct extensive experimental evaluations on real data we extracted from financial dataset comprising Yahoo Finance data, and non-financial data consisting of Google Trends data and approximately 30 million related Bitcoin and Ethereum. Our experimental results show that the hybrid models involving LSTM/1D-CNN with ARIMA/ARIMAX outperformed the individual models for the long-term prediction of cryptocurrency prices.

Keywords: Deep Learning · Machine Learning · Sentiment Analysis · Cryptocurrency · Price · Predictive Models

1 Introduction

Cryptocurrency or digital currency has gained a lot of confidence from savvy and sophisticated investors. The primary reason for its growing prominence is the ease of transaction facilitated by decentralized operational control [18]. However, the success of any cryptocurrency largely depends on whether people around the globe have started considering it to be useful and productive or not [18]. This was clearly evident when the price of Bitcoin and Ethereum reached an all-time high in 2021 after their adoption as a payment means after the announcement from PayPal and the investment arm of big public companies [14].

© The Author(s), under exclusive license to Springer Nature Singapore Pte Ltd. 2024
D. Benavides-Prado et al. (Eds.): AusDM 2023, CCIS 1943, pp. 177–191, 2024.
https://doi.org/10.1007/978-981-99-8696-5_13

Of course, the price of cryptocurrencies has been quite volatile and there are many factors, from internal to external and political, that have a direct or indirect bearing on the price of cryptocurrencies [12]. Internal factors comprise the Demand and Supply of Coins Circulation, Transaction Costs, Rewards System, Forks (Rule Changes), and Mining Difficulties. These can be further conflated with technical features and blockchain-based features. The technical features include financial data of the cryptocurrency, such as returns, while the blockchain-based features include the number of cryptocurrency transactions [11]. The external factors consist of Cryptomarket (Market Trend, Attractiveness or Popularity, Speculations) and Macroeconomic (Stock Markets, Interest Rate, Exchange Rate, Gold Price) components. These factors are largely dependent on sentiment data and asset-based features. Furthermore, political factors also impact the price and include Restrictions (Ban) and Legalization (Adaptation) related information. The internal factors and the political factors are very difficult to measure. In account of this, many of the studies done to predict the price of these cryptocurrencies have used one or many external factors, including historical transactions and social media data, as input to machine learning, deep learning, or statistical models in order to extract the pattern from these data and predict the price, both short-term and long-term.

Since many of the factors influencing the price of cryptocurrencies are very hard to measure and precisely annotate in a way that can be used as input data for the models, identifying the influential features becomes substantially more crucial for price prediction. Thus having the right dataset becomes the key to any kind of model training. In many of the studies done in this area, the scale of data used for social media has been on the lower side [13]. Therefore, in this research paper, we analysed the sentiments of around 30 million real-time tweets related to two leading cryptocurrencies, namely Bitcoin and Ethereum. We proposed deep learning and hybrid models for price prediction using various combinations of datasets, such as historical transactions, sentiment data, and Google Trends data for long-term prediction.

The research contributions of this paper are threefold:

- The performance of a machine learning model is correlated with the quality of the input data it receives. Accordingly, as part of this research, a comprehensive data set for sentiment scores of approximately 30 million real-time tweets related to Bitcoin and Ethereum was created. We created the dataset by mining and pre-processing these tweets before analyzing the sentiments in a way that could be fully utilized by the models to understand the impact of social media sentiments on cryptocurrency's price prediction.
- An approach to derive the weighted average of the Twitter sentiments for cryptocurrency's price on a day using Twitter sentiment for predicting stock price movement [8] showed that the metadata from highly influential Twitter accounts could skew the result. We addressed the skewness problem as the simple average method would only consider the compound score of the sentiment and ignore the information of the Tweet's metadata.

– Statistical models and deep learning models together can handle the specifics of the dataset, like seasonal part and trend, respectively. This can be improved when the models are trained using various combinations of datasets. So we benchmarked a few existing models on all combinations of datasets on a very large scale. Along with that we also proposed a new hybrid model using 1D-CNN along with ARIMA and ARIMAX. We were able to establish that the hybrid models with LSTM were slightly better suited for long-term prediction of the price of cryptocurrencies than 1D-CNN based models.

The rest of this paper is organized as follows. Related studies are discussed in Sect. 2. The proposed approach and algorithms are discussed in Sect. 3. The experiments and results related to the approaches are described in Sect. 4, followed by a detailed discussion of the results in Sect. 5. The paper concludes with the conclusion of the research and scope of future work in Sect. 6.

2   Related Work

ARIMA or Auto Regressive Moving Average Model has been reported as the best statistical model to predict the price of cryptocurrencies [12]. A moving average method was used to predict the price of Bitcoin Cash taking into account the moving average for 2, 3, 4 and 7 days and it was observed that the two-day moving average showed the best performance in terms of the mean absolute error percentage [3]. Another study [16] explored an interesting angle of establishing a correlation between trend and volume for the Bitcoin market, but it was concluded that trends did not depend on volume change. In another research [15], the study tried to compare the performance of ARIMA with LSTM for Bitcoin. The outcome showed that ARIMA performed better than LSTM in terms of Root Mean Square Error (RMSE).

When it comes to time-series data, ARIMA is considered to be an excellent method to handle the seasonal part, while RNN is very good at forecasting the trend of it. This led to a study that developed a hybrid model combining RNN with ARIMA [5]. It was observed that Artificial Neural Networks perform better when it comes to long-term predictions. It was also noted in the same research that if the model is parameterized properly, it can improve the result significantly.

Another research [7] tried to predict the price of three cryptocurrencies, namely Bitcoin, Litecoin and Ethereum, using three different models LSTM (Long Short Term Memory), GRU (Gated Recurrent Unit) and bi-LSTM (bidirectional LSTM). The study found that the GRU model showed the best performance in terms of Mean Absolute Percentage Error (MAPE). It was highlighted that if sentiment data is combined with transactional data, then there is great potential to explore new dimensions in such research.

To understand the behaviour of Bitcoin for short-term price prediction, Jaquart et al. [11] analyzed approximately 9 months of data for minute-based duration using six machine learning models and four types of datasets including technical, blockchain-based, sentiment and asset-based. A similarity was observed between the prediction of the price of the stock market and the prediction of the price of Bitcoin, but the limitation of the general prediction (accuracy 50.9% to 56%) was mainly due to the number of features that cannot be included in the research. It was also mentioned that if the duration was increased from one minute to a larger value, the performance might have improved. Another observation in the same research was that the RNN model along with gradient boosting classifiers was found to be more suitable for the prediction task with the technical features having more influence on the price prediction than blockchain-based features and sentiment data.

In another study, historical data was combined with sentiments from Chinese social media, and the LSTM model was used to forecast the price of cryptocurrency [9]. It was found that the model performed better than the autoregressive model by 18.5% in the case of precision and 15.4% in the case of recall. Similarly, Twitter and Google Trends data were used to predict the price of Bitcoin and Ethereum [2], and the results showed that the volume of tweets contributed more to the prediction of price change than the sentiments captured on Twitter.

Another research used news data and historical data to predict the price of Ethereum using the LSTM model and it was observed that sentiment did play a major role in the forecast process [17]. A novel approach using the 1D-CNN model was also used on sentiment data, blockchain transaction history, and financial indicators [4]. The research observed that 1D-CNN outperformed the LSTM model when tried on Bitcoin. In the same research, a trading strategy was devised so that the loss can be kept to a minimum when the Bitcoin market is down and the profit can be increased when the market is up. In fact, 1D-CNN has gained a lot of popularity these days in time-series forecasting, in addition to its established task of image recognition.

3   Methods

3.1   Price Prediction Models

Deep Learning Models. Two Deep Learning models, LSTM and 1D-CNN, were used to evaluate the long-term forecasting process. RNN has been used in past studies as described in Sect. 2. However, it suffers from the problem of vanishing gradients. LSTM (Long Short-Term Memory) addresses this problem and can forecast based on various sequences of data, especially long-term sequences of data, leading to its popularity growth as a model to forecast time-series data. 1D-CNN is also emerging as a popular model to predict time-series data [4], especially since it behaves more like computing the moving average in this price prediction problem.

Hybrid Models. Four hybrid models were used in the research to understand how statistical and deep learning models perform when combined together on different types of datasets. As described in Sect. 2, it was observed that hybrid models, by combining Statistical and Deep Learning models, can perform better than individual models alone. This is because, in hybrid mode, each type of model can handle specifics of the dataset including seasonal part and trend. Therefore, we proposed four types of hybrid models in this research: (1) LSTM and ARIMA, (2) LSTM and ARIMAX, (3) 1D-CNN and ARIMA, (4) and 1D-CNN and ARIMAX. The models consisting of LSTM or 1D-CNN that are combined with ARIMA were evaluated on historical data, while the same models that are combined with ARIMAX were used on historical data along with a combination of Twitter sentiment data, Tweet volume, and Google Trends data. The residual of ARIMA or ARIMAX was passed as input to the LSTM or 1D-CNN model. The final prediction of the residual was then converted back to the predicted value to compare it against the test value. Figure 1 explains this process through a flow chart.

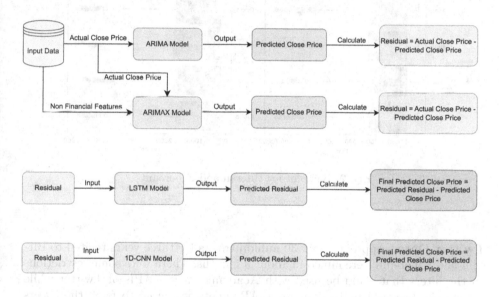

Fig. 1. Hybrid Model Workflow

3.2   Evaluation

The evaluation metrics are an important part of the forecasting process. They tell the quality of the prediction through error measures. The evaluation metrics chosen for this study are RMSE (Root Mean Square Error), loss based on MSE (Mean Square Error) and MAPE (Mean Absolute Percentage Error). Accuracy was not used as a metric since the target variable was a continuous value. In general, these metrics are used together to better analyze the result of a model since no one metric is perfect in all terms.

3.3 Data Collection and Pre-processing

Historical Data. The historical datasets for both Bitcoin and Ethereum for the last 5 years were downloaded from Yahoo Finance [1]. Initial data analysis was performed on this dataset. It did not require pre-processing. The standard deviation was high for both Bitcoin and Ethereum supporting the volatile trend of these cryptocurrencies. The data being positively skewed is also supported by the fact that the median value is less than the mean value. This is displayed in the box plots shown in Fig. 2 where the boxes are divided into unequal parts by the median line, the left part being smaller than the right part. The box plot also indicates outliers in the price for both Bitcoin and Ethereum.

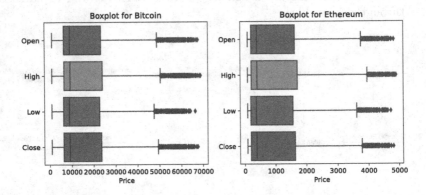

Fig. 2. Box plot for Bitcoin and Ethereum

Twitter Data. Approximately 30 million tweets for three years related to Bitcoin and Ethereum were mined in about 3 months. There were some restrictions on the filters that could be used with Academic Access APIs of Twitter while mining the tweets. Due to that, more API quota was used to fetch the tweets and then the noise was removed as part of pre-processing for better results of Sentiment Analysis. The Tweets were changed to lowercase for further filtering. Hashtags were removed from bitcoin, btc, ethereum, eth and cryptocurrency words to keep them in the Tweets while removing other hashtags. Bot accounts, hashtags, user mentions, newline, http, https links, emojis and & symbol were removed from the Tweets to have only relevant information in the Tweets.

Google Trends Data. Google Trends data comprising interest data, related query data, and related topics data were mined for the same duration as the Twitter data for Bitcoin and Ethereum. This data provides information about the interest of people across the globe through the means of their search on Google.

3.4   Sentiment Analysis

After the pre-processing step, sentiment analysis was performed on the Twitter data. Based on the literature review, Vader (Valence Aware Dictionary and sEntiment Reasoner) [10] was selected to implement sentiment analysis as it is specifically tuned for sentiments in social media. Vader provided a positive, negative, neutral and compound score for every Tweet. Thus, the sentiment of each Tweet was calculated as Positive, Negative or Neutral using Vader analysis. The next step was to calculate the weighted average of the sentiments for a day. Calculating simple average would have just considered the compound score of the sentiment, but the technique used to calculate the weighted average takes all the metadata related to the Tweets into account. In one of the research done using Twitter sentiment for predicting stock price movement [8], the idea of weighted average was dropped as the metadata from very influential accounts could have skewed the result. We addressed this problem in our research by normalizing all the metadata between 0 and 1 before taking the weighted average as described in Algorithm 1. Finally, based on this sentiment score, Tweets were classified as Positive, Negative or Neutral [10]. Daily Tweet volume was also calculated along with Positive, Negative and Neutral Tweet volume.

Algorithm 1. Sentiment score by weighted average method

$norm\_retweet\_count \leftarrow norm(retweet\_count)$

▷ Normalize metadata

$norm\_reply\_count \leftarrow norm(reply\_count)$
$norm\_like\_count \leftarrow norm(like\_count)$
$norm\_quote\_count \leftarrow norm(quote\_count)$
$norm\_followers\_count \leftarrow norm(followers\_count)$
$norm\_following\_count \leftarrow norm(following\_count)$
$norm\_tweet\_count \leftarrow norm(tweet\_count)$
$norm\_listed\_count \leftarrow norm(listed\_count)$

$weight \leftarrow norm\_retweet\_count \times norm\_reply\_count \times norm\_like\_count \times norm\_quote\_count \times norm\_followers\_count \times norm\_following\_count \times norm\_tweet\_count \times norm\_listed\_count$

▷ Calculate weight

$sentiment\_score \leftarrow \frac{sum(compound\_score \times weight)}{sum(weight)}$

▷ Calculate weighted average

---

4   Experiments and Results

Extensive experiments were conducted and analyzed using deep learning and hybrid models to understand which financial and non-financial indicators were best suited to predict the price of cryptocurrencies. The models were also fine-tuned using random hyper-parameter tuning. Various scaling techniques like Normalization and Standardization along with different batch sizes, sequence length, learning rates, configuration of the LSTM model or 1D-CNN model, and regularization like dropout were tried to explore the model's output. The data was split into 70% training, 15% test, and 15% validation. With this split, the prediction was done for a duration of around 3 months to 6 months depending on the dataset as shown in Table 1 and Table 2.

4.1 Deep Learning Models

LSTM and 1D-CNN are state-of-the-art models for time-series forecasting. Both of them were used to understand how these deep learning models performed with financial and non-financial indicators for long-term prediction.

LSTM. The first set of experiments was tried on the historical data by using various scaling types such as normalisation and standardisation along with different batch sizes, sequence lengths, and learning rates. It was observed that the normalized data had a better result than the standardized data. The batch size of 64 had an edge over other batch sizes and dropout layer also helped in regularization. So the model having this configuration data was chosen to experiment further using five different data sources: (1) historical, (2) historical and Twitter sentiment, (3) historical and Tweet count, (4) historical and Google Trends, (5) historical, Twitter sentiment, Tweet count and Google Trends. LSTM had the best outcome for long-term prediction using the financial indicator. The results using non-financial indicators were quite close. The pattern of the result for both cryptocurrencies was very similar, as can be seen in Table 1 and Table 2.

1D-CNN. Since 1D-CNN is also a deep learning model, the process and setup of the experiments were similar to that of LSTM. Different values of scaling type, 1D-CNN configuration, learning rate, and batch size were used in the experiments on all five types of data sources. In this case, also, Normalization was a better choice for scaling where batch size of 32 worked best for Bitcoin and 64 for Ethereum. 1D-CNN model showed the best output using the financial indicator in the case of Bitcoin whereas, in the case of Ethereum, the non-financial features outperformed the financial features. This can be observed in Table 1 and Table 2.

4.2 Hybrid Models

The deep learning models were combined with statistical models to explore if the hybrid models performed better than the base models. LSTM and 1D-CNN were combined with ARIMA for financial data and ARIMAX for financial and non-financial data together. The residual of the statistical model

Table 1. Result for Bitcoin sorted on RMSE

Model Name	Indicator Type	RMSE	MAPE
LSTM and ARIMA	Historical	1725.33	0.027
LSTM and ARIMAX	Historical, Twitter Sentiment, Tweet Count and Google Trends	1866.90	0.028
LSTM and ARIMAX	Historical, Twitter Count	1901.51	0.028
LSTM and ARIMAX	Historical, Google Trends	1903.28	0.028
LSTM and ARIMAX	Historical, Twitter Sentiment	1921.08	0.058
1D-CNN and ARIMAX	Historical, Twitter Sentiment, Tweet Count and Google Trends	2026.15	0.033
1D-CNN and ARIMAX	Historical, Twitter Sentiment	2066.43	0.033
1D-CNN and ARIMAX	Historical, Twitter Count	2090.57	0.033
1D-CNN and ARIMAX	Historical, Google Trends	2112.42	0.034
LSTM	Historical	2557.93	0.044
LSTM	Historical, Google Trends	2926.65	0.052
1D-CNN and ARIMA	Historical	2983.2	0.051
LSTM	Historical, Twitter Sentiment	3387.52	0.058
1D-CNN	Historical	3679.92	0.057
LSTM	Historical, Twitter Sentiment, Tweet Count and Google Trends	3801.79	0.066
LSTM	Historical, Twitter Count	4075.726	0.075
1D-CNN	Historical, Twitter Count	4151.62	0.07
1D-CNN	Historical, Twitter Sentiment	4520.34	0.077
1D-CNN	Historical, Twitter Sentiment, Tweet Count and Google Trends	4741.96	0.083
1D-CNN	Historical, Google Trends	5743.11	0.109

(ARIMA/ARIMAX) was passed as input to the deep learning model (LSTM/1D-CNN) to predict the final residual value, which was then converted back to the actual predicted value to compare it against the test value.

LSTM and ARIMA. The first hybrid model was configured using LSTM and ARIMA. In this model, the residual generated from ARIMA model using the Close value of financial data was fed to LSTM model for training. Standardized data with batch size of 32 worked better for Bitcoin whereas, for Ethereum, it was normalized data with batch size of 32. In general, the hybrid models outperformed the base models as can be seen in Table 1 and Table 2.

1D-CNN and ARIMA. The next hybrid model was configured using 1D-CNN and ARIMA. Standardized data with batch size of 64 was chosen for Bitcoin whereas, for Ethereum, it was normalized data with batch size of 64. This hybrid model outperformed its base models for Bitcoin but in the case of Ethereum, the performance of 1D-CNN base model was better than that of the hybrid model, as can be seen in Table 1 and Table 2.

Table 2. Result for Ethereum sorted on RMSE

Model Name	Indicator Type	RMSE	MAPE
LSTM and ARIMAX	Historical, Twitter Sentiment, Tweet Count and Google Trends	153.21	0.033
LSTM and ARIMAX	Historical, Google Trends	153.58	0.033
LSTM and ARIMAX	Historical, Twitter Count	153.70	0.033
LSTM and ARIMAX	Historical, Twitter Sentiment	154.41	0.033
1D-CNN and ARIMAX	Historical, Twitter Sentiment, Tweet Count and Google Trends	156.55	0.034
1D-CNN and ARIMAX	Historical, Google Trends	157.14	0.034
1D-CNN and ARIMAX	Historical, Twitter Sentiment	158.70	0.034
1D-CNN and ARIMAX	Historical, Twitter Count	161.45	0.035
LSTM and ARIMA	Historical	177.62	0.043
LSTM	Historical	212.39	0.051
1D-CNN	Historical, Twitter Sentiment	236.33	0.055
1D-CNN	Historical	262.33	0.065
1D-CNN and ARIMA	Historical	282.30	0.069
1D-CNN	Historical, Twitter Count	284.9	0.066
1D-CNN	Historical, Twitter Sentiment, Tweet Count and Google Trends	314.65	0.073
1D-CNN	Historical, Google Trends	437.42	0.095
LSTM	Historical, Twitter Sentiment, Tweet Count and Google Trends	1293.12	0.306
LSTM	Historical, Twitter Sentiment	1402.95	0.342
LSTM	Historical, Twitter Count	1495.91	0.37
LSTM	Historical, Google Trends	1504.02	0.373

LSTM and ARIMAX. LSTM and ARIMAX hybrid model was used to test the behaviour of non-financial indicators along with financial indicators. Standardized data with batch size 32 was used in the case of Bitcoin whereas normalized data with batch size 32 was chosen for Ethereum. The hybrid models outperformed the base models, as can be seen in Table 1 and Table 2.

1D-CNN and ARIMAX. The last set of experiments was conducted on the hybrid model configured using 1D-CNN and ARIMAX. Standardized data with batch size 64 was chosen for Bitcoin and normalized data with batch size 64 for Ethereum based on the results. In this case also, the hybrid models outperformed the base models as can be seen in Table 1 and Table 2.

Figure 3 and Fig. 4 show the visualization of the top 10 models for Bitcoin and Ethereum respectively.

4.3 Granger Causality

Granger Causality is a statistical test to determine whether one time series can help predict the value of another time series [6]. Twitter data and Google Trends data were tested with historical data to check if they were useful in

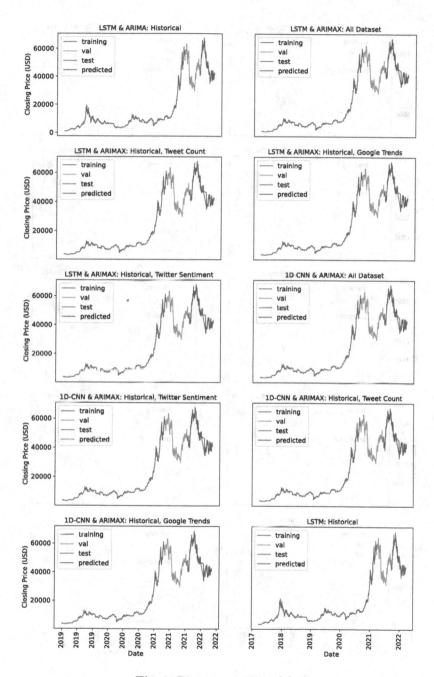

Fig. 3. Bitcoin top 10 models

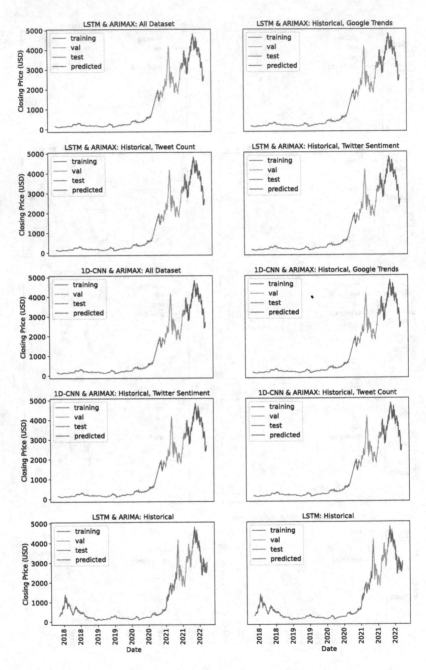

Fig. 4. Ethereum top 10 models

predicting the price. Based on the p-value, Twitter volume for Bitcoin showed that it significantly causes the close value of historical data in the first Lag itself. This means that Twitter volume has a strong impact on the price prediction, whereas for Ethereum it was after 5 lags. Twitter sentiment score did not indicate a strong correlation, whereas Google Trends data showed that it affects the close price after 7 lags in the case of Bitcoin and after 6 lags in the case of Ethereum. Twitter Volume having more impact on price prediction than Twitter sentiment was also indicated in one of the research [2] mentioned in Sect. 2. This can also be visually verified through the graph shown in Fig. 5 for Bitcoin Tweet volume.

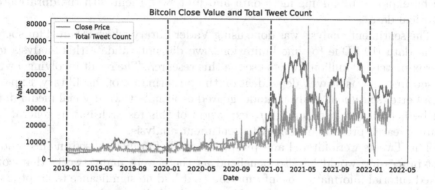

Fig. 5. Granger Causality Close Price and Tweet volume for Bitcoin

5 Discussion

The best result for our data for Bitcoin was for the hybrid model of LSTM and ARIMA based on the financial indicator. This model can handle specifics of the dataset including seasonal part and trend. This was followed by the hybrid models of LSTM and ARIMAX based on financial and non-financial data. The difference in result was very marginal. In the case of Ethereum, the best result was for hybrid model based on both financial and non-financial indicators. In theory, the non-financial data should have improved the performance of the Bitcoin hybrid model as observed in the case of Ethereum but it is possible that the noise present in the non-financial data, primarily tweets related to Bitcoin, could have affected the performance of the model. The hybrid model based on 1D-CNN and ARIMAX also had a very close prediction and very less differences in terms of error.

LSTM utilizes the long-term memory of the data, whereas 1D-CNN focusses on the spatial aspect of the data. The result of any model depends a lot on the data fed to it and that could be one of the reasons why the result of hybrid models of LSTM was slightly better than that of 1D-CNN. This result could have been improvised further by fine-tuning the model's configuration.

6 Conclusion and Future Work

Overall, the results indicate that the hybrid models perform better than the basic model itself. Hybrid model with LSTM is observed to have a slight edge over 1D-CNN. This could be due to the nature of the data including financial and non-financial both.

There are some areas that can be explored in future studies to improve the price prediction of cryptocurrencies. Due to the quota limit, approximately 3 years of Tweets were used in the research, whereas historical data was worth 5 years. Furthermore, the importance of Twitter data and Google Trends data can be explored by mining more data and matching them with the duration of historical data.

The sentiment analysis was done using Vader library which is tuned for social media data [10]. Due to time limitations, we did not validate this analysis for approximately 30 million Tweets used in this research. The result involving Twitter sentiments was heavily dependent on the performance of this library. Exploring a better sentiment analysis model geared especially toward social media data can be helpful not just as a future extension of this research but in general for many research projects dealing with sentiment analysis.

The Tweets were filtered and pre-processed before being used in this study. Post-processing would be also beneficial as there were many tweets that contained tutorial information or information that had no significance to cryptocurrencies. These tweets are difficult to filter out through a regular filtering process, since they also use the Bitcoin or Ethereum hashtags. If their weight is included in calculating the sentiment score or Tweet volume, they might skew the result. Future research may improve this filtering process, which can in turn help a lot of research projects involving Twitter data.

The news could also have a significant impact on the general sentiment of people towards cryptocurrencies. News data might be added to the datasets to explore whether it affects the price prediction performance. More external factors, such as stock market trends and gold prices, as described in Sect. 2 could also be included in the research to evaluate their impact on price prediction.

LSTM utilizes the long-term memory of the data whereas 1D-CNN focusses on the spatial aspect of the data. So a hybrid model using 1D-CNN and LSTM can be explored to evaluate if it is able to predict the price of the cryptocurrencies better than the other hybrid models explored in this paper.

References

1. Yahoo finance. https://finance.yahoo.com/cryptocurrencies/
2. Abraham, J., Higdon, D.W., Nelson, J., Ibarra, J.: Cryptocurrency price prediction using tweet volumes and sentiment analysis. SMU Data Sci. Rev. **1**(3) (2018). https://scholar.smu.edu/datasciencereview/vol1/iss3/1
3. Abu Bakar, N., Rosbi, S., Kiyotaka, U.: Forecasting cryptocurrency price movement using moving average method: a case study of bitcoin cash. Int. J. Adv. Res. **7**, 609–614 (2019). https://doi.org/10.21474/IJAR01/10188

4. Cavalli, S., Amoretti, M.: CNN-based multivariate data analysis for bitcoin trend prediction. Appl. Soft Comput. **101**, 107065 (2020). https://doi.org/10.1016/j.asoc.2020.107065
5. Fathi, O.: Time series forecasting using a hybrid Arima and LSTM model, pp. 1–7. Velvet Consulting (2019)
6. Granger, C.W.J.: Investigating causal relations by econometric models and cross-spectral methods. Econometrica **37**(3), 424–438 (1969). https://doi.org/10.2307/1912791
7. Hamayel, M.J., Owda, A.Y.: A novel cryptocurrency price prediction model using GRU, LSTM and bi-LSTM machine learning algorithms. AI **2**(4), 477–496 (2021). https://doi.org/10.3390/ai2040030
8. Hu, K., Grimberg, D., Durdyev, E.: Twitter sentiment analysis for predicting stock price movements (2018). http://cs230.stanford.edu/projects_fall_2021/reports/103158402.pdf
9. Huang, X., et al.: LSTM based sentiment analysis for cryptocurrency prediction. In: Jensen, C.S., et al. (eds.) DASFAA 2021. LNCS, vol. 12683, pp. 617–621. Springer, Cham (2021). https://doi.org/10.1007/978-3-030-73200-4_47
10. Hutto, C., Gilbert, E.: Vader: A parsimonious rule-based model for sentiment analysis of social media text. In: Proceedings of the International AAAI Conference on Web and Social Media, vol. 8, no. 1, pp. 216–225 (2014). https://doi.org/10.1609/icwsm.v8i1.14550
11. Jaquart, P., Dann, D., Weinhardt, C.: Short-term bitcoin market prediction via machine learning. J. Financ. Data Sci. **7**, 45–66 (2021). https://doi.org/10.1016/j.jfds.2021.03.001
12. Khedr, A.M., Arif, I., Raj, P.V.P., Bannany, M.E., Alhashmi, S.M., Sreedharan, M.: Cryptocurrency price prediction using traditional statistical and machine-learning techniques: a survey. Intell. Syst. Account. Finance Manag. **28**, 3–34 (2021). https://api.semanticscholar.org/CorpusID:234260503
13. Lamon, C., Nielsen, E., Redondo, E.: Cryptocurrency price prediction using news and social media sentiment (2017). https://api.semanticscholar.org/CorpusID:33873494
14. Nakamoto, S.: Bitcoin: a peer-to-peer electronic cash system (2008). https://bitcoin.org/bitcoin.pdf
15. Saxena, A., Sukumar, T.: Predicting bitcoin price using LSTM and compare its predictability with arima model (2018). https://api.semanticscholar.org/CorpusID:220374761
16. Szetela, B., Mentel, G., Bilan, Y., Mentel, U.: The relationship between trend and volume on the bitcoin market. Eurasian Econ. Rev. **11**(1), 25–42 (2021). https://doi.org/10.1007/s40822-021-00166-5
17. Vo, A.D.: Sentiment analysis of news for effective cryptocurrency price prediction. Int. J. Knowl. Eng. **5**, 47–52 (2019). https://doi.org/10.18178/ijke.2019.5.2.116
18. Wachter, J.: Cryptocurrency and blockchain: an introduction to digital currencies. https://www.coursera.org/learn/wharton-cryptocurrency-blockchain-introduction-digital-currency/home/info

Application Track

Multi-dimensional Data Visualization for Analyzing Materials

Amit Vurgaft(✉) (iD)

Department of Materials Science and Engineering and the Solid-State Institute
Technion — Israel Institute of Technology, Haifa, Israel
vurgaftamit@gmail.com

Abstract. High-throughput chemical synthesis is extensively used for analyzing materials since it allows a systematic probing of a large span of three space parameters: initial conditions, final products, and the characteristics of the obtained products. High-dimensional data visualization is required to fully understand the relations between these three spaces. However, most methods are limited to up to three dimensions (3D) at the time per graph. Here, we show how correlation analysis and parallel coordinate plots reveal the relations in a multidimensional space. This representation technique is general and may serve many synthetic and fabrication related processes. We demonstrate the power of our approach on a specific chemical colloidal synthesis of $CsPbBr_3$ nanocrystals, which results in highly emissive semiconductor nanocrystals, relevant for photonic applications. The colloidal synthesis of these nanocrystals is a multi-parameter process, with complex inter-relationships between the parameters, such as precursors and organic ligands concentrations and reaction temperature. The resulting nanocrystals present distinct morphological differences, resulting in a detectable shift of their emission spectrum. We use a dataset of 1351 samples to investigate the relations between and within these three spaces. In our case study, we have identified trends and anomalies in the data that provide directions for further research and illustrated thereby the potential of correlation analysis and parallel coordinates to explore patterns and relationships.

Keywords: High-dimensional Data · Data Visualization · Interdisciplinary Applications of Data Science

1 Introduction

In the last decade, data-driven methodology has gained immense popularity in materials science [1]. Data science techniques have proven to be valuable in designing and optimizing synthesis parameters by visualizing parameter spaces. In the past, one of the main challenges was to create large, informative datasets to apply these techniques to. Today, thanks to high-throughput chemical synthesis, large experimental datasets could be created with high consistency in a short time frame [2].

© The Author(s), under exclusive license to Springer Nature Singapore Pte Ltd. 2024
D. Benavides-Prado et al. (Eds.): AusDM 2023, CCIS 1943, pp. 195–210, 2024.
https://doi.org/10.1007/978-981-99-8696-5_14

The high-throughput experiment provides multidimensional data: environmental conditions, such as temperature and humidity; chemical parameters, such as volumes and proportions of precursors; and physical parameters of the synthesis itself, such as stirring rate and cooling time—all these can be part of the input space. The output of such experiment is also multi-dimensional, comprising data from chemical or physical measurements, either directly obtained or derived quantities. All of these parameters can be assigned to one of three typical parameter spaces: the initial conditions, the final products and the characteristics of products, which will be noted as I, P and C respectively. Each of the relationships between these spaces has meaning in the eyes of a materials scientist, as shown in Fig. 1.

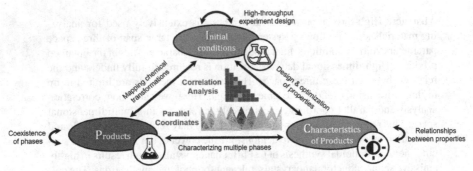

Fig. 1. In materials research, one typically deals with the following three parameter spaces: the initial conditions, the final products, and the properties of the products, which we refer to as I, P, and C, respectively. Correlation analysis and parallel coordinates are two multi-dimensional data visualization methods that can be utilized to understand the relations between and within these three spaces.

Many conventional methods of data visualization, such as bar graphs, line plots, histograms, and scatter plots, are limited to two or three dimensions that can be plotted against each other [3]. Such representations fail to represent the full parameter space of the system and therefore do not convey the underlying complexity of the full multidimensional relationships. Visualization of higher dimensional patterns and associations is possible but are more complex and the advancement of such techniques is still an active area of research [4].

To address the challenges posed by high dimensional data, we here employ two conventional methods of data visualization: correlation analysis and parallel coordinates. We demonstrate how these methods can be used for multi-dimensional visualization, discuss their applications in the context of materials analysis, and showcase how they complement one another.

Although high-throughput data and related methodologies surrounding it have been discussed extensively in the literature [5], the adoption of multidimensional visualization methods in materials science remains limited, partially due to scarcity of high-throughput data until recently and the unfamiliarity with the multi-dimensional visualization methods among the materials science community. Therefore, we aim to bring these multi-dimensional visualization methods to the attention of the material science community and to demonstrate their potential in materials analysis.

In the context of high-throughput synthesis, the methods of correlation analysis and parallel coordinates can be used in various ways for investigations. Firstly, the study of the physical laws and processes that are responsible for the specific spread of data, can be done by exploring the spaces of P vs. C, P vs. P and C vs. C. Secondly, these methods could serve as a tool for optimizing material manufacturing process, by distinguishing between the successful and problematic datapoints and presenting a roadmap for each case. This is done by focusing on the spaces of I vs. C. Finally, the methods could be used as a tool for design and validation of high-throughput experiments. By focusing on the spaces of I vs. I, the input features could be tested for their spread in the multi-dimensional space, and the methods could serve as a tool for anomaly detection and quality assurance beforehand or in real time.

2 Related Work

The utilization of the parallel coordinate method for materials analysis has been relatively limited. One of the notable works, by Rickman [6], included an analysis of multi-dimensional materials properties charts for 25 metallic and ceramic systems using parallel coordinate plot, followed by a principal component analysis (PCA) for dimensionality reduction. In another work by Kamath et al. [7], a parallel coordinate plot was used to obtain insights into the properties of additively manufactured stainless steel parts, based on 462 computer experiments. The work of Bhattarai et al. [8] has provided several insights about silicate melt viscosity values and their relationships with pressure, temperature and composition. Even less studies combined parallel coordinate plot with correlation analysis, like the work of Rickman et al. [9], exploring the characteristics of 82 experimentally fabricated high-entropy alloys.

Additionally, the method of parallel coordinates was implemented in TE Design Lab, a virtual laboratory for thermoelectric material design [10]. This interactive web-based implementation includes various functionalities to customize and analyze the added plot, including highlighting subrange of the full-scale values, and changing the order of axis.

Correlation analysis is a more popular method in the context of materials analysis, often applied as a supplementary step rather than a comprehensive and systematic procedure for obtaining insights. The most common example for application of correlation analysis is the use of correlation plots for investigating the relationships between different material characteristics, such as the emission intensity, photocurrent and lifetime [11], or the hydrogen absorption energy and electronegativity difference [12]. Another common example involved the use of correlation maps to examine the internal relation (autocorrelation) of a single continuous feature, such as one-dimensional X-ray scattering spectra [13] and surface enhanced Raman spectroscopy [14].

Furthermore, correlation maps are used extensively in the context of machine learning (ML), one of the largest branches of data science, which is greatly applied in the field of materials science [15, 16]. ML models can be used to predict the target feature by utilizing a set of given input features. During the preliminary steps of the ML workflow, correlation maps are integrated to the model in order to explore the relations between the characteristics of the material, which will be used later on as the input features to the models [17, 18].

While ML gained significant popularity in the last years, visualization techniques, and particularly those shown in this work, offer several advantages over ML techniques in certain contexts. Firstly, many of ML techniques perform as "black box" algorithms, offering limited explainability. In contrast, the data analysis methods presented in this work provide a straightforward and intuitive way to explore and understand data, making them accessible to a broader audience, including non-technical stakeholders.

In addition, correlation plots and parallel coordinates require less computational effort and less data preprocessing than ML, making them a practical choice for smaller datasets or rapid exploratory analysis. Also, they do not require a specific knowledge for optimization, as some ML algorithms do. Another distinct advantage, further elaborated in this paper, pertains to the differential attention given to anomalies in the data: Since ML algorithms present an averaged output based on values of all input datapoints, the unconventional relationships might go unnoticed. In contrast, visualization methods enable quick identification of outliers and allow the user to decide whether a specific point is a "noise" to be disregarded or a unique novelty requiring further investigation.

3 Case Study

To illustrate the applicability of correlation analysis and parallel coordinates to materials research, we present a case study that investigates the interrelationships among different parameters in a colloidal synthesis, a popular method to synthesizing inorganic nanocrystals of cesium-lead-bromide perovskite ($CsPbBr_3$). These nanocrystals garnered a lot of attention from the material science community because of their high photonic yield and emission tunability [19], making them very appealing for numerous photonic applications like photovoltaic cells [20] and lasers [21]. In addition, $CsPbBr_3$ synthesis is ideal for high-throughput experiment: The synthesis is done in ambient condition and in the range of temperature close to room temperature, the precursors space is intermediate size, and the analysis of the samples is done by simple absorption-emission spectroscopy.

This work was based on a dataset that includes synthesis information, calculated products and spectroscopy measurements of 1351 samples, published as a companion dataset of a previous study [22]. In these experiments several precursor solutions – dilute oleylamine (OLA), oleic acid (OA), lead (Pb) and cesium (Cs) – were mixed in a vial and the reaction was initiated by adding bromide (Br) solution. The synthesis was carried out at different combinations on a Hamilton NIMBUS-4 MicroLab robot.

Both data analysis methods explored here are done by drawing the spaces of initial conditions (I), products (P) and the characteristics of products (C).

In our dataset, the I space consists of the concentrations of reagents in the solution: [OLA], [OA], [Pb], [Cs] and [Br] in mM, and the temperature of the synthesis T in °C. These initial conditions were originally specified in the dataset for all samples.

The P space is the product space of Cs-Pb-Br, consisting of the fractions of nine morphologies (also called phases): $PbBr_x$, Cs_4PbBr_6, $CsPb_2Br_5$, $PbBr_2$, 1 ML, 2 ML, 3

ML, 4 ML and $CsPbBr_3$, so that the sum of the nine fractions approaches 1. The fractions of products used for P space, were given in the dataset as well.

For the C space, we focused on optical characteristics extracted out of the emission spectrum of the samples. The emission spectrum is the electromagnetic radiation (light) frequencies being emitted by a material due to the absorption of external energy. The first characteristic is the photoluminescence quantum yield (PLQY), representing the number of photons emitted as a fraction of the number of photons absorbed, a critical character-istic for the utilization of these nanocrystals in photonic applications. The second one is the central wavelength (CWL) of the emission spectrum, signifying the wavelength at which the electromagnetic radiation with the highest intensity has been emitted. Lastly, we have the full width at half maximum (FWHM) of the emission spectrum, represent-ing the width of a spectrum curve measured between those points on the y-axis which are half the maximum amplitude, providing insights to the spectral broadening of the emitted radiation.

To compute these optical characteristics, we used the spectroscopy part of the dataset. The FWHM was calculated from the emission spectra data, and the CWL was defined as the central wavelength of the FWHM. The PLQY was calculated based on the area of the sum of the Gaussians, fitted semi-automatically to the emission spectra.

The pre-processing involved formatting data into a usable dataset. First, we created a tabular structure of a m × n matrix, with m rows corresponding to m samples, and n columns which contain the I, C and P values for each sample. To ensure accuracy and reliability, the raw data has been carefully screened for the optical characteristics, addressing out-of-range values, impossible data combinations, and missing values fre-quently contained in the raw data, in order to avoid misleading and non-physical results. Finally, we remained with m = 1136 samples in the dataset for the C space.

3.1 Correlation Analysis

Correlation analysis is a strong and well-established technique for data analysis [23]. It is based on a **correlation plot**, a scatter plot of two variables. The way to evaluate the association between the variables is the **correlation test**. The most common one is Pearson correlation [24], which measures both the strength and direction of the linear relationship between two continuous variables, and calculated as:

$$r = \frac{\sum_i (x_i - \bar{x})(y_i - \bar{y})}{\sqrt{\sum_i (x_i - \bar{x})^2 \sum_i (y_i - \bar{y})^2}} \tag{1}$$

where x and y are two vectors of length m, and $\bar{x}$ and $\bar{y}$ corresponds to the means of x and y, respectively. However, Pearson correlation accounts only for a linear dependence between the variables, and often it is necessary to define a more relevant metric, as will be demonstrated further in this paper. To investigate the dependence between multiple variables at the same time, a **correlation matrix** is used, consisting of ordered correlation plots in a tabular structure. **Correlation map**, or **Corrgram**, is a visual display technique of correlation matrix, which is very useful to highlight the most correlated variables in a data table. The idea is to display the pattern of correlations in terms of their signs and magnitudes.

3.2 Parallel Coordinates

Parallel coordinate plot is a long-standing and efficient technique for high-dimensional datasets [25, 26]. Its main advantage lies in its ability to simultaneously identify correlations across multiple dimensions, making it easier for researchers to find trends, identify outliers and perform quality checks in large multivariate data.

Brushing, an interaction technique particularly useful in parallel coordinates, enables the user to select a subset of a dataset, that is then highlighted [27]. In addition, using different colors can be used to facilitate the detection of patterns across the variables and discovery of the relationships between them. In our work we used Python's "Plotly" package [28] to create an interactive display and to highlight data lines that fall within a subrange of the full-scale axis values for one or more axes simultaneously.

4 Applications and Discussion

In the following, we demonstrate how the use of correlation analysis and parallel coordinates reveals distinct insights from our dataset. Each combination of the spaces (I, C and P) shows a specific pattern of the data and can serve a different purpose. We will explore all possible combinations of these spaces thoroughly, except I vs. P, which has been thoroughly investigated in a prior study by the creators of the dataset.

4.1 C vs. P: Characterizing Multiple Phases

We start with an investigation of the characteristics of the products versus the product fractions (C vs. P). In Fig. 2 we selected the PLQY as our C parameter, and then used the nine different phases of the Cs-Pb-Br space as our P parameters. Already at first glance, we can learn qualitatively from the data spread about the effect of each phase on PLQY: while in high fractions of some phases, a low PLQY can be seen, other phases show high PLQY in their high fractions. To distinguish between these trends quantitatively, we use Pearson's correlation coefficient, r. The correlation values evaluate the ratio of emitted photons per absorbed photons, with a positive r corresponding to emitting phases with high emission-to-absorption ratio. Hence, the method enables classification of phases according to the selected characteristic.

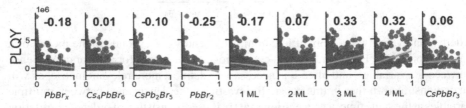

Fig. 2. Correlation plots of C-P, for PLQY vs. fraction of phases. From these correlation plots we can statistically evaluate how each phase affects the PLQY. In each figure, the number in bold is r, the Pearson's correlation coefficient, and the red line is the linear regression fit. The phases with negative r (e.g., PBr$_2$,) are non-emitting phases, while other phases with positive r (e.g., 3 ML), are the emitting phases. (Color figure online)

An additional example investigating C-P relationship with another characteristic of the products – CWL, presented in Fig. 3. This time, a parallel coordinate plot was chosen, due to its capability to visualize the wavelength by color. The purple-to-red color-bar corresponds intuitively to the physical emission wavelengths values of 400 to 550 nm. The plot reveals that there is a typical emission wavelength for each one of the emitting phases, which corresponds to the known values in the literature [29, 30]. Non-emitting phases can be noticed by multicolor lines. The reason for this is that the obtained CWL in these phases is affected by the existence of the other emitting phases.

Through the parallel coordinate method, the typical emission wavelength of each phase was found. The method could be utilized as a general and systematic way for finding a specific property of an unknown sample of mixed phases. Moreover, the main advantage of this analysis method is that it does not require a synthesis of a pure sample of each one of the phases of the material, to find their properties. Using the high-throughput data, both methods succeeded in finding the optical characteristic in all phases simultaneously based on multi-phased samples.

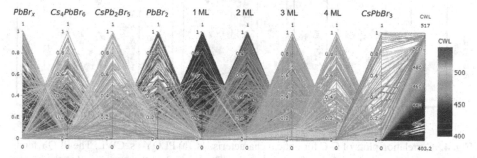

Fig. 3. Parallel coordinates of C-P, for CWL vs. fraction of phases. A typical emission wavelength is clearly noticed for each one of the emitting phases (1 ML, 2 ML, 3 ML, 4 ML and CsPbBr$_3$), which corresponds to the known values from literature. Non-emitting phases (such as PbBr$_x$) can be noticed by multicolor lines.

4.2 C vs. C: Relationships Between Properties

A different angle on the dataset comes through the relationships between the characteristics of the products (C vs. C). Figure 4 presents the correlations of the optical characteristics with themselves. In Figs. 4a and 4b, we can see that the shape of the data is different next to the emission wavelengths of the known morphologies of the 2ML, 3ML, 4ML and CsPbBr$_3$ phases. Around these wavelengths, the PLQY tends to increase, and the FWHM tends to decrease. This indicates that pure samples of the morphologies often have better optical characteristics then the mixed samples. For example, FWHM of 2 ML and 3 ML is less than 20 nm, while in mixed samples of these phases it reaches 30 nm. Figure 4c shows that high PLQY doesn't necessarily correspond to low FWHM, as there are high PLQY samples throughout the FWHM scale. We will note that while the original dataset included many samples of 1 ML, they are not reflected in the analysis due to filtering of the data.

From the spread of the data, it can be also understood that many of the samples are of the known emitting phases, but their emission is low with decreased PLQY values. However, their low values of FWHM indicate a well-defined sharp emission peak typical to the physical behavior of these materials.

One interesting anomaly in the data is a point with extremely high PLQY, noted with arrow in Fig. 4. At first sight, this point may be discarded as an outlier, as it inconsistent with the data set. However, examining the full parameter space reveals that the reason for this point to stand out is the lack of other samples with similar precursors properties. Later in-depth examination (Fig. 7), shows that this samples is indeed uniquely made. This data point serves as a notable example of an anomaly that could have been missed when using ML algorithms since it could have been interpreted as a "noise" in the data, and not as a local extremum in the multidimensional space.

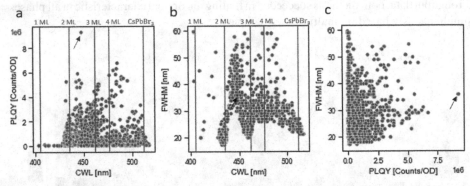

Fig. 4. Correlation plots of C-C for optical characteristics. (a) PLQY vs. CWL. The PLQY tends to increase at the emission of specific wavelengths. These wavelengths correspond to the known morphologies of 1 ML, 2 ML, 3 ML, 4 ML and $CsPbBr_3$ nanocrystals (marked in red lines). The arrow points to an interesting anomaly with extremely high PLQY. (b) FWHM vs. CWL. The FWHM drops when the CWL is of a specific morphology. However, in 4 ML it does not follow the same rule. (c) FWHM vs. QY. High PLQY doesn't necessarily correspond to low FWHM. (Color figure online)

Additional anomaly from this exact dataset led to a revolutionary insight, published by the author [31]. In that work, CWL was plotted against the Stokes Shift, another optical characteristic, and showed an unexpected non-monotonic trend.

These examples show the great capability of the correlation analysis in identifying trends and anomalies, to shed light on the physics of the investigated synthesis.

4.3   P vs. P: Coexistence of Phases

We will proceed further with the correlation analysis, towards investigation of the products, i.e., the P space. Here we present a tool for exploring the coexistence of phases in Cs-Pb-Br space. The first step is done by finding the coexistence of two-phases in a sample. In a correlation plot of two phases (Fig. 5a), a single point reflects the fractions of these two phases in the sample. For example, a point of (0.1, 0.7) is a synthesis which results in the following combination of phases: 10% of phase x, 70% of phase y, and 20% of other 7 phases.

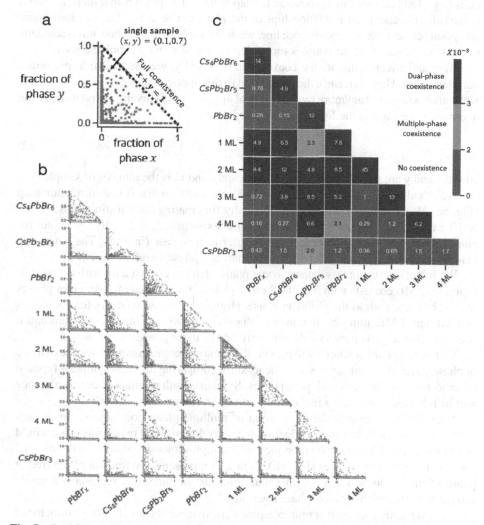

Fig. 5. Correlation analysis of P-P for dual phases coexistence. (a) Correlation plot of two phases. Each datapoint represent coexistence of two phases per synthesized sample. (b) Correlation matrix of nine phases. The collection of datapoints per graph represents the correlation of two phases of all the dataset. The full correlation matrix shows the coexistence of all possible phases. (c) Correlation map of nine phases. The values are the calculated value of the coexistence test.

There are 1351 points in this figure, representing the whole dataset. Datapoints along the line x + y = 1 represent samples in which there was an exact coexistence of the x and y phases. However, since we want to investigate the mutual dependence of all nine phases simultaneously, we shall use a correlation matrix (Fig. 5b). In this matrix, every cell is a correlation plot of two different phases and consists of the whole dataset.

Inspecting the correlation matrix, one can notice that two different patterns exist: some cells (e.g., 2 ML vs. 1 ML) have a data in a form of a triangle, while other cells (e.g., $CsPbBr_3$ vs. $PbBr_x$) have an L-shaped data. The physics that hides behind it is the different coexistence relationships of the phases. The triangular cells have more datapoints close to the full coexistence line, while the L-shaped cells show no coexistence of the two phases, since no samples show high values of x and y simultaneously.

To quantitatively estimate the coexistence of x and y, we would like to perform a correlation test. However, since the correlation in this application is not linear in essence, we cannot be assisted anymore by the classical approach of Pearson's correlation. As an alternative, we suggest the following x-y coexistence test:

$$A = \frac{1}{m} \sum_{i=1}^{m} x_i \cdot y_i \tag{2}$$

where x and y are the two phases in a specific cell, and m is the number of samples.

We calculate this value for each one of the cells and visualize it in a correlation map (Fig. 5c). We set numerical thresholds manually for creating three distinct cells: $A > 3 \times 10^{-3}$ (in red) for phases which satisfy dual-coexistence, $A \leq 2 \times 10^{-3}$ (in blue) for phases which show no coexistence, and an intermediate case (in pink). The latter case belongs to phases which coexist with more than one phase simultaneously.

We found out that in this dataset, some phases tend to coexist a lot with others. For example, 2 ML coexists with 5 of the 8 other phases. On the other hands, other phases tend to be pure, such as the $CsPbBr_3$ phase. Highest coexistence occurs for the phases of 1 ML and 2 ML, implying that their synthesis is sensitive to initial conditions and it is challenging to get a pure sample with only one of these products.

Different types of synthesis will provide different correlation matrices for coexistence of phases. Therefore, this analysis can be used for comparing different synthesis types in order to select a synthesis with pure phases. Synthesis with minimal phase coexistence will be reflected in blue cells in their respective correlation map.

Next, we will examine the third group of multiple-phase coexistence. Since each correlation plot is limited to investigation of a two-dimensional relationship, we would utilize our second data visualization method for high-dimensional data. In the parallel coordinates plot of the phases (Fig. 6b), each axis represents the fraction of a different phase of the products. Each polyline that connects the nine axes represents a single sample with its nine respective phase fractions.

Now, we study each case of multiple-phase coexistence separately. We go back to the correlation map and focus on one of these cells: 1 ML vs. $CsPb_2Br_5$ (green-black cell in Fig. 6a). Using brushing we select the subranges on the parallel coordinates plot to be (1 ML > 0.1) and ($CsPb_2Br_5 > 0.1$), which means sample that contain these both phases in a considerable amount (green-black dashed lines in Fig. 6b). After this selection, we see that samples that satisfy this condition contain two additional phases: Cs_4PbBr_6 and 2 ML, which were obtained as an output of the method (green solid lines in Fig. 6b). Going back to the correlation map, we can map all the other phases that correspond to this four-phase coexistence (green cells in Fig. 6c). This validates the result by fitting to other dual-phase coexistence cells. For completeness of the analysis, the examination of the other multiple-coexistence cells can be performed to find all combination of coexisting phases.

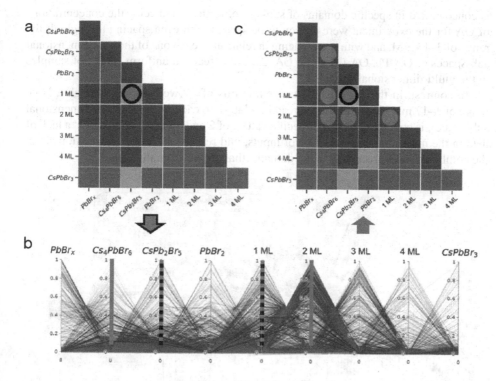

Fig. 6. Analysis of the products P-P for multiple phase coexistence. (a) The correlation map is first used to selected a cell of multiple-phase coexistence, consisting of $CsPb_2Br_5$ + 1 ML + unknowns (green-black cell). (b) Parallel coordinate plot reveals the coexistence of four phases simultaneously: $CsPb_2Br_5$ + 1 ML + 2ML + Cs_4PbBr_6. The original two phases are in dashed green axes, and the obtained phases are in solid axes. (c) Returning back to the correlation map validates the result by fitting to other dual-phase coexistence cells (green cells). (Color figure online)

In conclusion, we show here that we can methodically map the coexistence of multiple phases. Using correlation analysis, we classify each relationship of phases to three groups: no coexistence, dual-phase coexistence, and multiple-phase coexistence. Encountering a case of multiple-phase coexistence, the parallel coordinates plot enables the identification of all coexisting phases.

4.4 I vs. I: High-Throughput Experiment Design

Now we will focus on another space, not yet analyzed – the initial conditions space, I. When designing a high-throughput experiment, one is generally interested in achieving a uniform spread of the input features in order to cover a maximum volume in the multidimensional input space with a minimum number of samples.

In the correlation matrix of I-I (Fig. 7a), two types of cells can be noticed: those in which the data is spread uniformly throughout the sub-space, and those in which the data is concentrated in specific domains of sub-space. In the green cells, the concentrations of OA for the experiment were chosen to be spread with even spacing through all the range of 3–13 mM, and with even spacing in relation to each one of the two-dimensional sub-spaces of OA-Pb, OA-OLA and OA-Cs. This creates a uniform spread of samples in the multi-dimensional space.

In contrast, in the red cells the concentrations of Br were spread through a large range of 1–17 mM, however their spread in relation to each one of the two-dimensional sub-spaces, is concentrated in the specific values of 2 and 13 mM. This creates a lack of data in the multi-dimensional space of inputs, and results in lack of information about the combination of precursors in the domains that are far from this value.

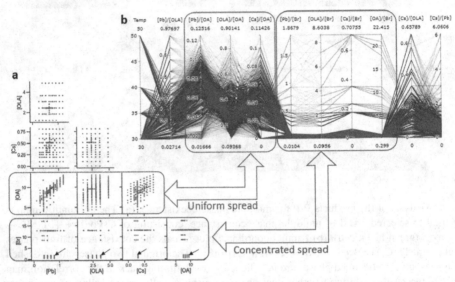

Fig. 7. Analysis of the input features I-I for precursors concentrations. Features with uniform spread of data are selected in the green frame, while features with concentrated spread are selected in red frame. (a) Correlation matrix of the precursors (all values in mM). The arrow points to the anomaly datapoint from Fig. 4. (b) parallel coordinates of the precursors ratios. (Color figure online)

As an example, the arrow is pointing to the anomaly datapoint with maximal PLQY from Fig. 4. This reveals that too few experiments were performed in the surrounding precursor domain, causing this sample to appear as an anomaly in PLQY. A similar conclusion can be drawn using the parallel coordinate method (Fig. 7b), where uniform spread forms a cluttered plot, while concentrated spread characterized by a sparse plot.

As shown, our methods can be used to visualize the spread of the input features and to indicate "holes" in the multidimensional input space. Then, in turn, a plan for a high-throughput experiment could be validated or modified if necessary.

4.5 C vs. I: Design and Optimization of Properties

The last useful representation to be discussed here is the relationship between the characteristics of the products and the initial conditions of the synthesis (C vs. I). This is advantageous for optimizing a material property towards a desired value and for finding initial conditions that produce the optimal result in I-space.

To demonstrate functionality of this tool we optimize the parameter-space to maximize the photoluminescence quantum yield (PLQY) and suggest a roadmap for highly emitting samples. We plot the PLQY against the reagent relations and the temperature, using the parallel coordinate plot (Fig. 8). Coloring coded PLQY values, as defined in the color bar, allows for the discrimination of low- and high-emission samples. Furthermore, it can be seen that there is a specific path, which most of the high PLQY samples are concentrated along it. By following this theoretical path (marked in dashed orange polyline) as a roadmap, one can reproduce the reagents concentrations that are required, based on this high throughput synthesis.

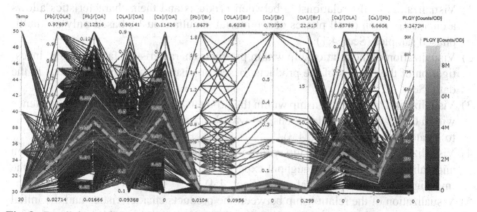

Fig. 8. Parallel coordinates C-I for the PLQY vs. initial conditions. A certain path with higher PLQY can be noticed through all samples. This path (marked in dashed orange polyline) can be used as an optimized roadmap for maximal PLQY. By following it, one can reproduce the rea-gent concentrations that are required. (Color figure online)

The strength of this method is that any other physical parameter can also be examined in the same manner, which allows moving the focus of the exploration and optimization to any other output feature.

Besides the roadmap for the desired property, by examining each axis separately, we can learn about the most important features for the property design. Good separation between different color ranges on a specific axis indicates a high importance of the respective feature, while overlap and disorder of colors indicate its insignificance. This can be used, on the one hand, to find irrelevant input features that can be ignored for future synthesis and, on the other hand, to indicate important parameters that may not have been considered significant in the first place.

5 Conclusion

In this work, we present a novel application of two visualization methods that can offer a practical and insightful alternative to current tools in the materials researcher's toolbox for exploring multivariate data.

We demonstrate the effectiveness of correlation analysis and the parallel coordinates plot for finding high dimensional patterns within three high-dimensional spaces commonly encountered in the context of materials synthesis: the initial conditions, the products and the characteristics of the products. To exemplify the approach's utility, we analyzed a dataset of a high-throughput experiment, with 1351 samples of cesium-lead-bromide nanocrystals synthesis, which involved multiple precursors and products.

This approach constitutes a powerful tool for exploring complex interrelationships that can guide materials research. For reference, a summary of the important aspects is given below:

1) Visualization of the relationship between products and their characteristics allows a characterization of all phases of a material simultaneously, based only on multi-phased samples (Sect. 4.1).
2) Visualization of the relationship within products characteristics facilitates the investigation of the physics of the products, based on visual trends and anomalies in the data (Sect. 4.2).
3) Visualization of the relationship within the products aids in finding phase coexistence within the material and allows a useful way the compare different types of synthesis to create a pure phase (Sect. 4.3).
4) Visualization of the relationship within the initial conditions can be used for design and validation of a high-throughput experiment, ensuring sufficient coverage of the investigated multidimensional space (Sect. 4.4).
5) Visualization of the relationship between the products characteristics and the initial conditions enables the optimization of a desired property. In addition, it points to the more significant initial conditions, which should be more precisely controlled to reach a desirable goal (Sect. 4.5).

It is essential to emphasize that multidimensional synthesis like the one provided in our case study, is common in the materials science research. Therefore, we strongly encourage a broader implementation of these multi-dimensional visualization methods in the application of materials analysis.

Lastly, our work showcases the insightful and practical application of data visualization technique in material science. We believe that a closer collaboration between data science researchers and the materials science community, with the aim of bringing more modern data analysis techniques to bear in materials science, holds great promise.

Acknowledgment. The author wishes to thank Yehonadav Bekenstein (Department of Materials Science and Engineering and the Solid-State Institute, Technion — Israel Institute of Technology, Haifa, Israel) for his support. This research did not receive any specific grant from funding agencies in the public, commercial, or not-for-profit sectors.

References

1. Himanen, L., Geurts, A., Foster, A.S., Rinke, P.: Data-driven materials science: status, challenges, and perspectives. Adv. Sci. **6**, 1900808 (2019). https://doi.org/10.1002/advs.201 900808
2. Chighine, A., Sechi, G., Bradley, M.: Tools for efficient high-throughput synthesis. Drug Discov. Today **12**, 459–464 (2007). https://doi.org/10.1016/j.drudis.2007.04.004
3. Ward, M.O., Grinstein, G., Keim, D.: Interactive Data Visualization: Foundations, Techniques, and Applications. AK Peters/CRC Press (2010)
4. Heinrich, J., Weiskopf, D.: State of the Art of Parallel Coordinates. Eurographics (State of the Art Reports). Eurographics Assoc. 95–116 (2013). http://dx.doi.org/10.2312/conf/EG2013/stars/095-116
5. Liu, Y., et al.: High-throughput experiments facilitate materials innovation: a review. Sci. China Technol. Sci. **62**, 521–545 (2019). https://doi.org/10.1007/s11431-018-9369-9
6. Rickman, J.M.: Data analytics and parallel-coordinate materials property charts. NPJ Comput. Mater. **4**, 1–8 (2018). https://doi.org/10.1038/s41524-017-0061-8
7. Kamath, C., El-Dasher, B., Gallegos, G.F., King, W.E., Sisto, A.: Density of additively-manufactured, 316L SS parts using laser powder-bed fusion at powers up to 400 W. Int. J. Adv. Manuf. Technol. **74**, 65–78 (2014). https://doi.org/10.1007/s00170-014-5954-9
8. Bhattarai, D., Karki, B.B.: Parallel coordinates-based visual analytics for materials property. In: VISIGRAPP 2019 - Proceedings of the 14th International Joint Conference on Computer Vision, Imaging and Computer Graphics Theory and Applications, vol. 3, pp. 83–95 (2019). https://doi.org/10.5220/0007375400830095
9. Rickman, J.M., et al.: Materials informatics for the screening of multi-principal elements and high-entropy alloys. Nat. Commun. **10**, 1–10 (2019)
10. Gorai, P., et al.: TE Design lab: a virtual laboratory for thermoelectric material design. Comput. Mater. Sci. **112**, 368–376 (2016). https://doi.org/10.1016/j.commatsci.2015.11.006
11. Draguta, S., et al.: A quantitative and spatially resolved analysis of the performance-bottleneck in high efficiency, planar hybrid perovskite solar cells. Energy Environ. Sci. **11**, 960–969 (2018). https://doi.org/10.1039/c7ee03654j
12. Xu, L., Jiang, D.-E.: Understanding hydrogen in perovskites from first principles. Comput. Mater. Sci. **174**, 109461 (2020). https://doi.org/10.1016/j.commatsci.2019.109461
13. Franke, D., Jeffries, C.M., Svergun, D.I.: Correlation Map, a goodness-of-fit test for one-dimensional X-ray scattering spectra. Nat. Methods **12**, 419–422 (2015). https://doi.org/10.1038/nmeth.3358
14. Dieringer, J.A., et al.: Surface enhanced Raman spectroscopy: new materials, concepts, characterization tools, and applications. Faraday Discuss. **132**, 9–26 (2006). https://doi.org/10.1039/b513431p
15. Wei, J., et al.: Machine learning in materials science. InfoMat **1**, 338–358 (2019). https://doi.org/10.1002/inf2.12028
16. Liu, Y., Zhao, T., Ju, W., Shi, S., Shi, S., Shi, S.: Materials discovery and design using machine learning. J. Materiomics. **3**, 159–177 (2017). https://doi.org/10.1016/j.jmat.2017.08.002
17. Im, J., Lee, S., Ko, T.W., Kim, H.W., Hyon, Y.K., Chang, H.: Identifying Pb-free perovskites for solar cells by machine learning. NPJ Comput. Mater. **5**, 1–8 (2019). https://doi.org/10.1038/s41524-019-0177-0
18. Pilania, G., Mannodi-Kanakkithodi, A., Uberuaga, B.P., Ramprasad, R., Gubernatis, J.E., Lookman, T.: Machine learning bandgaps of double perovskites. Sci. Rep. **6**, 1–10 (2016). https://doi.org/10.1038/srep19375
19. Sutherland, B.R., Sargent, E.H.: Perovskite photonic sources. Nat. Photonics **10**, 295 (2016)

20. Park, N.-G.: Perovskite solar cells: an emerging photovoltaic technology. Mater. Today **18**, 65–72 (2015)
21. Yakunin, S., et al.: Low-threshold amplified spontaneous emission and lasing from colloidal nanocrystals of caesium lead halide perovskites. Nat. Commun. **6**, 1–9 (2015)
22. Dahl, J.C., Wang, X., Huang, X., Chan, E.M., Alivisatos, A.P.: Elucidating the weakly reversible Cs-Pb-Br perovskite nanocrystal reaction network with high-throughput maps and transformations. J. Am. Chem. Soc. **142**, 11915–11926 (2020). https://doi.org/10.1021/jacs. 0c04997
23. Gogtay, N.J., Thatte, U.M.: Principles of correlation analysis. J. Assoc. Physicians India **65**, 78–81 (2017)
24. Freedman, D., Pisani, R., Purves, R.: Statistics (international student edition). In: Pisani, R. Purves, 4th edn. WW Norton & Company, New York (2007)
25. Heinrich, J., Weiskopf, D.: Parallel coordinates for multidimensional data visualization: basic concepts. Comput. Sci. Eng. **17**, 70–76 (2015)
26. Inselberg, A.: The plane with parallel coordinates. Vis. Comput. **1**, 69–91 (1985)
27. Kosara, R.: Indirect multi-touch interaction for brushing in parallel coordinates. In: Visualization and Data Analysis 2011, pp. 78–84 (2011)
28. Inc., P.T.: Collaborative data science (2015). https://plot.ly
29. Bohn, B.J., et al.: Boosting tunable blue luminescence of halide perovskite nanoplatelets through postsynthetic surface trap repair. Nano Lett. **18**, 5231–5238 (2018). https://doi.org/ 10.1021/acs.nanolett.8b02190
30. Bekenstein, Y., Koscher, B.A., Eaton, S.W., Yang, P., Alivisatos, A.P.: Highly luminescent colloidal nanoplates of perovskite cesium lead halide and their oriented assemblies. J. Am. Chem. Soc. **137**, 16008–16011 (2015). https://doi.org/10.1021/jacs.5b11199
31. Vurgaft, A., et al.: Inverse size-dependent Stokes shift in strongly quantum confined CsPbBr 3 perovskite nanoplates. Nanoscale **14**(46), 17262–17270 (2022). https://doi.org/10.1039/D2N R03275A

Law in Order: An Open Legal Citation Network for New Zealand

Tobias Milz[✉][iD], Elizabeth Macpherson[iD], and Varvara Vetrova[iD]

University of Canterbury, Christchurch 8041, New Zealand
tobias.milz@pg.canterbury.ac.nz,
{elizabeth.macpherson,varvara.vetrova}@canterbury.ac.nz

Abstract. The potential of leveraging network science in the area of law has long been advocated and highlighted through case and legislation networks. Yet this particular subdomain of data management and information retrieval is still heavily underutilised in both practice and research. One of the contributing factors to this problem is the lack of openly available legal data. This paper describes the development of a legal citation network for New Zealand. In contrast to traditional case citation networks, this data repository also includes legislation and court data. Our network provides the data and references from over 300,000 decisions, 10,000 legislations and 115 courts from all levels of jurisdiction. Additionally, we present exemplifying network analysis results to reveal previously hidden information about the New Zealand legal system and to motivate future research in this domain.

Keywords: Case Citation Network · Legal Tech · Neo4j · Network Science · PageRank · Data Mining

1 Introduction

The law is a fundamental pillar of human societies as it shapes, controls and governs how humans conduct business, behave and interact with each other. Recent advances in computer-assisted technologies such as NLP, data science and AI are creating opportunities to support the practice, research and study of this pervasive domain. It is therefore not surprising that there has been an increase in investments into supporting technologies for the legal industry (also known as "legal tech" or "law tech") over the last demi-decade [20]. A sub-discipline of particular appeal is concerned with the area of assisted legal data management and information retrieval.

Supporting law researchers and practitioners to retrieve information from the vast amount of ever-growing legal documentation is of natural interest to the legal research community. When researching a legal case, finding relevant similar cases, old and new precedents or applicable legislations are common tasks for legal practitioners. Case citation indices are a well-known and long-established tool for aiding tasks like these. Similar to the practice of quantifying

© The Author(s), under exclusive license to Springer Nature Singapore Pte Ltd. 2024
D. Benavides-Prado et al. (Eds.): AusDM 2023, CCIS 1943, pp. 211–225, 2024.
https://doi.org/10.1007/978-981-99-8696-5_15

the excellence of research publications, the number of references a case receives can indicate its importance or relevance to the research task at hand [26,27]. However, by converting these indices into citation networks we can utilise other network science and analysis tools to reveal additional "hidden" and novel information about the cases, laws and legal system.

Unfortunately, access to openly available legal data is still lacking in New Zealand and case citation indices are, for the most part, only commercially available via providers such as LexisNexis [6]. Although the potential of such networks has been highlighted by previous research from various countries and legal systems [13,18,21,22], to the best of our knowledge, no case citation network has been analysed nor published as open access data for New Zealand. Thus, we identified a need to develop and provide a case citation network that contains sufficient data to support the analysis of the New Zealand legal system and the development of future legal research applications. With that, we also aspired to extend the usual paradigm of case citation networks to include legislation and court data.

For this reason, we created a legal citation network for New Zealand containing over 300,000 court cases, 10,000 legislations and 115 courts from all areas of law and jurisdiction. Additionally, we demonstrate how this data can be used to reveal new information about the legal system and support common legal research tasks by applying network analysis algorithms and metrics.

The paper is structured as follows: Sect. 2 provides an overview of the related works and how this research extends previous studies in this domain, Sect. 3 describes the process of creating the legal citation network, Sect. 4 highlights and showcases selected network analysis results and lastly, Sect. 5 summarises this study and discusses opportunities for future research.

2   Related Work

Case citation indexing as a tool to identify potential precedent cases has been common practice since the early 19th century [14]. Hence, it is not surprising that the study of case citation networks has received some attention over the years and even more so with the increasing digitalisation of legal documents. Due to the diversity of the law domain, most research on case citation networks is focused on either a specific legal system, court or country such as the U.S. state and federal courts [26], the U.S. Supreme Court [11,13], the Austrian Supreme Court [14], the European Court of Justice and the General Court [27], the European Court of Human Rights [18], selected European courts [15], Canadian courts [22] and German courts [10,21,24].

Furthermore, these studies investigate different aspects or use cases of these networks. For instance, [18] use the case citation network of the European Court of Justice to compare the decision-making between domestic and international courts, [26] compares the case network to a research literature citation network and [27] analyses the use of centrality measures such as PageRank and degree centrality for case relevance ranking.

While the studies above mostly focus on court cases and their citations, [16,17,25] created and analysed legislation networks. However, we have not been able to identify any study that tried to connect more than one type of legal document other than [21]. In their study, the authors provide a German legal citation network that contains both laws and cases, but the citations are only extracted from cases (towards cases and laws) and not from the laws.

3 Creating the Legal Citation Network

We use the term "legal citation network" to describe a repository of legal documents that are interconnected based on the references that exist between them. Unlike the more commonly known case citation networks, a legal citation network is not limited to only one kind of legal record. Instead, the nodes of a legal citation network can represent a variety of legal resources including cases, courts, legislations, journals, publications, reforms, treaties and so on. In this section, we will describe the creation process of our legal citation network starting with the data acquisition, citation extraction and finally the database development.

3.1 Data Acquisition

Unfortunately, research in the legal tech domain can often be obscured by the lack of openly available legal data. Given the sensitive nature of legal documents and their intrinsic privacy concerns, this problem is not unique to New Zealand but common around the world. Consequently, efforts have been made to provide anonymised legal data for different countries [1–4,23]. However, in the case of New Zealand, all openly-available data sources are only providing read access without APIs. Furthermore, the data is stored in different formats and document types, including non-searchable or non-editable PDFs.

To overcome this problem, we assembled web scrapers, parsers and an OCR pipeline to collect and convert the contents from these sources into uniform and machine-readable text. The result is a new open legal dataset containing New Zealand case, legislation and court data.

Case and Court Dataset. A case or court decision is a document that contains the details of a court proceeding. This includes the information about the parties involved, the case description, the judge and the judgement. Using web scrapers we mined the data of 338,360 cases from 1842 to 2022 from the New Zealand Legal Information Institute's website (NZLII) [4]. About half (51.7% or 175,084) of these cases had their data already published in HTML format, while the rest (48.3% or 163,276) were only available as PDF documents. A large proportion of these PDF files are scanned documents that are not machine-readable. Hence, we created an OCR pipeline that converts each page into an image and then applies Google's OCR library Tesseract[1] to identify and extract the text. See Fig. 1 for an illustration of the case data mining process.

---

[1] https://github.com/tesseract-ocr/tesseract.

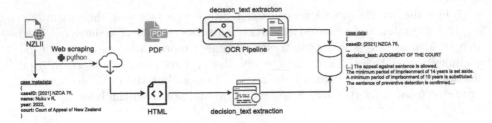

Fig. 1. Illustration of the case-data collection step. The data for this research was acquired from the NZLII database. About half of their cases are either in HTML or native and scanned PDF format. Depending on the document type, the decision texts are extracted using Python scrapers or an OCR pipeline.

For each case, we also parsed the metadata of the files to identify the year, title, neutral citation identifier (used as a unique caseID) and name of the corresponding court. The names of the courts and their shortcode were then also processed and added to a separate court dataset. In total we identified 115 unique courts from all levels of jurisdiction. Table 1 provides an example of the case and court data that we collected.

Table 1. Case, legislation and court dataset examples.

caseID	title	year	court	decision_text
[2021] NZCA 75	Hunter v R	2021	Court of Appeal	Judgement of the court [...] The appeals against conviction are dismissed
...	...	...	...	

legID	title	year	type	legislation_text
345447	Income Tax Act 2007	2007	Act	(1) A person who receives a refund for a tax year under section EK 12 [...]
...	...	...	...	

name	shortcode
High Court of New Zealand	NZHC
...	...

Legislation Dataset. The legislation data was scraped from the New Zealand Parliamentary Counsel Office website [5]. This open-access database contains most legislations for New Zealand in both HTML and PDF format. In total, we collected 10,402 unique legislations. Each legislation text was then supplemented with its year of proclamation, title and type (i.e. "Act", "Bill" or "Secondary Legislation"). These descriptors could be extracted and referred from their corresponding parsed HTML documents. Table 1 provides an example of the extracted legislation data.

3.2   Citation Extraction

The next step was to create automatic citation extraction algorithms to identify references towards other cases and legislations from either source. Related work has shown different approaches to this problem [19,21,23] depending on the corresponding citation conventions and their syntactic complexity.

In New Zealand, the syntax for citing legal resources can be more complex than compared to other countries like Germany [21]. However, using the guidelines provided in [9] and by investigating many examples with the help of law professionals, we determined some of the most common citation patterns. We won't be able to go into detail about the intricacies of these different citation conventions as it would be out of the scope of this publication, but we have highlighted the most common examples in Table 2.

Table 2. Case and legislation citation examples. Citation patterns and conventions differ depending on the year of the case, the responsible court and whether it was reported.

Citation Type	Example
Neutral Case Citation	Crawford v Phillips [2018] NZCA 208
Reported Case Citation	Taylor v NZ Poultry Board [1984] 1 NZLR 394 (CA)
Unreported Case Citation	R v Te Huia CA327/06
Legislation Citation	Income Tax Act 2007

Case Citations. We developed multiple complex regular expressions for each pattern and tested them on a small dataset of 100 randomly chosen court cases that we labelled manually. In total, we missed 8.7% (12 out of 138) of the manually identified citations. 16.7% of those were due to typographical errors in the original citation. The other mistakes were mostly due to imprecise citation behaviours or the use of abbreviations that we will include in the future. Citations that cannot be resolved into an actual reference were not counted. These are citations without any clear identification number, file number or report name. For example, the citation "Jackson v. Ward" can not be linked to a unique case especially since the names might have been anonymized to "J v. W" in the actual document.

Legislation Citations. For references towards laws, we can make use of string-matching methods as legislation citations are much less ambiguous and are always supposed to be cited by their exact name. During the evaluation process, only one (out of 71) citation was missed due to a typographical error in the original document. Here we also did not count citations to legislations that we cannot resolve into an actual reference (i.e. all references to bills and acts that we do not have in the dataset are ignored).

Table 3 summarises the result of applying these methods to our datasets. In total, we managed to identify 427,701 case and 312,016 legislation references. Self-citations from and to the same document were removed as they do not carry any significant legal meaning.

Table 3. Total number of identified references from cases and legislations using complex regular expression and string matching methods.

Source	Target	Number
Case	Case	427,639
Case	Legislation	244,159
Legislation	Case	62
Legislation	Legislation	67,857

3.3   Converting the Data into a Graph

Akin to the related works in [21,24,25], we interpret case citation networks as directed graphs where an edge from vertex (X) to vertex (Y) represents that (X) has cited (Y) at least once. In other words, multiple citations from one source to the same target still only count as one reference (i.e. no multiple edges). Consequently, a typical case citation graph can be described as a directed simple graph without loops:

$$G = (V, E) \tag{1}$$

with V representing the set of vertices (cases) of the graph and E the set of edges (references). In our graph, however, there are multiple vertex and edge types with different properties as we included the data and citations of courts and legislations as well. In this section, we will introduce our citation graph and its properties as it is represented in our database.

The Neo4j Graph. The next step was to convert the datasets and the extracted references into a network or graph-like data structure. Accordingly, we converted and imported our newly created datasets into a Neo4j database which we provide upon request[2]. Neo4j, as a graph-based database, supports the application of network algorithms directly on the database and allows for efficient querying of highly connected data like our citation network. Figure 2 illustrates the basic schema and the different types of relationships and nodes of the graph[3].

Graph Nodes. Our graph consists of the following node types: "Case", "Court" and "Legislation". Each node represents an entry of the corresponding dataset with the same properties as shown in Table 1. For example, a case node is a specific court case with a *title*, *year*, *court* and *decision_ text*.

---

[2] Please contact tobias.milz@pg.canterbury.ac.nz to gain access to the data.

[3] It shall be noted that we will adjust our terminology for this section and refer to vertices and edges as "nodes" and "relationships". This is to be consistent with the naming convention provided by Neo4j.

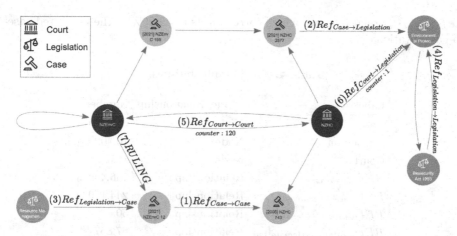

Fig. 2. Illustration of the relationship and node types in the Neo4j graph. Cases are connected to their corresponding court (7) and can cite other cases (1) or legislations (2). Legislations can cite each other (4) and cases (3). Courts can indirectly cite each other, themselves (5) and legislations (6) via their cases.

Graph Relationships. There are 7 relationship types in our graph. Relationships (1), (2), (3) and (4) (as shown in Fig. 2) are equivalent to the respective references that we identified in Sect. 3.2 and highlighted in Table 3. In other words, these relationships indicate that the source node has cited the target node at least once (as a reminder, duplicate citations were not counted as they do not carry additional legal meaning). Accordingly, these are directed unweighted relationships without loops (self-citations were removed). Relationship (5) is created between the corresponding courts of two connected cases. It represents an indirect citation between the two courts [21,23] and is therefore also directed. However, in contrast, this relationship can loop on the same node because cases can obviously reference other cases from the same court. Furthermore, it has a property that counts the number of references (i.e. number of unique cases) made from the source court to the target court. Similarly, relationship (6), describes and records the number of indirect citations (i.e. number of unique cases) from a court to a legislation. Lastly, relationship type (7) connects each case and its corresponding court.

Graph Summary. Table 4 provides an overview of the properties of the Neo4j graph. As the relationship numbers suggest, not all extracted case citations from Sect. 3.2 were added as relationships. This is due to two factors. First, as explained in Sect. 3.3, multiple citations from one case to the same target are interpreted as only one reference. Secondly, references to targets that are not in the database are excluded. This way, we ensure that the data in the database is consistent and does not contain "empty" nodes (i.e. nodes without property values). For example, we identified but omitted references to cases from outside New Zealand (e.g. the UK or Australia). In total, the database contains 236,458

$Ref_{Case \to Case}$ relationships, which corresponds to 55.29% of the extracted case-to-case citations.

Table 4. Neo4j graph statistics.

Label	Node/Relationship	Number
Case	Node	338,360
Legislation	Node	10,402
Court	Node	124
$REF_{case \to case}$	Relationship	236,458
$REF_{case \to legislation}$	Relationship	244,159
$REF_{legislation \to case}$	Relationship	50
$REF_{legislation \to legislation}$	Relationship	67,857
$REF_{court \to court}$	Relationship	761
$REF_{court \to legislation}$	Relationship	7,387
$RULING$	Relationship	338,360
Total	Node	**348.886**
Total	Relationship	**895,032**

4   Network Analysis

With the completion of the citation network[4], we can now utilise network science algorithms, metrics and visualisations to analyse its properties. The goal is to demonstrate how these tools can be leveraged to support common legal research tasks. Furthermore, we can compare our network's structure to similar studies from other countries and legal systems and make assumptions about the New Zealand legal system as a whole. However, for the most part, the results from this study are presented as a quantitative report as we will leave the in-depth legal interpretation for another study.

4.1   In-Degree Distribution

The distribution of all links across all nodes can give an insight into a network's behaviour and structure. In a random network, each node would have an equal chance of being linked to a new incoming node. Scale-free networks, on the other hand, contain "hubs" that have a higher chance of receiving new links. This property of preferential attachment [7] often leads to a "rich-get-richer"-like behaviour. Some examples are literature citation networks, actors starring in movies, the internet and cellular metabolism [8].

---

[4] In this section we consider the Neo4j graph as a network and adapt the appropriate terminology.

In case citation networks this principle would result in the effect that cases that are already heavily cited have a higher chance of getting cited again. Previous studies have found signs of this behaviour in the case citation networks of the Austrian Supreme Court [14], the U.S. Supreme Court [26] and German courts [21].

Figure 3 illustrates the in-degree distribution of case nodes in our network. As evident by this power-law degree distribution, most cases receive few to no incoming links. In fact, 95.32% of cases receive less than 4 citations, while the rest of cases (4.68%) receive 80.66% of all citations.

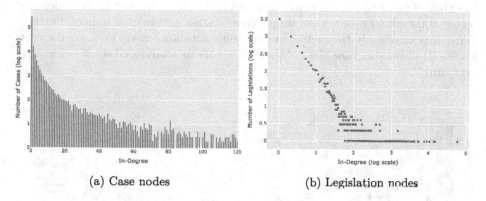

(a) Case nodes   (b) Legislation nodes

Fig. 3. In-degree distribution of the case and legislation nodes (tail-end of the horizontal axis is cut off for better visibility). In-degree value of a node is the number of incoming links the node receives (i.e. the number of references from other court cases or legislations). The course of the graphs resembles a characteristic power law distribution, signifying a scale-free network structure.

This supports the assumption that the New Zealand case citation network is also exhibiting a scale-free network structure, akin to the networks of the related works mentioned above. The in-degree distribution of the legislation network shows similar results with the majority of nodes (70.66%) receiving three or fewer incoming links. These results could indicate that there are few substantially influential agents (legislations and cases) that predominantly shape the landscape of New Zealand law.

4.2   In-Degree Centrality

Network science offers an array of centrality measures to quantify the influence and determine the role of important nodes in the network. In-degree centrality is the most elementary of these measures as it only reflects the number of links a node receives. It can reveal popular nodes that are likely to contain a lot of information, however, further interpretation can be limited.

For example, when looking for an appropriate precedent case as part of your legal research, a high in-degree centrality value can point to an influential case that is cited frequently and therefore very likely relevant to your case. On the other hand, it can merely highlight already well-known cases that established common procedural principles. That is to say, some references are made out of routine or for procedural purposes and do not always indicate a strong similarity or importance to your actual case. However, leveraging the fact that our network also contains legislations, we can provide additional functionality and improve the information gain of in-degree centrality.

For instance, Fig. 4a ranks the most cited cases with the condition that these also cite the "Accident Compensation Act". As cited legislations are often a much more reliable indicator of the underlying legal topic of a case (as compared to the area of the responsible court) our network allows for much more detailed search and ranking results than simple case citation networks or indices. Furthermore, we can reapply the metric to only count citations made in a specific year. Figure 4b illustrates how the network allows us to analyse citation behaviours over time.

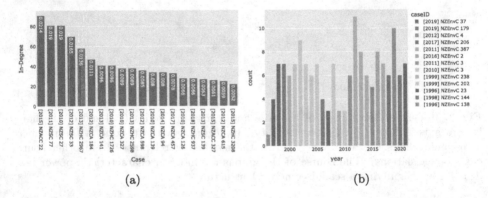

(a)   (b)

Fig. 4. Figure (a) shows a bar chart of the 20 most cited court cases that cite the Accident Compensation Act 2011. This is an example, of a targeted search for prevalent cases of a specific legal domain (accident compensation). Figure (b) illustrates the most cited environmental court decisions by year, showing the development of important court cases in this domain over time.

4.3 PageRank Centrality

As opposed to in-degree centrality, PageRank does not only consider the number of direct connections of a node but also the "quality" or importance of the nodes it is connected to. In citation networks, this has the effect that few citations from important nodes can be weighted more than many citations from less influential nodes. This way PageRank can help to identify less obvious influencers of information flow within the network and discover relevant nodes that might otherwise be overlooked due to a lower in-degree value.

Although the merit of leveraging PageRank in case citation networks has already been highlighted in previous works [27], we can take advantage of our legal citation network to increase its effectiveness. Unfiltered, we can use PageRank to highlight not only instrumental legal cases but also legislations and courts that significantly impacted the larger legal system. Furthermore, we can apply PageRank on a subset of cases filtered by their court or area of law (i.e. cases that cite specific legislations). This can provide a new avenue to discover very specific cases that had a smaller overall impact, but significant influence on their domain. As an example Fig. 5a shows the most important decisions of the family court, while Fig. 5b shows some of the currently most influential legislations.

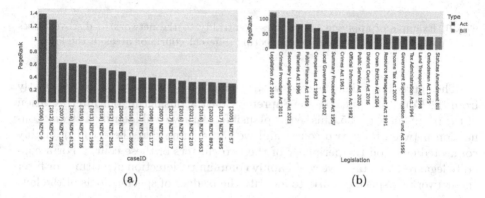

(a)   (b)

Fig. 5. Figure (a) shows a bar chart of family court decisions ranked by their PageRank value. This is another example, of a targeted search and ranking of cases. Figure (b) illustrates the most important legislations within the legislation network (i.e. we did not consider any of the links from the cases or courts). An analysis of the PageRank value over time could indicate the impact of new laws and how long it takes for these to affect the larger legal system,

4.4 Betweenness Centrality

Betweenness centrality is another example of how the legal citation network enables us to utilise network science metrics to discover influential cases that we cannot find with citation indices alone. Betweenness centrality compares the number of shortest paths that lead through each node. This has the potential to reveal bridges, which are nodes that connect otherwise less connected (or disconnected) clusters. In our network, these nodes could represent interdisciplinary court cases or legislations that link mostly unrelated domains of the law. Figure 6 shows the nodes with the highest betweenness value in our network.

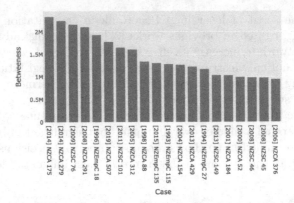

Fig. 6. Example of applying betweenness centrality to the network to discover cases that influence the legal system by connecting different courts or areas of the law.

The figure shows that the cases with the highest betweenness value are mostly from the Court of Appeal or the Supreme Court. Considering our interpretation of the results in Sect. 4.5, this seems plausible. Here we will identify a strong connection between these two courts and we can assume that there will be highly connected cases on the periphery of the two clusters of these courts. For a concrete legal research task, we could apply community detection algorithms or filter the network by specific courts to identify the bridges of specific areas of the law.

4.5   Court Citation Behaviour

Creating the network not only enables us to utilise network algorithms but we can use visualisation tools to investigate or confirm common citation patterns within the legal system. As an example, Fig. 7 illustrates the indirect citation behaviour between all courts. The width of a link represents the number of indirect references between the courts.

This visualisation indicates a high self-citation behaviour of the High Court (NZHC - centre node) and a strong connection to the Supreme Court (NZSC) and Court of Appeal (NZCA). This corroborates the expectations we had for the citation behaviour of the three highest courts in New Zealand. As is typical in common law countries (like New Zealand), the lower courts follow the decisions of the higher courts and cite them accordingly.

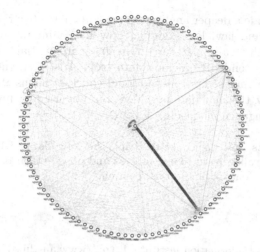

Fig. 7. This is an illustration of the court citation network within our legal citation network. The nodes represent the courts in our database abbreviated by their shortcode (e.g. NZHC - New Zealand High Court). The width of the links between the courts represents the number of references (or unique court decisions) made from the source court to the target court.

5 Conclusion and Future Research

In this study, we introduced and described the development and analysis of the first "legal citation network" for New Zealand. The primary goal of this study is to provide the network to the research community and demonstrate how it can be utilised to facilitate further research into legal tech-related information retrieval, NLP and data management technologies.

By scraping, parsing and converting legal documents from multiple sources, we managed to create the largest openly available legal data repository for New Zealand with over 300,000 court cases, 10,000 legislations and 115 courts. Using regular expressions, we detected and extracted references between the documents and converted them into links. The result is an interconnected citation network that we provide in the form of a Neo4j graph.

In the second part of this study, we showcased how the network can reveal new information about the New Zealand legal system and how it can support common legal research tasks by utilising network analysis algorithms and metrics. For instance, we illustrate how PageRank, in-degree and betweenness centrality can be used to identify relevant and influential cases, courts or legislations. Additionally, we discovered that the network shows signs of scale-free behaviour, similar to the case citation networks of other countries like the U.S., Austria and Germany [14,21,26]. These examples shall provide an understanding of the capabilities of the network and stir interest and motivation for researchers and law practitioners to leverage this data for future research.

This could include a deeper analysis and qualitative interpretation of specific areas of New Zealand law. For example, how has immigration law developed over time and which legal apparatuses were most influential? Additionally, the network could be extended to include data from Australia, the UK and other related countries to analyse how international court decisions affect the national legal system. Lastly, the legal data provided could be used for fine-tuning and/or training of legal language models like legal-BERT [12].

Acknowledgements. We would like to thank the New Zealand Legal Information Institute (NZLII) for allowing us to use, process and provide their data and would like to acknowledge their contribution to this research.

References

1. Australasian legal information institute. http://www.austlii.edu.au/. Accessed 20 Oct 2023
2. British and Irish legal information institute. https://www.bailii.org/. Accessed 20 Oct 2023
3. Free law project's courtlistener. https://www.courtlistener.com/. Accessed 20 Oct 2023
4. New Zealand legal information institute. http://www.nzlii.org/. Accessed 20 Oct 2023
5. New Zealand legislation website by New Zealand parliamentary counsel office/te tari tohutohu pāremata. https://www.legislation.govt.nz/. Accessed 20 Oct 2023
6. RELX group LexisNexis. https://www.lexisnexis.co.nz. Accessed 20 Oct 2023
7. Barabási, A.L.: The new science of networks. J. Artif. Soc. Soc. Simul. **6** (2003). https://doi.org/10.2307/20033300
8. Barabási, A.L., Bonabeau, E.: Scale-free networks. Sci. Am. **288**(5), 60–69 (2003). http://www.jstor.org/stable/26060284
9. Coppard, A., McLay, G., Murray, C., Orpin, J.: New Zealand Law Style Guide. 3rd edn. Thomson Reuters, Wellington (2018)
10. Corinna Coupette: Juristische Netzwerkforschung. Mohr Siebeck (2019). https://doi.org/10.1628/978-3-16-157012-4
11. Cross, F., Spriggs, J., Johnson, T., Wahlbeck, P.: Citations in the us supreme court: an empirical study of their use and significance. U. Ill. L. Rev., 489–575 (2010)
12. Elwany, E., Moore, D., Oberoi, G.: BERT goes to law school: quantifying the competitive advantage of access to large legal corpora in contract understanding. CoRR (2019). http://arxiv.org/abs/1911.00473
13. Fowler, J., Johnson, T., Spriggs, J., Jeon, S., Wahlbeck, P.: Network analysis and the law: measuring the legal importance of precedents at the U.S. supreme court. Polit. Anal. **15** (2007). https://doi.org/10.1093/pan/mpm011
14. Geist, A.: Using citation analysis techniques for computer-assisted legal research in continental jurisdictions. SSRN Electron. J. (2009). https://doi.org/10.2139/ssrn.1397674
15. Gelter, M., Siems, M.: Networks, dialogue or one-way traffic? An empirical analysis of cross-citations between ten of Europe's highest courts. Utrecht Law Rev. **8**(2) (2012). https://doi.org/10.18352/ulr.196

16. Katz, D.M., Coupette, C., Beckedorf, J., Hartung, D.: Complex societies and the growth of the law. Sci. Rep. **10**(1) (2020). https://doi.org/10.1038/s41598-020-73623-x
17. Koniaris, M., Anagnostopoulos, I., Vassiliou, Y.: Network analysis in the legal domain: a complex model for European union legal sources. J. Complex Netw. **6**, 243–268 (2018). https://doi.org/10.1093/comnet/cnx029
18. Lupu, Y., Voeten, E.: Precedent in international courts: a network analysis of case citations by the European court of human rights. Br. J. Polit. Sci. **42**, 413–439 (2012). https://doi.org/10.1017/S0007123411000433
19. de Maat, E., Winkels, R., van Engers, T.: Automated detection of reference structures in law, pp. 41–50. IOS Press (2006)
20. Mania, K.: Legal technology: assessment of the legal tech industry's potential. J. Knowl. Econ. (2022). https://doi.org/10.1007/s13132-022-00924-z
21. Milz., T., Granitzer., M., Mitrović., J.: Analysis of a German legal citation network. In: Proceedings of the 13th International Joint Conference on Knowledge Discovery, Knowledge Engineering and Knowledge Management - KDIR, pp. 147–154. INSTICC, SciTePress (2021). https://doi.org/10.5220/0010650800003064
22. Neale, T.: Citation analysis of Canadian case law. J. Open Access Law **1**(1), 1–51 (2013)
23. Ostendorff, M., Blume, T., Ostendorff, S.: Towards an open platform for legal information. In: Proceedings of the ACM/IEEE Joint Conference on Digital Libraries in 2020 (2020). https://doi.org/10.1145/3383583.3398616
24. Rönneburg, L., Fakultät, M.: Analyse eines juristischen Entscheidungscorpus mit Methoden der Netzwerkforschung und Sprachtechnologie. Ph.D. thesis (2021)
25. Sakhaee, N., Wilson, M.C., Zakeri, G.: New Zealand legislation network. In: Legal Knowledge and Information Systems: JURIX, vol. 294, p. 199 (2016)
26. Smith, T.A.: The web of law. SSRN Electron. J. (2005). https://doi.org/10.2139/ssrn.642863
27. Malmgren, S.: Towards a theory of jurisprudential relevance ranking Using link analysis on EU case law. Ph.D. thesis, Stockholm University (2011)

Enhancing Resource Allocation in IT Projects: The Potentials of Deep Learning-Based Recommendation Systems and Data-Driven Approaches

Li Xiao[1]([✉]), Samaneh Madanian[1]([✉]), Weihua Li[1]([✉]), and Yuchun Xiao[2]([✉])

[1] Auckland University of Technology, Auckland, New Zealand
{li.xiao,sam.madanian,weihua.li}@aut.ac.nz
[2] Zhejiang Gongshang University, Hangzhou, China
xyc@mail.zjgsu.edu.cn

Abstract. The dynamic landscape of Information Technology Project Management (ITPM) along with a recent emphasis on the concept of suitability, motivates organisations to better utilise their resources and improve current practices. This has been leading to explore the potential of deep learning-based Recommendation Systems (RecSys) in this domain. We focus on critical aspects of Agile Project Management, resource allocation, and performance monitoring. Our study also evaluates RecSys' effectiveness in enhancing project allocation and overall project success rates in ITPM. Analysing a diverse range of data, we observe a positive correlation between employee experience, skills, and project allocation ratings. The findings suggest that these variables exert a greater influence than traditional factors like age or educational background. We also demonstrate the benefits of leveraging historical performance data for future project planning and the utility of tracking ratings during project development. This paper contributes valuable insights for practitioners in IT project management, offering data-driven strategies to improve resource allocation and performance monitoring.

Keywords: IT Project Management · Resource allocation · Recommendation System

1 Introduction

The domain of Information Technology Project Management (ITPM) has attracted the great attention of both researchers and practitioners [9,12], mainly due to the dynamic transformations within the technology and software industry. This dynamic transformation also increases the IT projects' numbers and diversity. In the majority of instances, organizations encounter limitations that prevent the simultaneous execution of all envisioned projects. One of the major limitations primarily arises from the scarcity of resources. Consequently, organizations find themselves compelled to adopt the practice of assigning priority

© The Author(s), under exclusive license to Springer Nature Singapore Pte Ltd. 2024
D. Benavides-Prado et al. (Eds.): AusDM 2023, CCIS 1943, pp. 226–238, 2024.
https://doi.org/10.1007/978-981-99-8696-5_16

rankings to individual projects. This process, facilitated by diverse prioritisation methodologies, can ultimately lead to the de-prioritisation or virtual exclusion of projects deemed to possess lower precedence [5].

Therefore, in this research, we aim to pivot to two key areas. Firstly, we seek to explore ways to enhance ITPM and overall software development processes, and overall resource allocation. Secondly, we delved into the potential utility of deep learning-based recommendation systems (RecSys) within the purview of ITPM. The primary objective of these investigations is to address and resolve significant challenges prevalent in resource allocation and other pertinent sectors, with a view to augmenting the efficiency and efficacy of practices employed in IT project management.

Confronting the challenges within ITPM often necessitates examining the Agile Project Management (APM) paradigm, standard in contemporary software development since its inception in the 2000 s [9]. Initially gaining traction within small teams, APM broadened its scope to encompass larger organisations, thanks to scaled methodologies like the Scaled Agile Framework (SAFe) and Disciplined Agile Delivery (DAD). Further, the concept of IT portfolio management was introduced to harmonise with overarching business strategies to continuously define and deliver the right projects aligned with organisational business strategy in dynamic and uncertain environments [10]. Despite these strategic advancements, ITPM has frequently failed to consistently deliver the anticipated results, as evidenced by documented failure rates surpassing success rates [8].

On the other side, numerous successful software projects have embraced technology-centric methodologies, with RecSys acting as an important role. A few typical examples include Google App Store [3], Amazon, and Twitter, all benefiting from the implementation of RecSys. In contemporary software landscapes, RecSys has exhibited exceptional prowess in processing voluminous data sets, enabling efficient data filtration and seamless delivery of pertinent information. A comparison between project management data and the data generated on platforms such as Twitter [15] reveals that the former typically lags in both quantity and quality. Despite this, numerous research works are underway exploring the potential integration of RecSys to augment ITPM [8,14,16]. The integration of RecSys into ITPM holds promise for improving project allocation, resource management, and overall project success rates.

Additionally, integrating ITPM into resource allocation and performance monitoring, utilising performance ratings, remains a traditional yet crucial endeavour. The current challenges in this specific area are mostly related to subjectivity and personal preferences of decision-makers during the process that add to uncertainty around project prioritisation and resource allocations [13].

The incorporation of statistical elements into ITPM tools could yield significant benefits. In particular, detailed analysis of data distribution could unlock valuable insights, surpassing the limitations of specialised data. This research explores the potential advantages of applying statistical methods within ITPM tools, emphasising the impact of data distribution on performance ratings. Moreover, the investigation seeks to analyse the RecSys outcomes, thereby highlighting its relevance and significance within the sphere of ITPM. Through our investigations, we tend to enrich the knowledge domain of ITPM, thereby enhancing the efficacy of resource allocation and performance monitoring practices.

In this paper, we encountered scenarios within ITPM spanning diverse categories, including planning, resource allocation, monitoring, and risk management. We conducted an extensive study to ascertain the value of RecSys and IT project allocation data, integrating contemporary and classic scholarly articles. Additionally, we leveraged real-world data on ITPM allocation to assess its effectiveness via rating scores. In summary, the key findings and contributions are summarised as follows.

- We observed a positive correlation between employees' experience, skills, and corresponding ratings, as determined by ablation experiments. This suggests that employees possessing extensive experience and relevant skills tend to receive higher ratings in project allocation, challenging the deep-seated belief that factors such as age and educational background are the primary determinants.
- We analysed the correlation between language proficiency, name, and other factors with ratings. The findings underscore that these variables may not exert as significant an influence on project allocation ratings as experience and skills do.
- We explored how enhancements in resource allocation can be predicated on ratings during the planning phase. Understanding the correlation between ratings and employee allocation decisions can pave the way for more efficient resource distribution, bolstering project success rates.
- We investigated the practicality of tracking ratings using tools during the development phase. This approach offers a valuable mechanism for monitoring project progress and employee performance, facilitating timely adjustments and interventions.
- We delved into the potential advantages of leveraging historical performance data for future project development. By capitalising on past performance data, IT project managers can make more informed decisions, refining resource allocation strategies.

2   Related Works

Upon reviewing the existing relevant studies, we organised these studies into three key sections, i.e., prevailing challenges in ITPM, the importance of modern RecSys, and recent research exploring the integration of RecSys in ITPM.

Pacagnella and Da present a comprehensive literature review spanning two decades that delves into APM [9]. In their review, they reflected on the critical appraisal of APM, underscoring the necessity for tailoring policies, practices, and processes to suit the unique requirements of individual projects. Shastri et al. examine project managers' challenges in Agile teams [12]. Despite the Agile approach promoting roles such as product owner, scrum master, and coach, project managers continue to play a vital role in Agile projects.

Also to increase the ITPM visibility within organisations, portfolio management has been introduced as an organised and central approach for project selection, resource allocation, and actions to fund and sustain all IT projects in an organisation [11]. However, the concept itself has not resolved the issues of optimising resource allocation while having scheduling limitations. More specifically in IT organisations, there should be enough resources to deliver all of the proposed projects within the time allocated for the portfolio. Capacity planning achieves this balance, making sure resources are used as effectively and efficiently as possible, sequencing work that impacts the various work areas [2]. Different analytical and AI approaches have been introduced for these capacity planning and resource allocations [6]. Nevertheless, the focus of this research is on RecSys.

From a RecSys perspective, Cheng et al. introduce the concept of Wide and Deep learning, which amalgamates memorization and generalization benefits for recommender systems [3]. On the other hand, the deep component utilises deep neural networks to learn low-dimensional dense embeddings for sparse features, facilitating better generalisation to unseen feature combinations. Meanwhile, He et al. examine the use of deep neural networks for collaborative filtering in recommendation systems [7]. Both research works showcase the potential of deep learning techniques in enhancing RecSys' effectiveness.

Meanwhile, there have been several notable explorations of RecSys in the ITPM field. For example, Achrak and Chkouri recommend best practices, project manager behaviours, and organisational policies for Agile Methods [1], while Wei and Capretz discuss the use of open-source tools in Recommendation Systems in Software Engineering [16]. Sousa et al. propose a recommender system that suggests risks and response plans for a target project [14]. La and López outline the groundwork for developing an adaptive method for tool selection based on capturing information about project features and organisational capabilities [8].

Furthermore, Chiang and Lin argue that human resource allocation is critical for successful project outcomes in software development [4]. They propose a comprehensive framework to assist software companies in evaluating their resources and determining the feasibility of project estimations. This perspective aligns with exploring factors like experience, skills, and language proficiency.

Researches have revealed limitations and areas for improvement of RecSys, particularly in the use of the wide and deep model. Current trends in RecSys favor the adoption of Graph Neural Networks (GNNs) [15]. Zhang et al. proposed a motif-based graph attention network for web service recommendation (MGSR) that addresses the over-smoothing issue by incorporating network motifs in layer propagation [17]. GNNs offer the ability to learn more complex behaviors and diverse data connections, even mitigating the cold start issue. In our ITPM scenario, employee data may share more features, which can be better utilized without manually converting them into multi-hot and other features for the wide and deep. As GNNs show promise in RecSys, these challenges and incorporating network motifs and multi-source data connections can lead to more accurate and personalized recommendations, even for users or items with limited historical data. However, it is essential to acknowledge the potential limitations of GNNs and conduct rigorous experiments and comparative studies to understand their strengths and weaknesses in different scenarios. Exploring alternative approaches or hybrid models that combine GNNs with traditional models could also be fruitful in optimizing recommendation performance in ITPM tool applications.

3   Methodology

In this section, we elaborate on the methodology employed in this research work, which consists of three core components: data preparation, model architecture, and ablation study. Specifically, during the data preparation phase, we describe the features in the "IT Employee Data for Project Allocation" dataset, including employee attributes and skills. The section on model architecture introduces the wide and deep model, which combines categorical and numerical feature handling for precise project allocation. We systematically evaluate feature importance in the ablation study to gain insights into the model's rating predictions.

Data Preparation. The dataset utilised for this study, referred to as "IT Employee Data for Project Allocation," encompasses records for 1,000 employees dispersed across multiple CSV files. The "Employees" dataset provides comprehensive employee data, while the "Employee Skills" dataset augments the information suite for efficient project allocation management. This phase aims to tailor this data for input into our model within the RecSys for ITPM context. Initial preprocessing involves transforming the 1,000 employee data samples using LabelEncoder and StandardScaler functions from the sklearn library. This process efficiently translates categorical data into numerical features, which are partitioned into training and test datasets at a ratio of 7:3. Upon extraction from the CSV files, the dataset comprises 17 columns, encapsulating key attributes such as Employee ID, Employee Name, and Experience, among others. In addition, the "Employee Skills" CSV file contributes an additional 21 columns, capturing proficiencies such as Python, Machine Learning, Deep Learning, etc. Due to the high-dimensional feature space and the limited data points (1000), we employ a multi-hot manual feature representation for specific attributes like languages, interests, project counts, and skills, post their numerical transformation (refer to Table 1). This strategy facilitates encoding specific attributes' presence or absence, thereby preserving vital information encapsulated in the data. Following the consolidation and merging of columns, the dataset is segregated into training and test subsets, thereby ensuring it is adequately prepped for input into the model.

Table 1. Allocation Data Features

Feature	Type	Original Columns
Rating	Float	Rating
ID	integer	Eid
Experience	Integer	Experience
First_name	String	Ename
Surname	String	Ename
Total_projects	Integer	*Total_projects*
Languages	Strings to Multi-hot	Language1, Language2, Language3
Interests	Strings to Multi-hot	*Area_of_Interest_1,*
		Area_of_Interest_2,
		Area_of_Interest_3,
Projects_count	Multi-hot	*AI_project_count, ML_project_count,*
		JS_project_count, Java_project_count
		DotNet_project_count, Mobile_project_count
Skills	Multi-hot	Python, Machine Learning,
		Deep Learning, Data Analysis
		Asp.Net, Ado.Net, VB.Net
		C#, Java, Spring Boot,
		Hibernate, NLP, CV,
		JS, React, Node, Angular,
		Dart, Flutter, Vb.Net

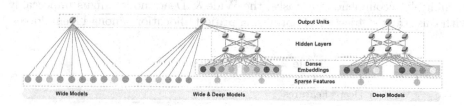

Fig. 1. Wide and deep architecture [3]

Model Architecture. The implementation of the RecSys within this study necessitates the use of the Wide & Deep model [3] for allocating employee data. The architecture of the model is demonstrated in Fig. 1. This model's effective training requires converting the training data frame into tensors, with the constitutive "wide" and "deep" elements depicted in Table 2.

The Wide & Deep model presents a hybrid architecture comprising distinct "wide" and "deep" components. The "wide" component, proficient in managing categorical features (data with numerous discrete values), oversees attributes like $Area_o f_I nterest$ and Languages. By leveraging feature crosses and transformations, this component discerns interactions between diverse categorical features, thereby learning high-order correlations and patterns. Conversely, optimised for manipulating numerical features (continuous-valued data), the "deep" component handles parameters like $Total_p rojects$, Experience, and Rating. The deep component, constituted by numerous neural layers, is adept at decoding intricate and nonlinear representations of the numerical input, thereby uncovering complex relational dependencies within the data.

The wide and deep model combines the strengths of a linear model (wide part) and a deep neural network (deep part). The wide component is a generalised linear model and the deep component is a feed-forward neural network. Each hidden layer is formulated in Eq. 1.

$$a^{(l+1)} = f\left(W^{(l)}a^{(l)} + b^{(l)}\right),$$ (1)

where l denotes the layer number and f refers to the activation function ReLUs.

$$P(Y) = \sigma\left(w_{\text{wide}}^T \cdot x + w_{\text{deep}}^T \cdot (\mathbf{a}^{(lf)}) + b\right),$$ (2)

where $P(Y)$ is the prediction rating score, w_{wide} is the vector of all wide model weights and w_{deep} are the weights applied on the final activations $a(lf)$.

In the final stage, these components converge to generate predictions. This amalgamation leverages the unique strengths of each component. The "wide" component excels at learning from sparse, broad signals, while the "deep" component focuses on deciphering complex patterns in dense, localised features. By integrating wide and deep architectures, the model addresses the heterogeneity of employee data obtaining relevant information from various features to generate precise, personalised project allocation recommendations. With proven efficacy in multiple recommendation tasks, the Wide & Deep model aligns impeccably with this study's objective of optimising project allocation among IT employees.

Table 2. Wide and Deep inputs

Wide or Deep	Features
Wide	Eid, Interests, Languages, Projects_count, Skills
Deep	First_name, Surname, Total_projects, Experience

To set up the experiment, we adopt the Adam optimiser with an initial learning rate of 0.005 and a weight decay parameter of 0.0001. We partitioned the dataset into training and test subsets, with 7000 instances allocated for training and 3000 for testing. Furthermore, we selected a batch size of 100 to balance computational efficiency and gradient estimation accuracy.

Ablation Study. The third component of our research methodology involves conducting an ablation study, predicated on the model's training and evaluation using the Mean Squared Error (MSE) loss function. The primary objective here is not to develop novel algorithms or models but to investigate the connections within the ITPM allocation data. More specifically, ablation methods are employed to investigate the relationship between the ratings and various aspects of the employee data. This strategy systematically removes each dimension (i.e., feature or variable) to observe the relationships between the rating and particular employee data attributes. Ablation experiments contribute to identifying the significance of individual features on the model's ability to predict ratings. We can determine the relative importance of specific employee data features on rating prediction by eliminating different features and comparing performance alterations. Such an analysis is indispensable for discerning the components of employee data exerting the most substantial influence on ratings and hence, critical for accurate rating predictions. This experiment is carried out over 300 epochs.

Upon completion of the ablation study, the final results must be evaluated from a statistical point of view. Our approach involves analysing the distribution of the data and leading discussions based on the ratings. The Shapiro-Wilk test and histograms will be used for this analysis. Given data limitations, complete knowledge of the ratings' quality or potential to increase data quantity might not be available. Nonetheless, by examining the data distribution and leveraging prior experience, we can employ reduction or summarization techniques to construct a robust ITPM tool. This approach will involve assessing the statistical properties of the rating data, which will reveal insights about the central tendency, spread, and shape of the data. Even without full information on the ratings' quality, prior experiences and domain expertise can inform the ITPM tool's development.

Potential data reduction techniques or summary methods might be necessary to compensate for the limited data quantity. Data reduction could involve aggregating ratings at different levels (e.g., team or department level), thereby providing a wider perspective. Summary statistics, such as mean, median, or percentiles, can offer valuable information without necessitating a vast dataset.

Using statistical analysis, prior experience, and data reduction strategies, we aim to construct a reliable and effective ITPM tool. Despite the constraints imposed by limited data availability, this tool will assist in resource allocation, performance monitoring, and decision-making within the organisation. Regular updates and stakeholder feedback will be essential to the tool's refinement and improvement.

4   Experimental Results and Discussion

The ablation tests for each feature are presented in a graphical representation of results, as depicted in Fig. 2. As can be seen from the figures, all the MSE losses quickly fit the data within the first 2 to 100 epochs. Notably, the variance in each ablation test can be attributed to the standard scaling of features, Adam optimiser properties, and log scaling in loss score. For example, the absence of "experience" saw the loss plummet to nearly 1 within the first ten epochs, yet this elevated MSE value persisted up to 300 epochs. This observation suggests that in this dataset, "experience" turns out to be a critical feature; its absence results in maintaining a high loss value. In contrast, the "skills" feature witnessed a slower drop, reaching a similar loss level after 200 epochs, possibly hinting at some dependency across its dimensions. However, our experimental results conclusively reveal that "experience" and "skills" significantly influence the ratings within the ITPM context.

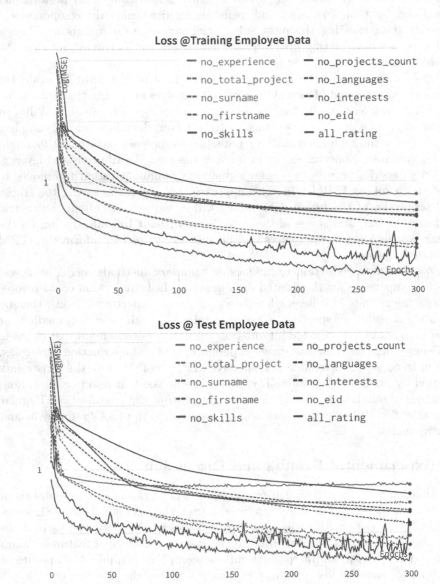

Fig. 2. Ablation experiments results.

Other features such as "speaking languages", "ID", and "name" (including first and last name) seemed to have minimal impact on the loss curve. Unexpectedly, despite the perceived importance of communication, the influence of speaking languages on ratings in software development appears insignificant. Other dimensions like project counts and interests also demonstrated minimal influence on the loss results. Based on these observations, when building a rating tool using RecSys, "ID" and "name" could be eliminated first.

A critical aspect of our research is to examine the original rating data, given that RecSys is optimally fitted to the rating data. In our analysis, we found that the mean rating is 1.29, the median rating is 1.17, and the 25th and 75th percentile ratings are 0.87 and 0.88, respectively. The raw ratings are a reflection of an organisation's standards and culture. Therefore, discerning what constitutes quality ratings and their measure is crucial. An ideal rating system must avoid excessively high ratings that could compromise the rating system and potentially lower the rating standards. A normal distribution could indicate quality results, with most people performing averagely in a fair and inclusive environment and only a few obtaining very high or low ratings. We analysed the existing data by plotting the ratings and performing a Shapiro-Wilk test to ascertain this.

The Shapiro-Wilk test yielded a statistic value of 0.88 and a p-value of 4.47e-27. Although the statistic value being close to 1 suggests that the data is relatively normally distributed, the p-value less than 0.05 indicates a deviation from a normal distribution. On examining the histogram (refer to Fig. 3), outliers around rating values 1 and 2 are noticeable. Despite these outliers, the overall distribution appears right-skewed, leaning towards a normal distribution. The ratings are generally lower than in a perfect normal distribution, with a higher concentration of scores around 1 and fewer scores above 3. In summary, while the data deviates from a perfect normal distribution, its shape is relatively close to normal. This finding is crucial for understanding the distribution of ratings and evaluating the impact of potential outliers on the dataset.

Based on the ablation test and distribution histogram, we infer that IT project allocation can significantly benefit from using RecSys with appropriate data. The tool's relevance and promise for optimising IT project allocation are evident from its widespread use in various industries like advertising, online shopping, and video platforms. However, the growing reliance on RecSys has given rise to challenges, particularly fraud, which can undermine the system's integrity and trustworthiness.

RecSys in advertising can have a significant impact, but it also comes with the risk of fraud. For example, in the context of ITPM, manipulations of key dimensions like "skills" can lead to inaccurate project allocations and unfair advantages. Robust security measures and careful validation of data inputs in RecSys are thus essential to prevent fraudulent activities and ensure the system's effectiveness.

Potential solutions to achieve a normal distribution of rating results and mitigate fraud include implementing objective rules and enhancing data diversity. Diversity in data dimensions is vital, as ITPM involves complex manage-

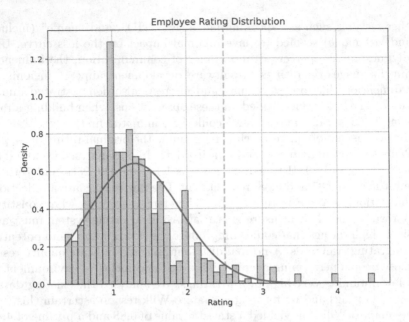

Fig. 3. Employee rating distribution

ment tasks that cannot be fully automated. Hence, future tools should consider diverse factors like "age", "educational background", "previous working experience", "market influence", etc. This approach not only increases the cost of fraud but also aids in detecting fraud patterns, thus enhancing the system's resilience against fraudulent activities.

5   Conclusion and Future Work

In conclusion, our investigation into RecSys application within the ITPM domain encompassed comprehensive literature reviews and empirical experiments utilising public datasets. Our literature review unearthed prevalent challenges in ITPM, especially within the context of APM, and the state-of-the-art research on RecSys within ITPM.

Through our empirical experimentation, which involved ablation tests on IT project allocation data, we underscored the crucial role of features such as employee experience and skills. We also noted the minimal influence of dimensions such as ID and name. It is imperative to acknowledge the limitations of our research, primarily arising from constrained data dimensions, limited data size, and the inherent variations in enterprise culture. These constraints imply that the universality of our findings might be restricted. Nonetheless, our research significantly contributes by setting a benchmark for understanding the interplay between ITPM dimensions. Additionally, it validates a process in resource allocation within the ITPM landscape, paving the way for future studies to extend and refine this body of knowledge.

Moving forward, several intriguing aspects remain to be explored. Firstly, conducting a more extensive investigation into other influential features not considered in this study, such as an employee's education, market influence, and background in different project roles, would be valuable. This could provide a more holistic view of the factors influencing project allocation. Secondly, introducing additional methods to handle fraud in RecSys, particularly within the ITPM domain, would be a beneficial future endeavour. This would aid in maintaining the integrity of the recommendation system and ensuring fair resource allocation. Furthermore, expanding the dataset with more diverse and extensive data from different organisations could refine the model and lead to a more generalised solution applicable across a wider range of contexts. Lastly, as machine learning and artificial intelligence evolve rapidly, integrating advanced models into RecSys for ITPM could provide more sophisticated and accurate recommendations, enhancing overall project success.

References

1. Achrak, E.M., Chkouri, M.Y.: Integrate and apply the recommendation system of agile methods. In: Ezziyyani, M. (ed.) AI2SD 2019. AISC, vol. 1105, pp. 287–297. Springer, Cham (2020). https://doi.org/10.1007/978-3-030-36674-2_30
2. Blichfeldt, B.S., Eskerod, P.: Project portfolio management - there's more to it than what management enacts. Int. J. Project Manage. **26**(4), 357–365 (2008). https://doi.org/10.1016/j.ijproman.2007.06.004
3. Cheng, H.T., et al.: Wide & deep learning for recommender systems. In: Proceedings of the 1st Workshop on Deep Learning for Recommender Systems, pp. 7–10 (2016)
4. Chiang, H.Y., Lin, B.M.: A decision model for human resource allocation in project management of software development. IEEE Access **8**, 38073–38081 (2020)
5. Dutta, R., Burgess, T.: Prioritising information systems projects in higher education. Campus-wide Inf. Syst. **20**(4), 152–158 (2003)
6. Ha, H., Madanian, S.: The potential of artificial intelligence in it project portfolio selection. In: International Research Workshop on IT Project Management. Asocciation of Information Systems (2020). https://aisel.aisnet.org/irwitpm2020/10
7. He, X., et al.: Neural collaborative filtering. In: Proceedings of the 26th International Conference on World Wide Web, pp. 173–182 (2017)
8. La Paz, A.I., López, R.I.: Recommendation method for customized it project management. Procedia Comput. Sci. **219**, 1938–1945 (2023)
9. Pacagnella Junior, A.C., Da Silva, V.R.: 20 years of the agile manifesto: a literature review on agile project management. Manage. Prod. Eng. Rev. 14 (2023)
10. Petit, Y.: Project portfolios in dynamic environments: organizing for uncertainty. Int. J. Project Manage. **30**(5), 539–553 (2012). https://doi.org/10.1016/j.ijproman.2011.11.007, special Issue on Project Portfolio Management
11. Rajagopal, S., McGuin, P., Waller, J.: Project portfolio management: leading the corporate vision. Palgrave Macmillan (2007). https://ezproxy.aut.ac.nz/login?url=https://search.ebscohost.com/login.aspx?direct=true&db=cat05020a&AN=aut.b1120574x&site=eds-live

12. Shastri, Y., Hoda, R., Amor, R.: The role of the project manager in agile software development projects. J. Syst. Softw. **173**, 110871 (2021)
13. Shaygan, A., Testik, Ö.M.: A fuzzy AHP-based methodology for project prioritization and selection. Soft. Comput. **23**, 1309–1319 (2019)
14. Sousa Neto, A., et al.: Towards a recommender system-based process for managing risks in scrum projects. In: Proceedings of the 38th ACM/SIGAPP Symposium on Applied Computing, pp. 1051–1059 (2023)
15. Twitter: the algorithm. https://github.com/twitter/the-algorithm (2023)
16. Wei, L., Capretz, L.F.: Recommender systems for software project managers. In: Evaluation and Assessment in Software Engineering, pp. 412–417 (2021)
17. Zhang, Y., Yu, J., Ruan, J., Wang, N., Madanian, S., Wang, G.: Motif-based graph attention network for web service recommendation. In: Proceedings of the 2023 Australasian Computer Science Week, pp. 143–146. Association for Computing Machinery (2023)

A Comparison of One-Class Versus Two-Class Machine Learning Models for Wildfire Prediction in California

Fathima Nuzla Ismail[1]([✉])(iD), Abira Sengupta[2](iD), Brendon J. Woodford[2](iD), and Sherlock A. Licorish[2](iD)

[1] Department of Biochemistry, University of Otago, Dunedin, New Zealand
nuzla.ismail@otago.ac.nz
[2] School of Computing, University of Otago, Dunedin, New Zealand
{abira.sengupta,brendon.woodford,sherlock.licorish}@otago.ac.nz

Abstract. Due to climate change, forest regions in California are increasingly experiencing severe wildfires, with other issues affecting the rest of the world. Machine learning (ML) and artificial intelligence (AI) models have emerged to predict wildfire hazards and aid mitigation efforts. However, the wildfire prediction modelling domain faces inconsistencies due to database manipulations for multi-class classification. To help to address this issue, our paper focuses on creating wildfire prediction models through One-class classification algorithms: Support Vector Machine, Isolation Forest, AutoEncoder, Variational AutoEncoder, Deep Support Vector Data Description, and Adversarially Learned Anomaly Detection. To minimise bias in the selection of the training and testing data, Five-Fold Cross-Validation was used to validate all One-class ML models. These One-class ML models outperformed Two-class ML models using the same ground truth data, with mean accuracy levels between 90 and 99 percent. Shapley values were used to derive the most important features affecting the wildfire prediction model, which is a novel contribution to the field of wildfire prediction. Among the most important factors were the seasonal maximum and mean dew point temperatures. In providing access to our algorithms, using Python Flask and a web-based tool, the top-performing models were operationalized for deployment as a REST API, with the potential to strengthen wildfires mitigation strategies.

Keywords: One-class SVM · ANN-AutoEncoder · ANN-Variational Auto-Encoder · Isolation Forest · `scikit-learn` · `PyOD`

1 Introduction

Wildfires have become a significant issue, destroying thousands of square kilometres of forest yearly. This type of disaster has a global impact on environments, the economy, and health. Natural wildfires are caused primarily by lightning, volcanic eruptions, dry climate, and vegetation. However, it has been documented

© The Author(s), under exclusive license to Springer Nature Singapore Pte Ltd. 2024
D. Benavides-Prado et al. (Eds.): AusDM 2023, CCIS 1943, pp. 239–253, 2024.
https://doi.org/10.1007/978-981-99-8696-5_17

that at least 90% of wildfires are caused by human behaviour, such as smoking in public, camping fires, and garbage burning [28]. As a result, continuous monitoring is required to address this serious issue and, more importantly, to forecast the possibility of widespread and intense wildfires. This brings us to the fundamental challenge that the public and fire management authorities inevitably face: the possibility of predicting wildfires well in advance to take timely action to mitigate damages.

ML and AI methods may aid researchers in developing models for monitoring and predicting wildfire anomalies in advance. However, technical limitations and environmental issues impede the process of monitoring and detecting wildfire occurrences and spread. Furthermore, the specific characteristics that may influence wildfire ignition remain as a research gap. This is due primarily to significant changes in atmospheric conditions, which frequently include air temperature, relative humidity, wind speed and direction, and spatial and temporal time-bounded features [21].

Many ML solutions for wildfire prediction have been developed by researchers, but only a few solutions make it to the deployment stage when it comes to practical use. Incorporating ML models into an Application Programming Interface (API) to develop user-friendly applications would improve the wildfire prediction domain. We look to address this opportunity, and provide the following contributions:

1. Using a fire incidence data set, we demonstrate how the application of appropriate One-class classification algorithms are better suited towards fire risk prediction than Two-class models.
2. The use of Shapley values identify features from the One-class ML models that significantly influence the risk of a wildfire event, providing explainability for our models.
3. A proposed architecture for the development and deployment of a web-based wildfire prediction tool that adopts the best One-class ML model.

For this study, the state of California was selected as the context for predicting the occurrence of wildfires. The experiment generated a set of historical fire data from California (2012 to 2016). Multiple One-class ML algorithms: Support Vector Machine (SVM), Isolation Forest (IF), Autoencoder (AE), Variational AutoEncoder (VAE), Deep Support Vector Data Description (DeepSVDD), and Adversarially Learned Anomaly Detection (ALAD) were investigated in these experiments. Repeated Five-Fold Cross-Validation (CV) was applied to the training data set to generate these models, yielding accuracy ranging from 90% to 99%.

The rest of the paper is organized as follows. In Sect. 2 we provide the study background and describe the One-class ML algorithms used in our experiments. Next, in Sect. 3 we describe the data set for the Californian case study. Our methodology is then provided in Sect. 4. The results of applying our methodology are presented in Sect. 5. In Sect. 6, the deployment of the ML models is discussed, followed by the web-based prototype evaluation in Sect. 7. Finally, in Sect. 8 we summarise our findings and outline opportunities for future work.

2   Background

Defining a negatively unbiased sample data set for complex events such as wild-fires is difficult. Without properly validating these data points, a slew of non-fire data points could be generated for a given location, date, and time. This problem can be solved by using a One-class classification model that defines a class boundary based on positive data labels [15].

When the model outcome probability is greater than the threshold value in One-class binary classification, it is labelled as an inlier (•) and when the model outcome probability is less than the threshold value in One-class binary classification, it is labelled as an outlier (?), which are based on the model output probability and the threshold value, as shown in Fig. 1a. Choosing an accurate threshold is critical for correctly classifying inliers and outliers. In principle, the classification boundary of One-class learning accepts many positive data labels while rejecting only a few outliers (see Fig. 1a). Positive data labels are used to train the model in One-class learning, whereas outliers are considered negative data labels or non-fire events.

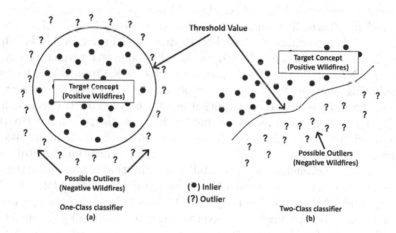

Fig. 1. The distinction between One-class classification (a) and Two-class classification (b). Compared with a One-class classification model, a Two-class classification model accepts inlier (positive) data labels but rejects outlier labels.

As noted above, the model's outcome probabilities, as well as its inlier and outlier predictions, are affected by the threshold value. When the threshold is greater than a certain value for a given prediction instance [10, p. 159], it will be detected as a fire (inlier). For our experiments, each One-class ML algorithm's functionality is described below.

In terms of our approaches, the Support Vector Machine (SVM) is a super-vised learning model that analyses data and identifies patterns for both classification and regression tasks [6]. The One-class variant refers to two types of One-class SVM (OCSVM). The standard OCSVM uses a sphere of minimal volume to contain a specified proportion of training instances [31]. The other OCSVM trains objects using a hyperplane in a kernel feature space. Data is transformed to a higher dimensional space in order to investigate the possibility of constructing a hyperplane decision boundary, with the assumption that all training points belong to one class and all non-training points belong to another. To generate the ML model in the experiments, the latter OCSVM algorithm was used. The One-class OCSVM operates in such a way that standard data clustered into a single region has a high density, while outliers are detected as low-density regions. New data points can be tested based on density regions to detect normal or outlier cases.

Isolation Forest (IF) is a binary random forest approach in which each node randomly chooses a dimension, and then a splitting threshold [17]. It will keep going until each node has a single sample. This method is used to build an ensemble of trees. A sample with exceptional values has a higher chance of being isolated early in the growth of the tree by chance than samples in clusters; as a result, the average depth of samples in the ensemble of trees directly affects the abnormality score [17].

A recent overview of ML and AI algorithms used in wildfire prediction are summarised in [2] which discussed models that are based on Artificial Neural Networks (ANN) including ones based on Radial Basis Function ANNs [23]. For our experiments we similarly adopted an AutoEncoder (AE) which is a type of multi-layer ANN for unsupervised learning that copies input values to output values, allowing mapping from high-dimensional space to lower-dimensional representation [26]. To reduce reconstruction errors, input data is encoded in the hidden layers. This method forces the hidden layers to learn the most patterns in the data while ignoring the "noise". Anomalies are defined as input data with a high reconstruction error. In contrast to an AE, the Variational AutoEncoder (VAE) learns the parameters of a probability distribution representing the data, which could make the model more adept at spotting anomalies [26].

Like AE and VAE, the goal of DeepSVDD [29] is to learn network parameters collaboratively while minimising the average distance from all data representations to the center for this algorithm. Normal data are closely mapped to the center for this algorithm, whereas anomalous data are mapped farther from the centre or outside a hypersphere [16]. In DeepSVDD, ANN are used as One-class classifiers, where any data points which the neural network rejects is categorised as an outlier. Network weights are derived from the training data. These trained network weights are then used in the process of testing new data instances. We have selected DeepSVDD [29] and ALAD [34] due to their popularity in performing prediction domains.

Finally, ALAD [34] is a reconstruction-based anomaly detection technique that assesses how well a sample is reconstructed by a Generative Adversarial Network (GAN). GANs are adopted as they can model complex high-dimensional distributions of real-world data, implying that they could be useful in anomaly detection. ALAD is a promising approach in complex, high-dimensional data.

ALAD is based on bidirectional GANs and contains an encoder network that maps data samples to latent variables. During training, this learns an encoder from the data space to the latent space, making it significantly more efficient at test time. ALAD assesses how far a sample is from its reconstruction by the GAN, where normal samples should be accurately reconstructed while anomalous samples are likely to be poorly reconstructed.

The OCSVM algorithm from the Python `scikit-learn` package [27] was used for the experiments. All the remaining methods including an alternative implementation of the OCSVM algorithm were taken from the Python `PyOD` package [35].

3   Data Set

The case study is based on California in the United States of America, which spans a land area of 423,970 square km[1]. From 2012 to 2016, 7,335 wildfire events were recorded in California by US Federal land management agencies, NOAA, the American Scientific Agency, MODIS 500m resolution satellite images, and the US Census Bureau[2]. The variables for the Californian data set were acquired accordingly, and are listed in Table 1. The collected data were combined into a data set that was geolocated and transformed into an appropriate format for further analysis[3]. These procedures were followed for the implementation of the use case in California.

4   Methodology

As demonstrated in Fig. 2, developing a decision support system for wildfire prediction involves a number of steps, including data preparation, processing, modelling, validation of ML models, and the potential for deployment of ML models.

Wildfire features, weather features, Live Fuel Moisture Content (LFMC) features, and social features are the four input categories that are used. The data set was encoded and scaled to test the ML models based on One-class classification. Below is a more thorough explanation of these steps.

The relevant classifier function calls were used during model training to fit the model to the data. Hyper-parameter tuning was used to configure the function's hyper-parameters, eventually producing one ML model for each classifier type that performed the best. During the tuning process, the hyper-parameters of the models were adjusted to achieve the best accuracy based on the most

---

[1] https://www.fire.ca.gov/our-impact/statisticsStatistics on CA wildfires and CAL FIRE activity.

[2] A different case study with 2.2 million acres burned in Western Australia was conducted as the second case study. However, due to page limitations, we are unable to discuss this data set and its associated results in this paper.

[3] This thesis provides more detail on the steps involved in data pre-processing [10].

Table 1. Variables used for ML models - Californian data set (7,335 Events)

No.	Feature	Description	Prior Research
1	IDATE	Fire Occurrence Date (Month & Date as an Integer)	[1]
2	LAT	Fire location latitude (degrees)	[1,11,33]
3	LON	Fire location longitude (degrees)	[1,11,33]
4	ELEVATION_m	Fire location elevation (in meters)	[1,7,11]
5	ACRES	Acres burnt (in acres)	
6	PPT_mm	Precipitation (in mm for the fire incident date)	[1,11,13]
7	TMIN_c	Minimum temperature (in Celsius for the fire incident date)	[11,13]
8	TMEAN_c	Mean temperature (in Celsius for the fire incident date)	[11,13]
9	TMAX_c	Maximum temperature (in Celsius for the fire incident date)	[1,11,13]
10	TDMEAN_c	Mean dew point temperature (in Celsius for the fire incident date)	[11,13]
11	VPDMIN_hpa	Minimum vapor pressure (in hectopascals) - Californian use case	[7]
12	VPDMAX_hpa	Maximum vapor pressure (in hectopascals) - Californian use case	[7]
13	lfmc_mean	Mean fuel moisture for a particular day (numeric)	[11]
14	lfmc_stdv	Standard deviation of fuel moisture for a particular day (numeric)	[11]
15	Mean_Sea_Level _Pressure	Mean sea level pressure of the nearest weather station to the wildfire event (in hectopascals) - (Universal Kriging)	[25]
16	Mean_Station _Pressure	Nearest mean weather station pressure to the wildfire event (in hectopascasl) - (Universal Kriging)	[25]
17	Mean_Wind _Speed	Mean wind speed for a given location (numeric mph) - (Universal Kriging)	[1,7,11]
18	Maximum_sustained _wind_speed	Maximum sustained wind speed for a given location (numeric MPH) - (Universal Kriging)	[7,11]
19	NAMELSAD	County name (string)	[13]
20	Population	Number of residents living in the respective county (numeric)	[13,24]

significant features determined by the ML algorithm. This procedure used the Python **hyperopt** package [4]. To elaborate, the first step was to specify relevant hyper-parameters for the ML models with predefined options and a range of values. The ML models were then trained for 80 iterations using various combinations of those hyper-parameters. Within each iteration, each model was trained using Five-Fold CV, and the average performance of that model was used to tune the hyper-parameters for the following model. Target values were predicted using testing data and then on the entire data set using the best-performing ML model via 20 times Five-Fold CV. This process produced mean Accuracy, Precision, Recall, and F1-Score classification metrics.

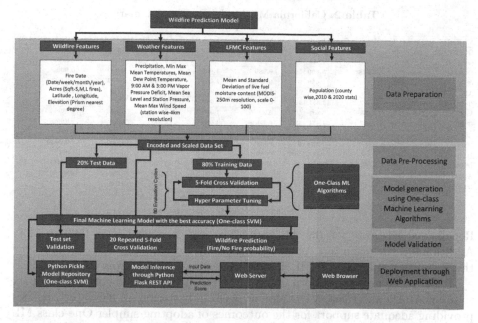

Fig. 2. The process of building a wildfire prediction model involves various steps from data preparation to deploying ML models through a web-based tool

5   Results

Here we compare the results of the ML models on the California wildfire data set as described in Sect. 5.1 with Two-class classification problems, which are described in the earlier research cited in Sect. 5.2. Additionally, Sect. 5.3 illustrates the most important features and their impact on One-class ML models for the two subtypes of OCSVM.

5.1   One-Class Machine Learning Model Results

Table 2 summarises the performance of the One-class ML models for the Californian data set. The number of inliers (fire positive) predictions, which correspond to the number of actual wildfire events, and the number of outliers (fire negative) predictions are used to assess the effectiveness of the applied ML approaches. The results from Table 2 highlight that the OCSVM (PyOD) model was the best performing One-class classifier, achieving a mean test Accuracy of 0.99, mean Precision of 1.00, mean Recall of 0.99, and mean F1-Score of 0.99. A more objective assessment of the OCSVM (PyOD) model through 20 × Five-Fold CV resulted in its performance being higher than the other ML models observed.

With mean test Accuracy of 0.99, Precision of 1.00, Recall of 0.99, and F1-Score of 1.00, the IF model produced results comparable with the other ML models validating its outstandng performance. Mean test Accuracy, Precision,

Table 2. California ML Model Results Summary

ML Technique	Data set Type	Data set Count	Inliers	Outliers	Mean Accuracy	Mean Precision	Mean Recall	Mean F1-Score	20 × Five-Fold CV
OCSVM	Train (80%)	5,868	5,806	62	0.989	1.000	0.989	0.994	0.990
(sklearn)	Test (20%)	1,467	1,443	24	0.983	1.000	0.983	0.991	±0.0030
OCSVM	Train (80%)	**5,868**	**5,809**	**59**	**0.989**	**1.000**	**0.990**	**0.990**	**0.990**
(PyOD)	Test (20%)	**1,467**	**1,458**	**9**	**0.993**	**1.000**	**0.990**	**1.000**	**±0.0028**
AE	Train (80%)	5,868	5,809	59	0.989	1.000	0.990	0.990	0.989
(PyOD)	Test (20%)	1,467	1,454	13	0.991	1.000	0.990	1.000	±0.0030
VAE	Train (80%)	5,868	5,809	59	0.989	1.000	0.990	0.990	0.989
(PyOD)	Test (20%)	1,467	1,454	13	0.991	1.000	0.990	1.000	±0.0028
IF	Train (80%)	**5,868**	**5,809**	59	0.989	**1.000**	**0.990**	**0.990**	**0.989**
(PyOD)	Test (20%)	**1,467**	**1,458**	**9**	**0.993**	**1.000**	**0.990**	**1.000**	**±0.0030**
DeepSVDD	Train (80%)	5,868	5,281	587	0.899	1.000	0.900	0.950	0.897
(PyOD)	Test (20%)	1,467	1,316	151	0.897	1.000	0.900	0.950	±0.0101
ALAD	Train (80%)	5,868	5,281	587	0.899	1.000	0.900	0.950	0.900
(PyOD)	Test (20%)	1,467	1,272	195	0.867	1.000	0.870	0.930	±0.0081

Recall, and F1-Score for the AE and VAE models ranged from 0.99 to 1.00. Additionally, the mean test Accuracy, Precision, Recall, and F1-Score values for the DeepSVDD and ALAD ML models were lower ranging from 0.87 to 1.00.

It should be noted that both OCSVM ML models, despite being less complex than an ALAD and DeepSVDD model, perform better on all mean test metrics providing adequate support for the outcomes of adopting simpler One-class ML models.

5.2 Two-Class Machine Learning Outcomes for the Same Ground-Truth Data

In assessing the One-class ML approach using the same ground truth data and a randomly generated equal amount of false data [32], created by applying Two-class ML models for the California region. Sayad [30] used a similar approach in representing negative samples using random timestamps and locations. Hence, the same approach was followed in creating a false data set. Furthermore, commonly used wildfire prediction models using Two-class ML models were investigated and chosen for this use case. As shown in Table 3, the Two-class ML algorithms were used with supporting literature for predicting wildfires.

Table 3. Performance of Two-class ML models

ML Algorithm	Supporting Literature	Mean Accuracy	Mean Precision	Mean Recall	Mean F1-Score
SVM	[8,11,22]	0.628	0.657	0.763	0.706
RF	[8,11,20]	0.679	0.664	0.724	0.693
Logistic Regression	[3,9,22]	0.676	0.651	0.756	0.697
XGBoost Regression	[19,20]	0.675	0.660	0.717	0.688
ANN	[8,9,22]	0.682	0.665	0.732	0.697

The outcome shows that similar Two-class-based ML models achieved mean test Accuracies from 0.63 to 0.68 for the test data set. Mean test Precision recorded values from 0.65 to 0.66 and average mean Recall values ranged from

0.73 to 0.76. Mean test F1-Score values recorded a range from 0.69 to 0.72. Hence, these results suggest that the Two-class models exhibit reduced performance for the selected data sets compared to One-class ML models using the same ground truth data. Therefore, One-class ML models can serve as good alternatives in prediction models such as wildfire risk, which has limited ground truth data over the period in question.

5.3   Feature Importance Derived Using Shapley Values

This section examines the results obtained through the application of Shapley values, which emphasize the most crucial features and their impact on One-class ML models. These values are obtained by using game theory principles and coefficients from the internal linear regression [18].

The Shapley value is a metric used to determine the average marginal contribution of each feature when considering all possible combinations (coalitions) of features [18]. To illustrate, to calculate the Shapley value of mean wind speed, one needs to evaluate all possible combinations of mean wind speed observations. For each combination, the marginal contribution of ignition probability will be assessed. By aggregating all the marginal contributions to ignition probability, the mean marginal contribution of ignition probability can be determined as the Shapley value's outcome.

Using Shapley values [18], the plot on the left in Fig. 3 shows the average impact of the features on the One-class OCSVM PyOD ML models' outputs. The most influential attributes included the maximum and average dew point temperatures associated with different seasons. Then Mean_Sea_Level_Pressure, PPT_mm, and lfmc_mean are the second set of essential features that influence wildfire prediction. For example, the temperature variables and lfmc_mean have a more significant impact on the model output for the risk of wildfire than does the population. Also, high LFMC is more susceptible to ignition and can signal more fire spread [11]. Mean_Sea_Level_Pressure is the average level of one or more bodies of water on Earth from which elevation can be calculated. With increasing elevation, sea level pressure decreases. Wind speed and direction are both factors in the wind effect. The dry wind is one of the primary causes of wildfire spreading. The rate of wildfire spread has been estimated to be around 8% of wind speed, regardless of fuel type, especially in dry fuel moisture conditions [12]. It can be noted that these same features are ranked highly across all the models, and hence, these top-ranked features should be given more importance in the modelling process. Furthermore, the Shapley value impact has been investigated in Fig. 3. The result of testing the features and models informed a web-based tool, which is presented in the following section to showcase the efficiency and practicality of One-class ML models.

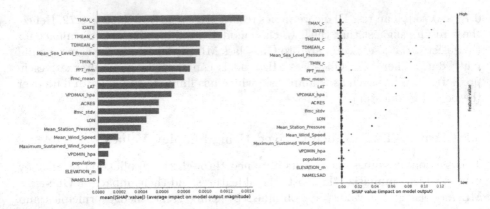

Fig. 3. Shapley values generated from the OCSVM PyOD model (right) shows that the mean and temperature values are high when the model output is predicting a positive fire occurrence

6   Deployment of Machine Learning Models

The web-based tool[4] is presented as a contribution to the state-of-the-art ML-based wildfire prediction domain (see Fig. 4). The main objective of this phase of research is to deploy the results of the selected ML model using a REST API, which will then be fed into a web-based tool. The web-based tool's goal is to provide long-term wildfire predictions based on One-class classification-based ML models that can predict the start of a wildfire one week in advance for any given location in California. This can also help international wildfire management authorities test wildfire prediction models across multiple geographies. The web-based tool is also useful for countries that do not have access to wildfire forecasting systems. However, this is not meant to replace current regional wildfire forecasting systems.

The ML model is fed with 20 features (see Sect. 3) from four categories and six probability rates of danger levels, which are mapped by the decision scores (d) of the One-class ML models: No Danger ($d \leq 0$), Low ($0 < d \leq 60$), Moderate ($60 < d \leq 80$), High ($80 < d \leq 90$), Very High ($90 < d \leq 97$) and Severe ($97 < d \leq 100$). These fire danger rating breakpoints used were similar to fire spread probabilities modelled by the US Wildland Fire Decision Support System [14] to create these threshold classes. The selection of these fire danger rating class thresholds was informed by historical fire danger outcomes, as documented in [5]. The Flask REST service for the web-based tool uses three other external REST APIs for generating wildfire prediction outcomes:

1. The publicly available free Open Topo Data elevation REST API which gives elevation data of any location when latitude and longitude are given.
2. The OpenWeather REST API provides historical, current and forecasted weather details through REST APIs for any point on the world.

---

[4] https://www.bushfirepredict.com.

3. USGS Earth Explorer Website hosts LFMC data as different vegetation indexes carry a file format of all locations of California based on a MODIS grid.

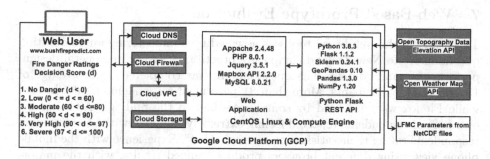

Fig. 4. The deployed ML model's architecture deriving wildfire prediction outcomes using six fire danger rating levels

The four main functionalities listed below were thus identified as the outcome of this web-based tool:

1. Choose some historical wildfire events to train the ML models and validate the model output. Users can also alter the input parameters and analyse and explore the most important features of the ML models.
2. Select any location in California using the map, manually enter the input features, and use a probability to predict wildfire susceptibility.
3. Search all input features for the next 7 d.
4. View historical yearly wildfire heat-maps based on ML model training and testing data.

Several technological advancements in the field of wildfires have emerged in recent years as a result of the high costs and practical difficulties of fighting wildfires. Technology training is at the top of the list because it is crucial for sharing common resources and standards for information on fire danger and communication between fire authorities. The use of technological systems to forecast and predict the occurrence of wildfires has enabled emergency response teams to plan ahead and take preventative action. Firefighters must receive the necessary training to deal with such emergencies. Implementing a straightforward, inexpensive prototype can significantly lower the price of intricate training. Large financial budgets can also be set aside for public education campaigns about wildfire prevention and natural wildfire occurrences. The infrastructure cost of the Google Cloud Platform (GCP) (virtual machine and domain name) constitutes the sole cost element for the implementation of this web-based tool. The remaining programmes and services are either open source, free, or have a free usage tier. Hosting this in an on-premise local area network may thus result in an infrastructure cost saving in terms of infrastructure. The web-based tool has a monthly fee of \$42 NZD and can forecast wildfires up to a week in advance[5]. To assess the utility of this tool, a user-based questionnaire evaluation has been carried out, the results of which are reported in the next section.

7   Web-Based Prototype Evaluation

The utility of the web-based tool was assessed by administering an 18-question questionnaire to New Zealand computing practitioners covering general design, performance, and content. The questionnaire was completed in an average of 15 min by 11 participants. More than 81 percent of respondents used Chrome, while Firefox was also used by some, according to the findings. Additionally, over 63% of respondents reported feeling extremely satisfied, with the remaining respondents rating their satisfaction as "somewhat". Experience with the mobile phone view using different browsers produced mixed results, with the majority of respondents being satisfied, 18% being neither satisfied nor dissatisfied, and the remaining 9% being extremely dissatisfied. The mobile phone version needs to be enhanced further as a result. The overall design of the tool resulted in an above-average ranking for all feedback. Performance-wise, the speed of information retrieval from the input feature fields, the speed of ML prediction, and the response time were all very quick. Additionally, the overall performance was rated as being better than satisfactory, with 80% stating that it performed excellently and with positive feedback exceeding 63% for the tool's content when measuring the understandability of input and output features of the accuracy of wildfire prediction outcomes. This results in a high-performance rating for the web-based tool.

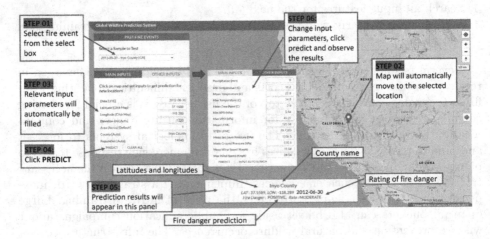

Fig. 5. Inyo 2012-06-30 Fire Event with main input features

---

[5] More information on the cost calculation can be found on [10, pp. 167–168].

8   Conclusion and Future Work

In summary, historical wildfire events in California were represented using a total of 20 features. Seven different One-class ML algorithms were used order to train multiple models. After the hyper-parameters of the ML models were tuned, the models were validated using repeated 20 × Five-Fold CV. The average test Accuracy of each ML model ranged from 0.90 to 1.00, demonstrating the ML models' high generic performance for the California data set. In addition, Precision, Recall, and F1-Score values were used to evaluate the effectiveness of the ML models.

Not only does our study address the need to create ML-based wildfire prediction models, but, more importantly, it identifies key features from these models that could influence wildfire ignition. As well as our findings being consistent with the outcomes of previous research we also showed the degree to which these identified features contribute to the risk of a wildfire event.

Finally, we described development of a web-based prototype that integrates the best performing ML algorithms and model of the sequence of wildfire events for wildfire occurrence mapping. The intended audiences for this tool are the general public and wildfire authorities.

However, as we only used one data set for this study, future work will involve the creation of more wildfire data sets from other countries, potentially using different features. Top-ranked features extracted from these ML models using Shapley values may be compared and contrasted with the findings from our existing work to show how the contribution of different features influence the risk of wildfire depending on the location.

References

1. Abdollahi, A., Pradhan, B.: Explainable artificial intelligence (XAI) for interpreting the contributing factors feed into the wildfire susceptibility prediction model. Sci. Total Environ. **879**, 163004 (2023). https://doi.org/10.1016/j.scitotenv.2023.163004
2. Alkhatib, R., Sahwan, W., Alkhatieb, A., Schütt, B.: A brief review of machine learning algorithms in forest fires science. Appl. Sci. **13**(14) (2023). https://doi.org/10.3390/app13148275, https://www.mdpi.com/2076-3417/13/14/8275
3. de Bem, P., de Carvalho Júnior, O., Matricardi, E., Guimarães, R., Gomes, R.: Predicting wildfire vulnerability using logistic regression and artificial neural networks: a case study in Brazil. Int. J. Wildland Fire **28**(1), 35–45 (2018). https://doi.org/10.1071/WF18018
4. Bergstra, J., Yamins, D., Cox, D.D.: Making a science of model search: hyperparameter optimization in hundreds of dimensions for vision architectures. In: Proceedings of the 30th International Conference on International Conference on Machine Learning, vol. 28, pp. I-115–I-123. ICML 2013, JMLR.org (2013)
5. Center, N.I.F.: National Wildfire Coordinating Group (NWCG). Interagency Standards for Fire and Fire Aviation Operations. Createspace Independent Publishing Platform, Great Basin Cache Supply Office: Boise, ID, USA (2019)

6. Cortes, C., Vapnik, V.: Support vector machine. Mach. learn. **20**(3), 273–297 (1995)
7. Donovan, G.H., Prestemon, J.P., Gebert, K.: The effect of newspaper coverage and political pressure on wildfire suppression costs. Soc. Nat. Resour. **24**(8), 785–798 (2011)
8. Ghorbanzadeh, O., et al.: Spatial prediction of wildfire susceptibility using field survey GPS data and machine learning approaches. Fire **2**(3), 43 (2019)
9. Goldarag, Y., Mohammadzadeh, A., Ardakani, A.: Fire risk assessment using neural network and logistic regression. J. Indian Soc. Remote Sens. **44**, 1–10 (2016). https://doi.org/10.1007/s12524-016-0557-6
10. Ismail, F.N.: Novel machine learning approaches for wildfire prediction to overcome the drawbacks of equation-based forecasting, Ph. D. dissertation, University of Otago (2022)
11. Jaafari, A., Pourghasemi, H.R.: 28 - Factors influencing regional-scale wildfire probability in Iran: an application of random forest and support vector machine. In: Pourghasemi, H.R., Gokceoglu, C. (eds.) Spatial Modeling in GIS and R for Earth and Environmental Sciences, pp. 607–619. Elsevier (2019)
12. Jain, P., Coogan, S.C., Subramanian, S.G., Crowley, M., Taylor, S., Flannigan, M.D.: A review of machine learning applications in wildfire science and management. Environ. Rev. **28**(4), 478–505 (2020)
13. Jiménez-Ruano, A., Mimbrero, M.R., de la Riva Fernández, J.: Understanding wildfires in mainland Spain. a comprehensive analysis of fire regime features in a climate-human context. Appl. Geogr. **89**, 100–111 (2017)
14. Jolly, W.M., Freeborn, P.H., Page, W.G., Butler, B.W.: Severe fire danger index: a forecastable metric to inform firefighter and community wildfire risk management. Fire **2**(3), 47 (2019). https://doi.org/10.3390/fire2030047
15. Khan, S.S., Madden, M.G.: One-class classification: taxonomy of study and review of techniques. Knowl. Eng. Rev. **29**(3), 345–374 (2014)
16. Kim, S., Choi, Y., Lee, M.: Deep learning with support vector data description. Neurocomputing **165**, 111–117 (2015)
17. Liu, F.T., Ting, K.M., Zhou, Z.: Isolation forest. In: 2008 Eighth IEEE International Conference on Data Mining, pp. 413–422. IEEE (2008)
18. Lundberg, S.M., Lee, S.I.: A unified approach to interpreting model predictions. In: Proceedings of the 31st International Conference on Neural Information Processing Systems, pp. 4768–4777. Curran Associates Inc. (2017)
19. Ma, J., Cheng, J., Jiang, F., Gan, V., Wang, M., Zhai, C.: Real-time detection of wildfire risk caused by powerline vegetation faults using advanced machine learning techniques. Adv. Eng. Inform. **44**, 101070 (2020). https://doi.org/10.1016/j.aei.2020.101070
20. Michael, Y., Helman, D., Glickman, O., Gabay, D., Brenner, S., Lensky, I.M.: Forecasting fire risk with machine learning and dynamic information derived from satellite vegetation index time-series. Sci. Total Environ. **764**, 142844 (2021). https://doi.org/10.1016/j.scitotenv.2020.142844
21. Miller, C., Hilton, J., Sullivan, A., Prakash, M.: SPARK – a bushfire spread prediction tool. In: ISESS 2015. IAICT, vol. 448, pp. 262–271. Springer, Cham (2015). https://doi.org/10.1007/978-3-319-15994-2_26
22. Nhu, V.H., et al.: Shallow landslide susceptibility mapping: a comparison between logistic model tree, logistic regression, naïve bayes tree, artificial neural network, and support vector machine algorithms. Int. J. Environ. Res. Public Health **17**(8), 2749 (2020)

23. Ntinopoulos, N., Sakellariou, S., Christopoulou, O., Sfougaris, A.: Fusion of remotely-sensed fire-related indices for wildfire prediction through the contribution of artificial intelligence. Sustainability **15**(15), 1–24 (2023). https://doi.org/10.3390/su151511527

24. Nunes, A., Lourenço, L., Meira Castro, A.C.: Exploring spatial patterns and drivers of forest fires in Portugal (1980–2014). Sci. Total Environ. **573**, 1190–1202 (2016). https://doi.org/10.1016/j.scitotenv.2016.03.121

25. Papadopoulos, A., Paschalidou, A., Kassomenos, P., McGregor, G.: On the association between synoptic circulation and wildfires in the Eastern Mediterranean. Theoret. Appl. Climatol. **115**(3), 483–501 (2014)

26. Patterson, J., Gibson, A.: Deep Learning: A Practitioner's Approach. O'Reilly Media, Inc. (2017)

27. Pedregosa, F., et al.: Scikit-learn: machine learning in Python. J. Mach. Learn. Res. **12**, 2825–2830 (2011)

28. Reisen, F., Duran, S.M., Flannigan, M., Elliott, C., Rideout, K.: Wildfire smoke and public health risk. Int. J. Wildland Fire **24**(8), 1029–1044 (2015)

29. Ruff, L., et al.: Deep one-class classification. In: Dy, J., Krause, A. (eds.) Proceedings of the 35th International Conference on Machine Learning. Proceedings of Machine Learning Research, vol. 80, pp. 4393–4402. PMLR (2018)

30. Sayad, Y.O., Mousannif, H., Al Moatassime, H.: Predictive modeling of wildfires: a new dataset and machine learning approach. Fire Saf. J. **104**, 130–146 (2019). https://doi.org/10.1016/j.firesaf.2019.01.006

31. Tax, D.M., Duin, R.P.: Support vector domain description. Pattern Recogn. Lett. **20**(11–13), 1191–1199 (1999)

32. Tien Bui, D., Bui, Q.T., Nguyen, Q.P., Pradhan, B., Nampak, H., Trinh, P.T : A hybrid artificial intelligence approach using GIS-based neural-fuzzy inference system and particle swarm optimization for forest fire susceptibility modeling at a tropical area. Agric. For. Meteorol. **233**, 32–44 (2017)

33. Tonini, M., D'Andrea, M., Biondi, G., Degli Esposti, S., Trucchia, A., Fiorucci, P.: A machine learning-based approach for wildfire susceptibility mapping. the case study of the liguria region in Italy. Geosciences **10**(3), 105 (2020)

34. Zenati, H., Romain, M., Foo, C.S., Lecouat, B., Chandrasekhar, V.: Adversarially learned anomaly detection. In: 2018 IEEE International Conference on Data Mining (ICDM), pp. 727–736. IEEE (2018)

35. Zhao, Y., Nasrullah, Z., Li, Z.: PyOD: a python toolbox for scalable outlier detection. J. Mach. Learn. Res. **20**(96), 1–7 (2019)

Skin Cancer Detection with Multimodal Data: A Feature Selection Approach Using Genetic Programming

Qurrat Ul Ain[(⊠)] [iD], Bing Xue [iD], Harith Al-Sahaf [iD], and Mengjie Zhang [iD]

Center for Data Science and Artificial Intelligence and School of Engineering and Computer Science, Victoria University of Wellington, P.O. Box 600, Wellington 6140, New Zealand
`{Qurrat.Ul.Ain,Bing.Xue,Harith.Al-Sahaf,Mengjie.Zhang}@ecs.vuw.ac.nz`

Abstract. Melanoma is the most deadly form of skin cancer and can be treated if detected at an early stage. This study develops a skin cancer image classification method using the feature selection ability of genetic programming and multi-modal skin cancer data. This study utilizes suitable feature descriptors to extract informative features that incorporate the scale, color, local, global, and texture information from the dermoscopic images, as well as effectively utilize domain knowledge to enhance the performance of binary and multiclass classification tasks. Existing approaches mainly rely on grayscale and texture information to classify skin cancer images. Designing an effective way to combine multi-channel multi-resolution spatial/frequency information has not been well explored to improve the classification performance of complex skin cancer images. To preserve all local, global, color, and texture information simultaneously, we extract Local Binary Patterns and wavelet decomposition features from multiple color channels. The proposed method is evaluated using a dermoscopic image dataset and compared to existing deep learning and GP methods. The results conclude that the proposed method outperformed the other methods in this study. With the interpretability of GP models, the proposed method highlights important domain-specific features with high discriminating ability between different types of skin cancers. This discovery validates the potential of the proposed method to improve dermatologists' real-time diagnostic ability.

Keywords: Cancer detection · Multimodal data · Image classification · Genetic Programming · Feature selection

1 Introduction

The occurrence of melanoma, the most lethal form of skin cancer, has seen a rapid rise in the last three decades [19]. Detecting skin cancer at its initial stages offers a high likelihood of recovery, with a five-year survival rate of 92% [19]. Recent advancements in computer-aided diagnostic (CAD) systems have enabled earlier detection of diverse skin cancers. Dermatologists examine several crucial visual attributes to formulate diagnoses using dermoscopy criteria; these include

© The Author(s), under exclusive license to Springer Nature Singapore Pte Ltd. 2024
D. Benavides-Prado et al. (Eds.): AusDM 2023, CCIS 1943, pp. 254–269, 2024.
https://doi.org/10.1007/978-981-99-8696-5_18

Asymmetry, Border irregularity, Color variation, and Diameter of the lesion, collectively referred to as the ABCD rule [20]. Another visual screening technique for skin cancer detection is the 7-point checklist method which includes regression areas, pigment network, streaks, asymmetry, blue-whitish veil, dots/globules, and presence/absence of six colors: black, white, red, dark-brown, light-brown, bluish-gray [5]. These fundamental medical characteristics aid dermatologists in precisely identifying various types of skin cancers.

Genetic programming (GP) is a bio-inspired technique that employs a process of genetic evolution to generate a population of computer programs (referred to as models or trees) aimed at solving a specific task [12]. GP incorporates genetic operators like reproduction, crossover, and mutation, applied iteratively to transform the existing generation of programs into a new one [12]. The resulting evolved program exhibits a tree-like structure comprising terminal nodes and internal nodes. Terminal nodes are features, while internal nodes are functions. GP has its inherent feature selection capacity by designating the most significant features as terminals, which often possess a greater discriminatory capability between classes compared to the original features. This particular attribute significantly impacts the achievement of robust performance. Beyond its classification applications, GP has been extensively explored for its utility in feature selection as well [21].

For effective classification of skin cancer images, the classification algorithm requires a range of distinct features with discriminative qualities, encompassing attributes such as local and global aspects, as well as scale, color, and texture characteristics, to achieve optimal performance. Features can be extracted from multiple color channels and scales of an image to incorporate texture and color information. Insights derived from research into the human visual system emphasize that the spatial/frequency representation contains both localized and overarching information. This revelation has driven researchers to formulate multi-scale texture models for image classification [7]. The wavelet decomposition applied to images captures multi-scale properties rendering itself as a valuable technique for texture analysis, facilitating the creation of informative features [9]. This served as the impetus for our decision to extract wavelet-based features in this study. In addition, it is essential to develop accurate classification methods or diagnostic systems that can be used in real-life situations to classify a particular type of skin cancer. In the medical field, correctly identifying a diseased image is more crucial than correctly identifying a non-diseased one [3,11]. Along with correctly classifying, dermatologists need to recognize significant features that can assist them in visual pattern analysis [9,22].

Convolutional neural networks (CNNs) have been increasingly used in dermoscopy image analysis during the past decade. Esteva et al. [8] trained an Inception network on clinical and dermoscopy images and achieved human-level classification performance. Liu et al. [13] developed a skin image classification model by combining domain knowledge and deep learning with clinical images. However, CNNs typically require data augmentation and appropriate model architecture to train effectively from scratch due to the limited size of medical datasets [6]. Using CNNs requires a significant amount of time and resources due to their high computational cost.

Numerous methodologies have been devised for extracting diverse feature types from lesion images [9,18]. These techniques assess the efficacy of these features in identifying melanoma within a binary image classification framework. However, they lack integration of multimodal data (including tabular and image data) and fail to amalgamate domain expertise with image-extracted features to differentiate between various cancer types. Furthermore, these approaches confine their applicability solely to binary classification tasks, specifically melanoma detection, without addressing the more intricate challenge of multiclass skin image classification. In contrast, this endeavor centers on devising a potent classification approach that leverages domain knowledge and features extracted from skin cancer images, showcasing competence across diverse scenarios.

Goals: This research introduces a novel skin cancer image classification approach that employs genetic programming for feature selection, effectively merging insights from dermatology and computer vision to address binary and multiclass image classification. In contrast to previous methodologies, the proposed method uniquely employs GP to select features by capitalizing on diverse attributes such as texture, color, border shape, and geometrical information. These attributes are derived from domain-specific knowledge and image-based features, enhancing the precision of skin cancer detection. This information is provided by multimodal data, i.e., domain (specialized) knowledge (7-point checklist) and image features extracted by image descriptors from multiple color channels and multiple resolutions. The method adopts a wrapper approach to pick informative features from these two sets of features. These selected features are subsequently fed into a classification algorithm, such as a support vector machine, to learn a classifier. Through this strategy, the proposed method aims to autonomously identify and utilize the most informative domain-specific and image-derived features. This research tackles the following key research questions:

- Which types of features are positively contributing to providing good classification performance for binary and multiclass classification tasks?
- Can the proposed approach provide better discriminating ability when a combination of feature sets is used than utilizing a single feature set?
- Can the proposed GP method outperform the existing deep learning and GP methods for skin cancer image classification?

2   Background

The primary objective of feature extraction in skin cancer imaging is to derive image features that closely align with those perceptible to dermatologists and can consequently be employed for identifying distinct forms of skin cancer [9]. In the current study, we extend this process by integrating domain-specific information in the form of tabular data, which is curated by expert dermatologists utilizing the 7-point checklist method. In addition, we encompass features extracted from

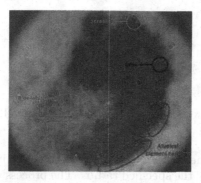

Fig. 1. Identification of some dermoscopic criteria based on 7-Point Checklist (7PC) with blue-whitish veil, atypical pigment network, dots, and streaks [15] (Color figure online).

dermoscopic images via widely recognized image descriptors suitable for skin cancer detection, notably the three-level pyramid-structured wavelet decomposition (3LWD) [7] and the Local Binary Pattern (LBP) [16]. These distinct categories of features are amalgamated within the framework of our study with the following aims: 1) to furnish the requisite discriminative information to the GP process, facilitating the effective selection of features during the course of evolutionary refinement, and 2) to explore which particular combinations of features exert greater prominence in the classification of dermoscopy images.

2.1   7-Point Checklist Domain Specific Features (7PC)

In recent years, a plethora of analytical methods based on scored algorithms have emerged with the dual aim of simplifying dermoscopic learning and enhancing the early detection of melanoma. Among these methods, the 7-point checklist [5] stands out as one of the most current and rigorously validated dermoscopic algorithms. This is attributed to its commendable balance between high sensitivity and specificity, even when employed by individuals without specialized expertise. Of the seven criteria in the checklist, three are major (atypical network, blue-whitish veil, and atypical vascular pattern) and four are minor (irregular streaks, irregular dots/globules, irregular blotches, and regression structures). Figure 1 shows the presence of pigment network, dots, streaks, and blue-whitish veil in a dermoscopic image. Initially chosen for their association with melanoma, these criteria also prove informative for other skin cancer types [5].

2.2   Local Binary Patterns (LBPs)

LBP serves as a widely employed image descriptor for feature extraction, developed by Ojala et al. [16]. This method undertakes a pixel-wise scan across an image, employing a sliding window with a predetermined radius. The central

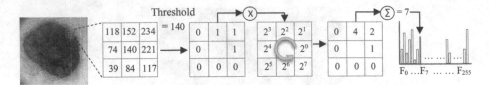

Fig. 2. Feature extraction with LBP and creating a LBP histogram.

pixel's value is determined through an assessment of the intensities of surrounding pixels situated along the specified radius. The process generates a histogram, representing a feature vector, as depicted in Fig. 2. There are two kinds of LBPs: uniform and non-uniform. Uniform patterns have a maximum of two 0 to 1 or 1 to 0 transitions. Non-uniform patterns have more than two such transitions. For example, 01000000 and 1110000 are uniform LBPs, whereas 00110100 and 00100100 are non-uniform LBPs. To improve efficiency, the feature vector can be reduced to $p(p-1)+3$ bins by using only uniform LBP patterns. Non-uniform patterns are grouped into a single bin. Commonly, LBP uses a window size of 3×3 pixels and a radius of 1 pixel ($LBP_{8,1}$).

Color, an integral facet of the ABCD rule [20], is pivotal for skin lesion classification. Color variations induce substantial diversity within the RGB (red, green, blue) space, potentially enabling features extracted from RGB color channels to effectively differentiate between classes. In the current study, uniform LBP features are extracted from each of the three RGB color channels to identify corners, blobs, and streaks in skin images, potentially improving performance.

2.3 Three Level Wavelet Decomposition (3LWD)

The analysis of texture plays a crucial role in identifying the visual characteristics of a lesion and is an essential component of clinical diagnosis. This is exemplified by the use of the ABCD rule in dermoscopy [7]. By employing pyramid-structured wavelet analysis, both the intricate details of local nuances, such as structure and texture and the broader global features that define the lesion's overall properties can be effectively captured [9]. In this study, a three-level pyramid-structured wavelet decomposition technique is opted, specifically applied to the red, green, blue, and luminance[1] color channels within the context of skin images. This way we include multichannel (four color channels) and multiresolution (three scales in 3LWD) image features.

To glean informative insights from wavelet coefficients, diverse statistical metrics are utilized, encompassing norm, skewness, energy, kurtosis, mean, entropy, average energy, and standard deviation. Further details can be referenced in [9]. Figure 3(a) illustrates a skin lesion image, and Fig. 3(b) depicts the application of pyramid-structured wavelet decomposition to the image in Fig. 3(a).

---

[1] luminance = $(0.299 \times R) + (0.587 \times G) + (0.114 \times B)$.

3   The Proposed Method

The proposed method, a wrapper-based feature selection approach using domain knowledge in GP (DKGP), for skin cancer image classification is described in this section. The overall structure is presented in Fig. 4. The dataset comes with images and tabular data based on 7PC. Each image in the dataset is given to the two image descriptors (3LWD and LBP) discussed in Sect. 2 to get two sets of feature vectors. These image feature vectors are then combined with domain-specific 7PC features to form two sets of feature vectors, namely 7PC+LBP, and 7PC+3LWD. The feature vectors of the combined dataset are partitioned into separate training and test sets. The training set is utilized to facilitate the evolution of individuals within the GP framework, equipped with informative features at their terminal nodes. Employing these selected features from the evolved GP individual, the original training and test sets undergo a transformation, resulting in reduced training and test sets. Subsequently, a classification algorithm, such as a support vector machine (SVM), is employed. This algorithm employs the transformed training set to build a classification model. The acquired classification model is then applied to the transformed test set to measure the performance of the trained model based on utilizing the extracted/selected features.

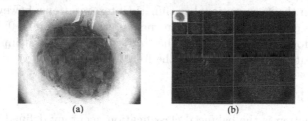

(a)   (b)

Fig. 3. Three-level pyramid-structured wavelet decomposition.

3.1   Terminal Set

The terminal set consists of five sets of feature vectors, extracted from the feature extraction methods discussed in Sect. 2.

1. 7PC: 12 dermoscopic features provided by expert dermatologists based on the seven-point checklist method as described in Sect. 2.1.
2. LBP: 59 LBP features extracted from each of the RGB channels and concatenated to make 177 (= 59 LBP features × 3 channels) features.
3. 3LWD: Wavelet-based texture features extracted from RGB and luminance color channels of the images to make a total of 416 (= 8 statistical measures × 13 nodes × 4 color channels) features.
4. 7PC+LBP: 7PC and LBP feature sets as mentioned above are concatenated to make a total of 189 (= 12 7PC + 177 LBP) features.
5. 7PC+3LWD: 7PC and 3LWD feature sets are concatenated to make 428 (= 12 7PC + 416 3LWD) features.

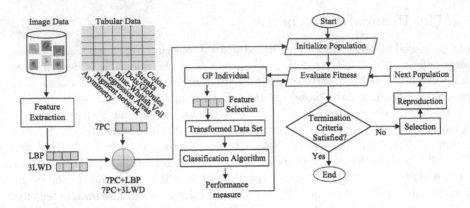

Fig. 4. The flowchart of the proposed DKGP method.

3.2 Function Set

The function set encompasses seven distinct operators, namely four arithmetic operators $(+, -, \times, /)$, two trigonometric operators (sin, cos), and one conditional operator (if). In the context of the arithmetic operators, the first three operators retain their conventional arithmetic interpretations; however, the division operator is guarded to yield 0 when encountering division by 0. The if operator demands four inputs and yields the third input if the first input surpasses the second input; otherwise, it yields the fourth input.

3.3 Fitness Function

The fitness function is the balanced classification accuracy defined as

$$fitness = \frac{1}{m} \sum_{i=1}^{m} \frac{TP_i}{TP_i + FN_i} \qquad (1)$$

where, m represents the number of classes, TP stands for true positive, FN signifies false negative, and the ratio $\frac{TP_i}{TP_i + FN_i}$ denotes the true positive rate of class i. In scenarios where class imbalances exist, i.e., leading to very different number of instances across classes, employing the standard overall accuracy, calculated by the ratio of correctly classified instances to the total number of instances, may generate results skewed towards the majority class. Consequently, adopting the balanced accuracy metric can mitigate this problem. This metric assigns equal importance to all classes within a dataset. By utilizing this fitness function, as shown in Equation (1), the GP process is guided towards achieving good performance across all classes.

3.4 Classification

Once the GP evolutionary process is complete, the best GP individual is identified, featuring selected features at its terminals based on the training data.

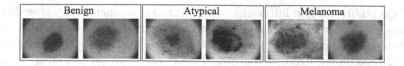

Fig. 5. Samples of the three classes in the PH2 dataset.

These chosen features play a pivotal role in transforming the original training and test data, resulting in new training and test sets. The modified training data is employed to train a classification method, such as SVM. Subsequently, the modified test data is used to evaluate the performance of the trained classification model on the test dataset.

4   Experiment Design

To conduct the experiments, we employ *10-fold cross-validation* to partition the datasets. This division assigns nine folds for training and one fold for testing, using stratified random sampling for balanced distribution. There are 30 independent runs of GP carried out, and the outcomes are presented as mean and standard deviation values for balanced accuracy and F1-score. In each GP run, an individual is evolved using the training data (9 folds), with the fitness function (Eq. 1) evaluating the accuracy of the DKGP method. The selected features within the evolved individual are then applied to adapt the test data (1 fold). This iterative procedure is repeated 10 times to yield results for 10-fold cross validation, ensuring each fold serves as a test set exactly once. Consequently, this entire process is replicated 30 times, employing distinct seed values, resulting in 30 sets of training and test accuracy results. The implementation leverages the Evolutionary Computing Java-based (ECJ) package version 27 [14].

For the wrapper classification methods, we utilize four techniques: Naive Bayes (NB), Support Vector Machines (SVMs), k-Nearest Neighbor (k-NN) with k=5 (balancing noise reduction and efficiency), and Random Forest (RF). Implementation is executed using the widely adopted Waikato Environment for Knowledge Analysis (WEKA) package [10]. Consistent with prior studies [1,2], of GP for skin cancer detection, this work opts for a Radial Basis Function kernel, which has shown superior performance compared to the default linear kernel. These settings have been adopted from previous studies [1,2] where they have shown the best performance amongst other settings.

4.1   Dataset

There are 200 images in the PH2 dataset [15]. These images are captured using a dermatoscope and measure approximately 768×560 pixels. They are divided into three categories: melanoma, common nevus, and atypical nevus. In dermatology, melanoma and common nevus refer to malignant and non-malignant lesions, respectively. Atypical nevus, on the other hand, is a non-malignant lesion that could potentially develop into tumor cells later. To conduct binary classification experiments, 40 instances of melanoma are designated as the "malignant" class. The "benign" class consists of a total of 80 common nevus and 80 atypical nevus, as per [2]. Exemplar samples from this dataset are depicted in Fig. 5.

4.2   GP Parameters

The GP parameters employed in the proposed method are detailed in Table 1. The evolutionary process persists until either a maximum of 50 generations is achieved or a flawless classification model with 100% accuracy is identified. These are commonly adopted GP settings [2].

Table 1. GP Parameter Settings.

Parameter	Value	Parameter	Value
Generations	50	Selection type	Tournament
Population Size	1024	Tournament size	7
Crossover Rate	0.80	Tree depth	2–6
Mutation Rate	0.19	Initial Population	Ramped half-and-half
Elitism	0.01		

4.3   Classification Methods for Comparison

Existing GP Method: Embedded GP (EGP-4) [1] is an existing GP approach that evolves four trees in its individual to perform melanoma detection in a binary classification task. It utilizes image features only extracted using LBP image descriptor and the ABCD rule of dermoscopy [20].

Existing Deep Learning Methods: Moreover, we compare DKGP with recently developed deep learning methods for the PH2 dataset as discussed below:

- Patino et al. [17] devised a multiclass classification approach that incorporates distinct morphological operations to encompass asymmetry, color, and border attributes. The method employs three classification techniques: a fully connected neural network, SVM, and logistic regression.
- Alkarakatly et al. [4] introduced a 5-layer Convolutional Neural Network (CNN) to address the 3-class classification task posed by the PH2 dataset.

5   Results and Discussions

The results of binary and multiclass classification are presented in Tables 2 and 3, respectively. The values of these results in terms of balanced accuracy and F1-score represent the mean and standard deviation among the 30 GP runs.

5.1 Binary Classification

The results for the task of binary classification are presented in Table 2. Horizontally, the table consists of five blocks and the first block lists the features used to perform classification. The second, third, fourth, and fifth blocks each of which show the results of using NB, SVM, k-NN, and RF classification methods, respectively, used in the wrapper approach in the proposed DKGP method. There are five rows that show the feature sets used to perform classification. Here, 7PC, LBP, and 3LWD show the three sets of features that are used individually to perform experiments. 7PC+LBP shows that the two sets of features, 7PC and LBP, are concatenated together to perform experiments. Similarly, 7PC+3LWD shows the 7PC feature set concatenated together with 3LWD feature set before being provided to the proposed DKGP method. The results are given in terms of balanced accuracy and F1-score. The bold sign shows the highest accuracy achieved among the four classifiers in a row using a feature set.

There are five sets of experiments using the five feature sets as mentioned in Sect. 3.1. Among the five feature sets (Table 2), DKGP achieved the highest performance with 88.69% average accuracy using the 7PC feature set and NB as a wrapped classifier. In addition, DKGP achieved the highest F1-score of 87.48% using the 7PC feature set and RF as a wrapped classifier. It has been observed that using the individual feature sets, i.e., 7PC, LBP, and 3LWD, 7PC has always performed better than the other two. For example, 7PC features provided 84.56% balanced accuracy and 85.89% F1-score using k-NN classifier, and outperformed the LBP feature set with 76.09% balanced accuracy and 77.53% F1-score, and also outperformed 3LWD with 79.69% balanced accuracy and 81.42% F1-score. There is another behavior seen using concatenated feature sets when provided to the SVM classifier. SVM provided the highest classification accuracy of 87.44% using the 7PC+LBP feature set and outperformed the individual 7PC and LBP feature sets with 83.88% and 74.81% accuracy. This shows that combining the two feature sets (7PC with domain-specific knowledge and LBP with computer vision knowledge) allows SVM to learn better and discriminate among the two classes compared to using a single set of features. However, this trend has not been observed in the other classification methods: NB, k-NN, and RF.

5.2 Multiclass Classification

The results of multiclass classification are presented in Table 3. Among the five feature sets, 7PC with domain knowledge has been prominent and provides the highest test accuracy of 88.33% with k-NN wrapped classifier. Using the two image feature descriptors, i.e., LBP and 3LWD, DKGP could not achieve good results for the multiclass classification task. For instance, LBP and 3LWD provided the highest test accuracies of 58.75% and 70.00% on average with RF and k-NN, respectively. However, when these image features are combined with domain-specific features, there is a massive improvement of around 32% in test classification performance. For example, SVM achieved 55.42% classification accuracy on average using the LBP features, and it increased to 87.25% accuracy on average using 7PC+LBP features. This shows that 7PC features have high discriminative knowledge to distinguish between the three classes.

Among the four wrapped classification methods, k-NN achieved the highest classification performance of 88.33% and 88.89% on average in terms of balanced accuracy and F1-score using 7PC features, respectively. RF also achieved the highest performance with 7PC features. SVM achieved the highest performance of 87.25% and 86.08% in terms of balanced accuracy and F1-score, respectively, using the combination of domain-specific and LBP (7PC+LBP) features. NB achieved the highest performance of 84.50% balanced accuracy and 84.59% F1-score using 7PC features. These results are very encouraging for a very complex real-world problem of skin cancer detection where most of the classifiers are performing well for the multiclass classification task.

5.3 Overall Results

The outcomes of both binary and multiclass classification substantiate that the integration of domain knowledge within the proposed DKGP approach substantially enhances the classification algorithm's capability to construct a more precise classifier, as opposed to relying solely on features extracted from skin images. We have found an interesting behavior among all classification methods that they achieve better classification performance when utilizing the combination of domain-specific (7PC), and domain-independent (LBP, and 3LWD) feature sets, as compared to utilizing a single feature set. For instance, in the binary classification task, RF achieved 75.16% accuracy using LBP features and 77.06% accuracy using 7PC+LBP features, as shown in Table 2. Similarly, RF achieved 73.72% accuracy using 3LWD features and 76.04% accuracy using 7PC+3LWD features, as shown in Table 2. This trend has been seen in both binary and multiclass classification results. For example, in the multiclass classification task, SVM provided 55.42% accuracy using LBP features and 87.25% accuracy using 7PC+LBP features, as shown in Table 3. Similarly, SVM achieved 64.79% accuracy using 3LWD features and 81.67% accuracy using 7PC+3LWD features, as shown in Table 2. From these binary and multiclass classification results, it is concluded that utilizing domain-specific knowledge helps achieve performance gains in complex real-world image classification tasks.

Table 2. Results of the proposed DKGP method for **Binary Classification**: Accuracy and F1-score (%) on the test set.

Feature Set	NB		SVM		k-NN		RF	
	accuracy	F1-score	accuracy	F1-score	accuracy	F1-score	accuracy	F1-score
7PC	**88.69 ± 1.40**	85.91 ± 1.42	83.88 ± 1.90	84.32 ± 2.09	84.56 ± 2.21	85.89 ± 2.16	87.31 ± 1.92	87.48 ± 1.98
LBP	**76.41 ± 1.78**	75.78 ± 2.33	74.81 ± 2.43	74.67 ± 2.70	76.09 ± 2.47	77.53 ± 2.64	75.16 ± 2.53	76.91 ± 2.83
3LWD	74.69 ± 3.56	71.34 ± 3.68	**75.08 ± 2.62**	75.86 ± 2.50	79.69 ± 2.90	81.42 ± 3.07	73.72 ± 2.38	75.33 ± 2.46
7PC+LBP	84.31 ± 1.86	84.07 ± 1.60	**87.44 ± 2.13**	87.13 ± 2.29	80.25 ± 1.46	81.97 ± 1.61	77.06 ± 2.81	79.74 ± 2.70
7PC+3LWD	**80.31 ± 2.51**	77.12 ± 2.14	76.98 ± 2.59	80.40 ± 2.52	80.00 ± 1.63	82.00 ± 1.42	76.04 ± 2.56	79.19 ± 2.17

Table 3. Results of the proposed DKGP method for **Multiclass Classification**: Accuracy and F1-score (%) on the test set.

Feature Set	NB		SVM		k-NN		RF	
	accuracy	F1-score	accuracy	F1-score	accuracy	F1-score	accuracy	F1-score
7PC	84.50 ± 1.13	84.59 ± 1.19	86.81 ± 1.71	86.79 ± 1.84	**88.33 ± 1.78**	88.89 ± 1.97	87.50 ± 1.25	87.81 ± 1.23
LBP	52.50 ± 2.83	50.69 ± 3.02	55.42 ± 3.83	53.75 ± 3.61	58.33 ± 2.42	59.08 ± 2.36	**58.75 ± 1.42**	59.57 ± 1.69
3LWD	65.63 ± 2.63	63.88 ± 2.55	64.79 ± 2.21	64.42 ± 3.41	**70.00 ±1.42**	70.69 ± 1.65	67.71 ± 2.21	68.49 ± 2.28
7PC+LBP	80.00 ± 1.44	80.43 ± 1.26	**87.25 ± 2.53**	86.08 ± 2.82	80.21 ± 2.63	80.59 ± 2.77	79.58 ± 3.42	80.06 ± 3.69
7PC+3LWD	72.08 ± 2.92	70.24 ± 3.15	**81.67 ± 1.66**	82.63 ± 1.34	77.50 ± 2.83	78.51 ± 2.78	76.67 ± 2.00	77.45 ± 2.22

Table 4. Comparison with existing deep learning and GP methods on the PH2 dataset for Multiclass classification.

Method	task	Strategy	Results
Patino et al. [17]	multi	neural network	86.50%
Alkarakatly et al. [4]	multi	5-layer CNN	90.00% (overall)
Ain et al. [1]	binary	embedded GP	78.17%
DKGP	binary	domain knowledge GP wrapper	**88.69%**
=	multi	=	**88.33%**

5.4   Comparison with Existing Methods

Table 4 compares DKGP with existing methodologies, both in terms of the strategies employed and the results achieved. Patino et al. [17] attained a balanced accuracy of 86.50% within a multiclass classification context, employing 10-fold cross-validation. In the domain of binary classification, Ain et al. [1] devised a multi-tree GP approach for melanoma detection, securing a balanced accuracy of 78.17%. The uniformity in experimental settings across both [1,17] and our proposed DKGP method facilitates direct comparisons.

In binary classification, DKGP exhibits a performance of 88.69%, outperforming the second approach [1] by a substantial improvement of nearly 10% accuracy. For multiclass classification, DKGP achieves 88.33% accuracy, surpassing the method in [17] with an enhancement of nearly 2% accuracy. A direct comparison between DKGP and the method introduced by Alkarakatly et al. [4] poses challenges, as the latter relies on the overall accuracy within an imbalanced classification setting, potentially skewing outcomes towards majority class instances. Conversely, DKGP embraces balanced classification accuracy (as defined in Equation (1)), ensuring equal consideration for all classes.

5.5   Computation Time

Generally, the time required to train a deep learning method is quite long and usually takes several days. However, the proposed DKGP method is very efficient in terms of time computation. For instance, the longest computation time that DKGP utilized is only 7.08 min with k-NN as a wrapped classifier using

7PC+3LWD features in the multiclass classification task, which is very fast. Similarly, DKGP provided the highest classification performance of 88.69% and 88.33% in the binary and multiclass classification tasks utilizing only 3.36 and 2.47 s on average, respectively, to train the NB classifier with 7PC features. This shows the effectiveness and efficiency of the proposed DKGP method for skin cancer binary and multiclass image classification.

5.6 An Evolved GP Individual

We further demonstrate the effectiveness of our proposed DKGP method by analyzing a good evolved GP program as shown in Fig. 6. This individual is taken from the multiclass experiments where SVM is provided with the domain-specific features (7PC) and domain-independent features (LBP) in the form of a single feature set, i.e., 7PC+LBP. This GP individual achieved 87.96% classification accuracy on the test data of the PH2 dataset. The colored nodes represent terminal nodes and the white nodes represent function nodes. Since LBP features are extracted from RGB color channels, the red, green, and blue colored nodes correspond to the LBP feature extracted from red, green, and blue color channels of the dermoscopy images, respectively. The grey-colored nodes represent the 7PC domain-specific features.

We can clearly observe that, among the LBP features, the GP individual has selected a combination of different color channels which shows the importance of color in skin cancer image classification. This individual selects eight LBP features from the red channel (F6, F8, F32, F48, F49, F53, F58, F62), and six features each from the green channel (F70, F80, F98, F99, F108, F104) and seven features from the blue channel (F132, F139, F151, F156, F160, F172, F173). It is interesting to note that F8 and F80 appear three and two times, respectively, in this individual, showing their high discriminating ability between classes. Moreover, the expression $(\cos(F8) - \cos(F181))$ appears twice. To dig further into the binary pattern of these features, we found that F8 and F80 represent line ends and edges, which corresponds to the presence of streaks and regression areas in the skin lesions. This shows that this GP individual has achieved high classification performance by selecting important LBP texture patterns that have significant information about identifying dermoscopic criteria such as streaks and regression areas.

Among the 7PC features, the GP individual, as shown in Fig. 6 selects three features (F180, F181, and F182), which correspond to the presence/absence of streaks, regression areas, and blue-whitish veil, respectively. In addition, F181 has been selected twice showing its effectiveness in discriminating between benign melanoma, and atypical nevi classes.

Another important aspect is dimensionality reduction. This GP individual has selected 24 features (21 LBP and three 7PC) from a total of 189 7PC+LBP features. This shows that DKGP method provides dimensionality reduction of many folds while achieving good classification performance for the complex skin cancer image classification problem. This also shows the effectiveness of incorporating knowledge from dermatology and computer vision domains to improve classification performance in skin cancer images.

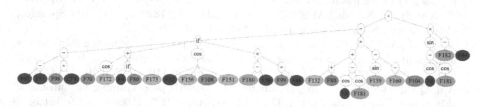

Fig. 6. A good evolved GP individual providing 87.96% balanced accuracy using SVM in the multiclass classification task on the unseen data.

6 Conclusions

This work has developed a GP-based wrapper approach utilizing domain knowledge in binary and multiclass skin cancer image classification. The proposed method incorporates various types of multi-channel and multi-resolution image features that possess important information related to RGB and gray-level pixel-based image properties, and combines them effectively with domain knowledge of dermatology to perform classification. The proposed DKGP method utilizes three types of features; domain knowledge extracted from 7PC, and domain-independent knowledge extracted from LBP, and 3LWD. The DKGP method efficiently employs diverse combinations of feature sets, offering them to GP for feature selection and subsequently employing a wrapper classification technique for accurate classification. This approach has demonstrated its efficacy in both binary and multiclass skin image classification tasks, surpassing the performance of existing GP and deep learning methods, thus affirming its capability to effectively discern differences between classes.

However, due to the constraint imposed by the availability of domain knowledge, the experiments are performed using only one dataset that comes along with this domain knowledge based on 7PC. We will investigate more datasets and suitable future extraction methods. Furthermore, our future pursuits will involve thoroughly exploring the integration of domain-specific knowledge with domain-independent image features, aiming to enhance classification performance.

References

1. Ain, Q.U., Al-Sahaf, H., Xue, B., Zhang, M.: A multi-tree genetic programming representation for melanoma detection using local and global features. In: Mitrovic, T., Xue, B., Li, X. (eds.) AI 2018. LNCS (LNAI), vol. 11320, pp. 111–123. Springer, Cham (2018). https://doi.org/10.1007/978-3-030-03991-2_12

2. Ain, Q.U., Al-Sahaf, H., Xue, B., Zhang, M.: Automatically diagnosing skin cancers from multimodality images using two-stage genetic programming. IEEE Trans. Cybern. **53**(5), 2727–2740 (2022). https://doi.org/10.1109/TCYB.2022.3182474

3. Ain, Q.U., Al-Sahaf, H., Xue, B., Zhang, M.: Genetic programming for automatic skin cancer image classification. Expert Syst. Appl. **197**, 116680 (2022). https://doi.org/10.1016/j.eswa.2022.116680

4. Alkarakatly, T., Eidhah, S., Al-Sarawani, M., Al-Sobhi, A., Bilal, M.: Skin lesions identification using deep convolutional neural network. In: Proceedings of the 2019 International Conference on Advances in the Emerging Computing Technologies, pp. 1–5. IEEE (2020)

5. Argenziano, G., Fabbrocini, G., et al.: Epiluminescence microscopy for the diagnosis of doubtful melanocytic skin lesions: comparison of the ABCD rule of dermatoscopy and a new 7-point checklist based on pattern analysis. Arch. Dermatol. **134**(12), 1563–1570 (1998)

6. Bisla, D., Choromanska, A., Berman, R.S., Stein, J.A., Polsky, D.: Towards automated melanoma detection with deep learning: data purification and augmentation. In: Proceedings of the IEEE Conference on Computer Vision and Pattern Recognition Workshops (2019). https://doi.org/10.1109/CVPRW.2019.00330

7. Chang, T., Kuo, C.C.J.: Texture analysis and classification with tree-structured wavelet transform. IEEE Trans. Image Process. **2**(4), 429–441 (1993)

8. Esteva, A., Kuprel, B., et al.: Dermatologist-level classification of skin cancer with deep neural networks. Nature **542**(7639), 115–118 (2017)

9. Garnavi, R., Aldeen, M., Bailey, J.: Computer-aided diagnosis of melanoma using border-and wavelet-based texture analysis. IEEE Trans. Inf. Technol. Biomed. **16**(6), 1239–1252 (2012)

10. Hall, M., Frank, E., Holmes, G., Pfahringer, B., Reutemann, P., Witten, I.H.: The WEKA data mining software: an update. ACM SIGKDD Explor. Newslett. **11**(1), 10–18 (2009)

11. Kawahara, J., BenTaieb, A., Hamarneh, G.: Deep features to classify skin lesions. In: Proceedings of the 13th International Symposium on Biomedical Imaging, pp. 1397–1400. IEEE (2016)

12. Koza, J.R.: Genetic Programming III: Darwinian Invention and Problem Solving, vol. 3. Morgan Kaufmann (1999)

13. Liu, Y., et al.: A deep learning system for differential diagnosis of skin diseases. Nat. Med. **26**(6), 900–908 (2020)

14. Luke, S.: Essentials of metaheuristics. Lulu, 2nd edn. (2013). http://cs.gmu.edu/sean/book/metaheuristics/

15. Mendonça, T., Ferreira, P.M., Marques, J.S., Marcal, A.R., Rozeira, J.: PH^2 - a dermoscopic image database for research and benchmarking. In: Proceedings of the 35th International Conference of the IEEE Engineering in Medicine and Biology Society, pp. 5437–5440. IEEE (2013)

16. Ojala, T., Pietikäinen, M., Harwood, D.: A comparative study of texture measures with classification based on featured distributions. Pattern Recogn. **29**(1), 51–59 (1996)

17. Patiño, D., Ceballos-Arroyo, A.M., Rodriguez-Rodriguez, J.A., Sanchez-Torres, G., Branch-Bedoya, J.W.: Melanoma detection on dermoscopic images using superpixels segmentation and shape-based features. In: Proceedings of the 15th International Symposium on Medical Information Processing and Analysis, vol. 11330, p. 1133018. International Society for Optics and Photonics (2020)
18. Satheesha, T., Satyanarayana, D., Prasad, M.G., Dhruve, K.D.: Melanoma is skin deep: a 3D reconstruction technique for computerized dermoscopic skin lesion classification. IEEE J. Transl. Eng. Health Med. **5**, 1–17 (2017)
19. Siegel, R.L., Miller, K.D., Wagle, N.S., Jemal, A.: Cancer statistics, 2023. CA: A Cancer J. Clin. **73**(1), 17–48 (2023)
20. Stolz, W., Riemann, A., et al.: ABCD rule of dermatoscopy: a new practical method for early recognition of malignant-melanoma. Eur. J. Dermatol. **4**(7), 521–527 (1994)
21. Tran, B., Xue, B., Zhang, M.: Genetic programming for multiple-feature construction on high-dimensional classification. Pattern Recogn. **93**, 404–417 (2019)
22. Xie, F., Fan, H., Li, Y., Jiang, Z., Meng, R., Bovik, A.: Melanoma classification on dermoscopy images using a neural network ensemble model. IEEE Trans. Med. Imaging **36**(3), 849–858 (2017)

Comparison of Interpolation Techniques for Prolonged Exposure Estimation: A Case Study on Seven Years of Daily Nitrogen Oxide in Greater Sydney

Prathayne Nanthakumaran[1](✉) ⓘ and Liwan Liyanage[2] ⓘ

[1] School of Computer, Data and Mathematical Sciences, Western Sydney University, Penrith, Australia
19625690@student.westernsydney.edu.au
[2] School of Computer, Data and Mathematical Sciences, Western Sydney University, Penrith, Australia
L.Liyanage@westernsydney.edu.au

Abstract. Continuous exposure to air pollutants over a long period of time adversely affects population health. Addressing this issue may help in reducing the disease burden. Thus, it is crucial to understand the spatial and spatiotemporal variation in this prolonged exposure to ambient air pollutants to make informed decisions. The objective of this study is to evaluate the performance of most commonly used spatial interpolation techniques in sparsely located real-world sensor data for the purpose of estimating the prolonged exposure to air pollutants. The secondary data obtained from NSW Air Quality Monitoring Network (AQMN) sites within Greater Sydney during 1st January 2011 - 31st December 2017 by considering the daily concentrations of Nitrogen Oxide (NO) were used for this study. Nearest Neighbour (NN) interpolation, Inverse Distance Weighted (IDW) interpolation without search radius and with search radius (10 km, 15 km, 20 km, 25 km, 30 km, 35 km, 40 km, 45 km and 50 km) were used to estimate the daily concentrations at unknown locations. The performance of these interpolation techniques was assessed based on leave location-out cross-validation (LLO-CV) using Root Mean Square Error (RMSE), Index of Agreement (d) and Coefficient of Determination (R^2). Results revealed that, IDW with search radius of 25 km and power value of one performed better for the given dataset. IDW outperformed NN interpolation technique. These findings may help policy makers to come up with strategies for disease management, control and mitigation.

Keywords: Prolonged exposure · Nearest Neighbour Interpolation · Inverse Distance Weighted Interpolation · Leave location-out cross-validation

1 Introduction

According to World Health Organization, in 2019 about 99% of the global population was exposed to air quality without adhering the WHO air quality

© The Author(s), under exclusive license to Springer Nature Singapore Pte Ltd. 2024
D. Benavides-Prado et al. (Eds.): AusDM 2023, CCIS 1943, pp. 270–283, 2024.
https://doi.org/10.1007/978-981-99-8696-5_19

guidelines [1]. Continuous exposure to air pollutants over a long period of time adversely affects population health. Diseases such as cataract [2–4], diabetes [5], cancer [6–9], cardio-vascular diseases [10–12] are linked with continuous exposure to air pollutants over a long period. Addressing this issue may help in reducing the disease burden. Estimating the prolonged exposure to air pollutants is a crucial step in identifying this association of the air pollutants with a given disease for disease management, control and mitigation. Thus, it is crucial to understand the spatial and spatio-temporal variation in the prolonged exposure to ambient air pollutants to make informed decisions.

The ambient air pollution levels are measured using the sensors located at monitoring sites across space at short time period such as hourly or daily records [13]. It is of interest to estimate the average exposure over a long span of time as several years. Estimating this prolonged exposure to ambient air pollutants using available sparsely located sensor data is to be handled spatially and temporally. The spatial aspect of the problem involves estimating the air pollutant concentrations at unknown locations using values at known locations. The temporal aspect of the problem involves estimating the average exposure to air pollutants at a given location over a span of time period.

When estimating the prolonged exposure using short-term measurements, majority of the techniques focus on either improving prediction for high resolution spatially while considering annual averages in time [14–18]; or considering a spatial point (i.e. centroid) to address the variation within the spatial polygon (i.e. spatial area) with or without considering high resolution in time [19]. This needs further exploration with respect to the selected data in analysing the efficiency of these techniques. There is a huge need in improving the estimation of prolonged exposure using short-term measurements addressing high resolution space and time especially due to the highly fluctuating nature of these data. Our research focuses on evaluating the performance of estimating prolonged exposure when considering high resolution spatially as well as temporally. The objective of this paper is to evaluate the performance of the most commonly used spatial interpolation techniques in real-world sparsely located sensor data for the purpose of estimating the prolonged exposure to air pollutants.

2   Methods and Methodology

2.1   Data Collection and Pre-processing

Data used for this study are Ambient Nitrogen Oxide (NO) Concentrations measured through an advanced Air Quality Monitoring System, the NSW Air Quality Monitoring Network (AQMN) [13]. These data were publicly available and downloaded by using the Data download facility provided under the *"Enhance air quality website and data delivery"* (EWADD) project [13]. Daily averages of ambient Nitrogen Oxide Concentrations at the sites located within or close to the Greater Sydney region and from 1st January 2011 - 31st December 2017 were considered for this study.

These data were found with a lot of missing values with gaps ranging from a few days to years. The missing values at each sites were explored using missing percentage and the longest gap [20]. The monitoring sites with missing per cent less than 20% and the longest gap of less than 30 consecutive days were selected [20]. These selected sites were imputed using Kalman Smoothing on Structural Time Series method in the imputeTS R package [21] as it has been proven efficient for relatively small missing values [22]. Missing values with large gaps were imputed using a bi-directional method based on regularized regression models [23]. The data thus prepared was used for further analysis. The sites selected for further analysis is as shown in Fig. 1.

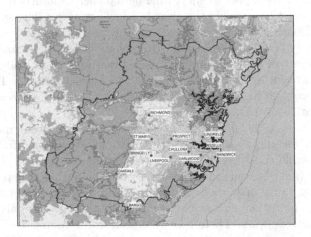

Fig. 1. NSW Air Quality Monitoring Network sites selected for further analysis of Nitrogen Oxide (NO)

2.2   Spatial Interpolation

It is of interest to estimate the daily NO concentrations at unknown locations using the available data. The idea is based on Tobler's first law of geography, *"everything is related to everything else, but near things are more related than distant things"* [24]. A variety of techniques are available for spatial interpolation. Some of them are non-geostatistical techniques such as Nearest neighbours, Natural Neighbours, Inverse Distance weighted, Trend surface analysis; geo-statistical techniques such as kriging and ensemble techniques [25]. While these methods are firmly established for spatial interpolation and temporal variability, accommodating daily fluctuations to estimate the prolonged exposure over several years becomes intricate due to the computational challenges posed by high-dimensional time points. Nearest Neighbour (NN) and Inverse Distance Weighting (IDW) interpolation technique were chosen for this study due to their simpler computational implementation for daily records of seven years of data with better results [19].

Nearest Neighbour Interpolation. The nearest neighbour interpolation is a straightforward interpolation technique, relying on the Voronoi diagram. This Voronoi diagram is constructed by assigning the unknown point location in space with the daily NO concentration of the nearest known point location from the set of point locations with known daily NO concentration [26].

Inverse Distance Weighted Interpolation. According to IDW interpolation technique, the daily NO concentration at an unknown location is calculated as,

$$u(\mathbf{x}) = \begin{cases} \frac{\sum_{d-1}^{N} w_i(\mathbf{x})u(\mathbf{x_i})}{\sum_{i=1}^{N} w_i(\mathbf{x})}, & \text{if } d(\mathbf{x}, \mathbf{x}_i) \neq 0 \text{ for all } i \\ u(x_i), & \text{if } d(\mathbf{x}, \mathbf{x}_i) = 0 \text{ for some } i \end{cases} \tag{1}$$

$$w_i(x) = \frac{1}{d(x, x_i)^p} \tag{2}$$

where $u(x)$ is predicted value at location x, w_i is the IDW weighting function assigned, $u(x_i)$ observed value at location x_i, $d(x, x_i)^p$ is the geographic distance between unknown location x and known location x_i and p is the power value [27,28]. The choice of p affects the influence of the weighting function. The most commonly used weighting function are reciprocal of distance (when $p = 1$) or squared distance (when $p = 2$) [27,28].

2.3   Performance Measures

The performance of these interpolation techniques was assessed using the following most commonly used performance measures.

The root mean square error (RMSE) is given by [29],

$$RMSE = \sqrt{[\frac{1}{n}\sum_{i=1}^{n}(O_i - P_i)^2]} \tag{3}$$

The index of agreement, proposed by [29], is given by

$$d = 1 - \frac{\sum_{i=1}^{n}(O_i - P_i)^2}{\sum_{i=1}^{n}(|P_i - \bar{O}| + |O_i - \bar{O}|)^2}, 0 \leq d \leq 1 \tag{4}$$

The coefficient of determination (R^2) [22] is given by,

$$R^2 = [\frac{\sum_{i=1}^{n}[(P_i - \overline{(P)})(O_i - \overline{(O)})]}{n.\sigma_p\sigma_o}]^2 \tag{5}$$

where O_i is the observed data point, P_i is the interpolated data point, $\bar{O}$ and $\bar{P}$ are the means of observed and interpolated data points respectively, σ_O and σ_P are the standard deviations of observed and interpolated data points respectively-and n is the total number of observations.

2.4 Approach

Daily NO concentrations are spatio-temporal in nature. It is of interest to estimate the daily NO concentrations at unknown locations over the period of time considered. Let $X_{S,T}$ be the daily NO concentration at site, S on a given day T. Here $S = 1, 2, 3, ..., s; T = 1, 2, 3, ..., t$ where s is the total number of sites in the study area and t is the total number of days in the study period. Different spatial models were obtained for each day T of the study period by considering only the spatial variation. The performance at a given site is calculated as the average of the performance measure over the study period for that site S. The performance at different sites were explored based on Leave-Location-out cross validation (LLO-CV) technique to avoid spatial over-fitting [30] as illustrated in Fig. 2.

3 Results and Discussion

3.1 Spatial Location of Sites

Figure 3 (A) and (B) summarises the elevation of the sites and the geographic distance between two sites respectively. It can be seen clearly that Bargo and Oakdale are located at the highest elevation compared to the other sites. Figure 1 and Fig. 3(B) depict that the sites are not evenly distributed across space. Oakdale is the closest site to Bargo while Bargo is located further away from the rest of the sites. On the other hand, sites such as Chullora, Liverpool, Prospect and Rozelle have more sites located closer to each other (Fig. 3(B)).

3.2 Comparison of the Performance of Spatial Interpolation Techniques

The overall performance of Nearest Neighbour (NN) interpolation and Inverse Distance Weighted (IDW) interpolation technique for the power of values one and two were compared as shown in Fig. 4. Results showed that IDW outperformed NN technique. This result was consistent for all three performance measures Root Mean Square Error (RMSE), Index of Agreement (d) and Coefficient of Determination (R^2) considered. Additionally, there was no significant difference in the overall performance between IDW with power values of one and two. The performance of these techniques was further explored in detail at each site as shown in Fig. 5.

Figure 5 showed that there was no significant difference in the performance of IDW techniques at sites for different power values of one and two. This result was consistent with the three performance measures considered. However, the performance of NN and IDW interpolation varied at different sites. According to RMSE, IDW performed better at Bargo, Bringelly, Liverpool, Prospect, Randwick and Rozelle compared to NN interpolation. However, NN interpolation performed better than IDW interpolation at Oakdale and Richmond. There

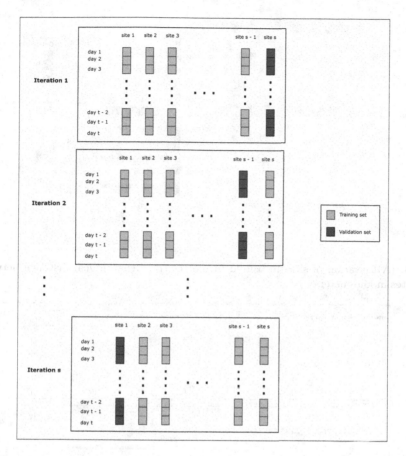

Fig. 2. Schematic representation of Leave-Location-out cross-validation

was no significant difference in the performance of these techniques at Chullora, Earlwood and Lindfield.

According to d, IDW interpolation outperformed NN interpolation at Bargo, Bringelly, Liverpool, Prospect, Randwick and Rozelle whereas vice-versa at Richmond. Additionally, there was no significant difference in the performance of these techniques at Chullora, Earlwood, Lindfield, Oakdale and St.Marys. According to R^2, IDW interpolation outperformed NN interpolation at Bargo, Chullora, Earlwood, Liverpool, Prospect, Randwick and Rozelle whereas no significant difference at Bringelly, Lindfield, Oakdale and St.Marys.

For a much more detailed understanding, the observed values at each site location were compared with the predicted values based on NN interpolation (Fig. 6) and IDW interpolation (Fig. 7).

Figure 6 and 7 highlighted that NN interpolation significantly underestimated the daily NO concentrations at Bargo, Liverpool and Prospect whereas significantly overestimated at Bringelly and Oakdale. IDW interpolation showed

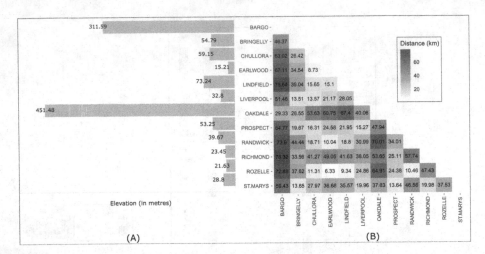

Fig. 3. (A)Elevation of sites measured in metres (B) Geographical distance between two sites in kilo-meters

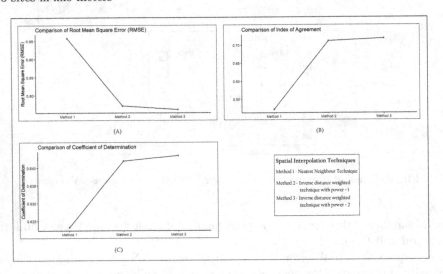

Fig. 4. Comparison of the performance of Nearest Neighbour interpolation, Inverse distance weighted interpolation with power = 1 and Inverse distance weighted interpolation with power = 2. (A) Based on Root Mean Square Error (RMSE) (B) Based on Index of Agreement (d) (C) Based on Coefficient of determination (R^2)

improved predictions at Bargo, Bringelly, Liverpool, Prospect and St.Marys. However, predictions at Liverpool, Earlwood and Prospect can be improved further.

Note that at Chullora, NN performed better than IDW interpolation. Additionally, both NN and IDW interpolation performed very poorly at Oakdale. Note that Oakdale is located at the highest elevation of approximately 451 m

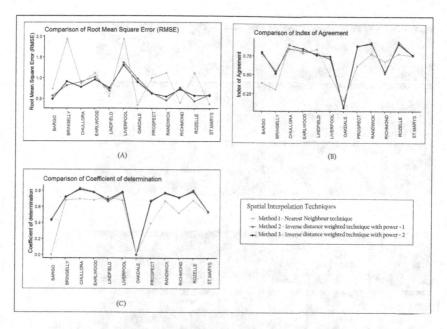

Fig. 5. Comparison of the performance of Nearest Neighbour interpolation, Inverse distance weighted interpolation with power = 1 and Inverse distance weighted interpolation with power = 2 at each site. (A) Based on Root Mean Square Error (RMSE) (B) Based on Index of Agreement (d) (C) Based on Coefficient of determination (R^2)

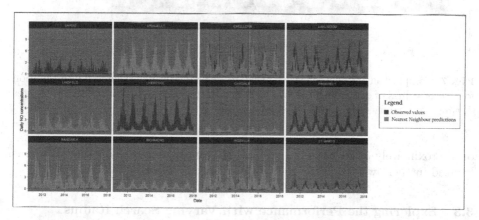

Fig. 6. Comparison of the predictions based on Nearest Neighbour interpolation with the observed values at different sites during 1st January 2011 - 31st December 2017

(as shown in Fig. 3) and further away from the rest of the sites. The poor results at Oakdale may be due to these geographic locations and further investigation is needed. However, the IDW prediction at Bargo, a site located at an elevation

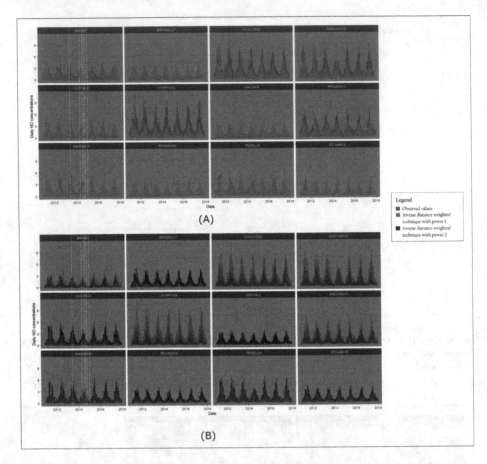

Fig. 7. Comparison of the predictions based on Inverse Distance Weighted interpolation for power values of one in (A) and two in (B) with the observed values at different sites during 1st January 2011 - 31st December 2017

of approximately 312 m, was not so bad. It is interesting to note that Bargo is located further away from the rest of the sites but closer to Oakdale.

3.3   Exploring the Performance with Varying Search Radius

The performance of IDW interpolation was explored by introducing a search radius. Figure 8 visualizes the selection of sites for interpolation based on search radius by using Earlwood as an example. This process is repeated for all the sites by varying search radius between 5 km - 50 km and the results were summarised in Table 1.

Within a 5 km radius, none of the sites have their respective neighbours. Only four sites (Chullora, Earlwood, Lindfield, Rozelle) have their respective neighbour within a 10 km radius. All sites except (Bargo, Oakdale and Richmond)

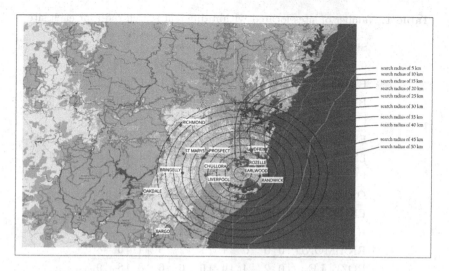

Fig. 8. A visual illustration of selecting search radius considering Earlwood as an example. Here black dotted circle of different radii (5 km to 50 km) highlight the monitoring sites within respective distances from Earlwood.

have at least one of their respective neighbour within a 15 km radius. The respective neighbour of Richmond is located within a 15 km–20 km search radius. It is worth noting that Bargo and Oakdale do not have any sites located within a 25 km radius. However, there exist one and two neighbours within 25 km–30 km of Bargo and Oakdale respectively.

Figure 9 clearly highlights the variation in performance for different search radii. Same as earlier, there was no significant difference in the performance of IDW interpolation for power values of one and two. Overall root mean square error was very low for a search radius of 10 km followed by a significant raise. The root mean square error value dropped to the search radius of 25 km and then increased. These findings were consistent with the Index of Agreement and Coefficient of Determination. It is important to note that only four sites had neighbours located within a 10 km radius. These sites had either only one or two neighbouring sites available for interpolation where as more neighbouring sites were available for interpolation when search radius was increased to 25 km.

Table 2 summarises the overall performance of NN interpolation technique, IDW interpolation utilizing search radius of 10 km and 25 km and IDW interpolation without utilizing search radius, based on performance measures RMSE, d and R^2. All the IDW interpolation outperformed the NN interpolation. The overall performance of IDW interpolation utilizing search radius performed better than that of, without utilizing search radius. Moreover, the IDW interpolation utilizing a search radius of 10 km performed better than that of the search radius of 25 km. However, it is important to note that the overall performance results were based on only four sites when considering search radius of 10 km, whereas the results were based on ten sites, when considering the search radius of 25 km.

Table 1. Number of sites located within different search radii for each site

Sites	Search Radius (in km)									
	5	10	15	20	25	30	35	40	45	50
BARGO	0	0	0	0	0	1	1	1	1	2
BRINGELLY	0	0	2	3	3	5	7	9	10	11
CHULLORA	0	1	3	6	6	8	8	8	9	9
EARLWOOD	0	2	3	4	6	6	7	8	8	9
LINDFIELD	0	1	1	4	5	6	6	8	9	9
LIVERPOOL	0	0	2	4	6	7	8	9	10	10
OAKDALE	0	0	0	0	0	2	2	3	4	5
PROSPECT	0	0	1	4	7	8	9	9	9	10
RANDWICK	0	0	2	4	4	4	6	6	7	8
RICHMOND	0	0	0	1	1	2	3	4	6	8
ROZELLE	0	2	4	4	6	6	6	8	8	9
ST.MARYS	0	0	2	4	4	5	5	9	9	10

Table 2. A comparison of the overall performance of different spatial interpolation techniques considered

Methods	RMSE	d	R^2
Nearest Neighbour Interpolation	0.951	0.623	0.521
IDW interpolation of power 1 without utilizing search radius	0.771	0.712	0.642
IDW interpolation of power 2 without utilizing search radius	0.761	0.719	0.644
IDW interpolation of power 1 utilizing search radius of 10 km	0.714	0.884	0.735
IDW interpolation of power 2 utilizing search radius of 10 km	0.749	0.874	0.730
IDW interpolation of power 1 utilizing search radius of 25 km	0.747	0.791	0.711
IDW interpolation of power 2 utilizing search radius of 25 km	0.749	0.792	0.708

Additionally, there was no significant difference in the power of IDW interpolation for this dataset. However, IDW interpolation based on the power value of one showed a slightly improved result despite the search radius.

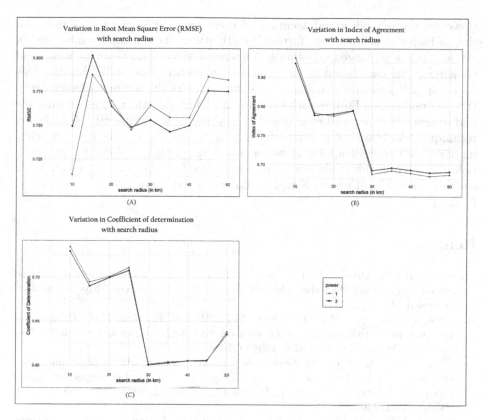

Fig. 9. Variation in the performance of Inverse distance weighted interpolation technique of power (p = 1 and 2) with search radius 5 km, 10 km, 15 km, 20 km, 25 km, 30 km, 35 km, 40 km, 45 km and 50 km. (A)based on root mean square error (RMSE) (B) based on Index of Agreement (C) based on Coefficient of determination

4   Conclusion

The conclusions obtained were based on the daily concentrations of ambient Nitrogen Oxide (NO) at sites within Greater Sydney during 1st January 2011 - 31st December 2017. The performance of spatial interpolation techniques Nearest Neighbour (NN) interpolation and Inverse Distance Weighted (IDW) interpolation was assessed using Root Mean Square Error (RMSE), Index of Agreement (d) and Coefficient of Determination (R^2) based on leave-location-out cross-validation (LLO-CV). IDW interpolation outperformed NN interpolation. Additionally, findings showed an improvement in results when using IDW interpolation with a search radius of 10 km and 25 km over the one without utilizing a search radius. To be more precise, IDW interpolation using a search radius of 10 km demonstrated slightly better performance compared to a radius of 25 km. However, the lack of monitoring sites still remains as an issue. Incorporating low-cost sensors in addition to these high cost sensors could be one of the solutions.

However, the reliability and accuracy of these low-cost sensor recordings with that of high-cost sensors need further investigation. These results can be used in estimating the prolonged exposure to Nitrogen Oxide at unknown locations and identifying the association of the prolonged exposure with public health. This may help policy makers to come up with strategies for disease management, control and mitigation. Future work involves extending the study by evaluating the performance of other well-established spatial and spatiotemporal interpolation techniques. Additionally, it is important to note that the results were considering daily NO concentrations. However, these results might differ for other air pollutants. Thus, this study is also to be extended for other major air pollutants such as ground level ozone, Nitrogen dioxide, Particulate matters (PM10 and PM2.5).

References

1. World Health Organization - Ambient (outdoor) air pollution. https://www.who.int/news-room/fact-sheets/detail/ambient-(outdoor)-air-quality-and-health. Accessed 24 Aug 2023
2. Choi, Y., Park, S.J., Paik, H.J., Kim, M.K., Wee, W.R., Kim, D.H.: Unexpected potential protective associations between outdoor air pollution and cataracts. Environ. Sci. Pollut. Res. **25**, 10636–10643 (2018)
3. Shin, J., Lee, H., Kim, H.: Association between exposure to ambient air pollution and age-related cataract: a nationwide population-based retrospective cohort study. Int. J. Environ. Res. Public Health **17**(24), 9231 (2020). https://doi.org/10.3390/ijerph17249231
4. Chua, S.Y.L., et al.: The association of ambient air pollution with cataract surgery in UK biobank participants: prospective cohort study. Invest. Ophthalmol. Vis. Sci. **62**(15), 7–7 (2021)
5. Janghorbani, M., Momeni, F., Mansourian, M.: Systematic review and metaanalysis of air pollution exposure and risk of diabetes. Eur. J. Epidemiol. **29**, 231–242 (2014)
6. Guo, Y., et al.: The association between lung cancer incidence and ambient air pollution in China: a spatiotemporal analysis. Environ. Res. **144**, 60–65 (2016)
7. Pope Iii, C.A., et al.: Lung cancer, cardiopulmonary mortality, and long-term exposure to fine particulate air pollution. JAMA **287**(9), 1132–1141 (2002)
8. Katanoda, K., et al.: An association between long-term exposure to ambient air pollution and mortality from lung cancer and respiratory diseases in Japan. J. Epidemiol. **21**(2), 132–143 (2011)
9. Chen, X., et al.: Long-term exposure to urban air pollution and lung cancer mortality: a 12-year cohort study in Northern China. Sci. Total Environ. **571**, 855–861 (2016)
10. Lee, B., Kim, B., Lee, K.: Air pollution exposure and cardiovascular disease. Toxicol. Res. **30**, 71–75 (2014)
11. Beelen, R., et al.: Long-term exposure to air pollution and cardiovascular mortality: an analysis of 22 European cohorts. Epidemiology, 368–378 (2014)
12. Meo, S. A., Suraya, F.: Effect of environmental air pollution on cardiovascular diseases. Euro. Rev. Med. Pharmacol. Sci. **19**(24), (2015)

13. Riley, M., Kirkwood, J., Jiang, N., Ross, G., Scorgie, Y.: Air quality monitoring in NSW: from long term trend monitoring to integrated urban services. Air Quality Climate Change **54**(1), 44–51 (2020)
14. Brauer, M., et al.: Ambient air pollution exposure estimation for the global burden of disease 2013. Environ. Sci. Technol. **50**(1), 79–88 (2016)
15. Liu, C., et al.: Associations between ambient fine particulate air pollution and hypertension: a nationwide cross-sectional study in China. Sci. Total Environ. **584**, 869–874 (2017)
16. Wang, Y., et al.: Spatiotemporal analysis of PM2. 5 and pancreatic cancer mortality in China. Environ. Res. **164**, 132–139 (2018)
17. Rushworth, A., Lee, D., Mitchell, R.: A spatio-temporal model for estimating the long-term effects of air pollution on respiratory hospital admissions in Greater London. Spatial Spatio-Temp. Epidemiol. **10**, 29–38 (2014)
18. Yazdi, M.D., et al.: The effect of long-term exposure to air pollution and seasonal temperature on hospital admissions with cardiovascular and respiratory disease in the United States: a difference-in-differences analysis. Sci. Total Environ. **843**, 156855 (2022)
19. Hanigan, I., Hall, G., Dear, K.B.G.: A comparison of methods for calculating population exposure estimates of daily weather for health research. Int. J. Health Geogr. **5**, 1–16 (2006)
20. Wijesekara, L., Nanthakumaran, P., Liyanage, L.: Space and time data exploration of air quality based on PM10 sensor data in greater Sydney 2015–2021. In: 15th International Conference on Sensing Technology, pp. 295–308. Springer, Sydney (2022). https://doi.org/10.1007/978-3-031-29871-4_30
21. Moritz, S., Bartz-Beielstein, T.: imputeTS: time series missing value imputation in R. R J. **9**(1), 207 (2017)
22. Wijesekara, W.M.L.K.N., Liyanage, L.: Comparison of imputation methods for missing values in air pollution data: case study on Sydney air quality index. In: Arai, K., Kapoor, S., Bhatia, R. (eds.) FICC 2020. AISC, vol. 1130, pp. 257–269. Springer, Cham (2020). https://doi.org/10.1007/978-3-030-39442-4_20
23. Wijesekara, L., Liyanage, L.: Air quality data pre-processing: a novel algorithm to impute missing values in univariate time series. In: 33rd IEEE International Conference on Tools with Artificial Intelligence, pp. 996–1001. IEEE, Washington DC (2021)
24. Tobler, W.R.: A computer movie simulating urban growth in the Detroit region. Econ. Geogr. **46**(sup1), 234–240 (1970)
25. Li, J., Heap, A.D.: Spatial interpolation methods applied in the environmental sciences: a review. Environ. Modell. Softw. **53**, 173–189 (2014)
26. Boots, B., Sugihara, K., Chiu, S.N., Okabe, A.: Spatial Tessellations- Concepts and Applications of Voronoi Diagrams, 2nd edn. John Wiley & Sons, West Sussex, England (2009)
27. Deligiorgi, D., Philippopoulos, K.: Spatial interpolation methodologies in urban air pollution modeling: application for the greater area of metropolitan Athens, Greece. Adv. Air Pollut. **17**(5), 341–362 (2011)
28. Shepard, D.: A two-dimensional interpolation function for irregularly-spaced data. In: Proceedings of 23rd ACM National Conference, pp. 517–524 (1968)
29. Willmott, C.J.: On the validation of models. Phys. Geogr. **2**(2), 184–194 (1981)
30. Meyer, H., Reudenbach, Hengl, T., Katurji, M., Nauss, T.: Improving performance of spatio-temporal machine learning models using forward feature selection and target-oriented validation. Environ. Model. Softw. **101**, 1–9 (2018)

Detecting Asthma Presentations from Emergency Department Notes: An Active Learning Approach

Sedigh Khademi[1,2]([✉]) [iD], Christopher Palmer[1] [iD], Muhammad Javed[1] [iD],
Gerardo Luis Dimaguila[1,2] [iD], Jim P. Buttery[1,2,3] [iD], and Jim Black[4,5] [iD]

[1] Health Informatics Group, Centre for Health Analytics, Melbourne Children's Campus, Melbourne, Australia
{sedigh.khademi,chris.palmer,muhammad.javed,
gerardoluis.dimaguil,jim.buttery}@mcri.edu.au
[2] Department of Paediatrics, University of Melbourne, Melbourne, Australia
[3] Infectious Diseases, Royal Children's Hospital, Melbourne, Australia
[4] Melbourne School of Population and Global Health, University of Melbourne, Melbourne, Australia
jim.black@health.vic.gov.au
[5] Department of Health, State Government of Victoria, Melbourne, Australia

Abstract. Emergency department (ED) triage notes contain valuable information for near real-time syndromic surveillance of emergent public health events. With the increasing use of machine learning algorithms for classification of ED triage notes, active learning (AL) offers a way to automatically obtain high-quality data for labelling, resulting in a reduced requirement for annotators to search for suitable data. Active learning involves selecting the most valuable data to enhance the model's learning process, guided by a query strategy that identifies samples with most impact on the model's training. The objective of this study was to assess the effectiveness of active learning in developing a high-performing model for ED syndrome detection. The research aimed to explore pool-based active learning and investigated various query strategies to improve the model's performance in identifying asthma presentations from ED triage notes. Our results showed that AL can be highly effective for reducing annotation effort while building reliable models. Uncertainty sampling strategy outperformed all other methods, achieving an F1 score of 0.91 to improve the baseline score by 7.6 percentage points.

Keywords: Active learning · large language models · emergency department · syndromic surveillance · natural language processing · chief complaints

1 Introduction

Emergency triage is the process of rapidly assessing and prioritizing patients based on their clinical presentations. Triage notes are the documents that are created during this process and contain a wealth of information about patients' medical conditions.

© The Author(s), under exclusive license to Springer Nature Singapore Pte Ltd. 2024
D. Benavides-Prado et al. (Eds.): AusDM 2023, CCIS 1943, pp. 284–298, 2024.
https://doi.org/10.1007/978-981-99-8696-5_20

These notes serve as valuable resources for identifying cases and syndromes in real-time, enhancing detection, and facilitating effective planning for public health events. In recent years, there has been an increase in the use of machine learning algorithms for surveillance and classification of ED triage notes [1]. These algorithms can be used to identify patients who are at risk for certain conditions, or to track the spread of disease.

Most modern Natural Language Processing (NLP) systems designed for detecting medical conditions rely on machine learning models trained using labelled data. The effectiveness of these models is determined by the amount of quality training data they have access to. Although these models can achieve remarkable results with sufficient supervision, gathering enough labelled data is a costly and manual process. The labelling process requires a detailed understanding of the problem domain, and the time that domain experts have available for labelling is limited. This limitation often becomes the primary factor that hinders the creation of new NLP models [2]. Active learning (AL) is an established approach to address the labelling bottleneck.

Active learning combines machine learning and human input to select the most valuable data to enhance the machine learning process. The machine learner is guided by "consulting the oracle" of a domain expert to identify information that yields the greatest understanding of the domain in the shortest amount of time [3]. Active learning is an iterative procedure that utilizes machine-driven algorithms to extract data from a pool of unlabeled data. Experts then annotate this data, which is used to enhance the machine learning model through training. This process is repeated until a stopping criterion is met. The goal is to build a high-performing model while minimizing the specialized human effort needed to locate records for annotation. Machine learning methods are employed to identify the most likely valuable data for annotation instead. The focus of this paper is to investigate whether the use of active learning can help train a high-performing model with reduced human annotation for clinical text classification tasks, specifically near real-time syndromic detection of asthma from emergency department notes. Early detection of changes in asthma presentation patterns, such as those that occur during thunderstorm asthma events, can inform staffing and resourcing and improve care [4].

1.1 Background

Active Learning (AL) comprises three main scenarios: (1) Pool-based AL, where the learner accesses a fixed set of unlabeled instances known as the pool; (2) Stream-based AL, where the learner receives instances one at a time; and (3) Membership query synthesis, where the learner generates new artificial instances to be labeled. When the pool-based scenario operates on a batch of instances rather than a single instance, it is referred to as batch-mode AL [5].

Pool-based sampling is a widely used approach to active learning [6]. To describe the algorithm of this approach, we have a small, labelled dataset L and a large unlabelled dataset U. The objective is to select the most suitable samples from U for labelling, in an iterative manner. At each step, the active learning algorithm leverages the information present in L to train a model M, then a query strategy is employed to identify the optimal candidates in U for further labelling, denoted as I. An oracle then assigns labels to I, which is subsequently incorporated into L. The process repeats until the stop criterion is met. Algorithm 1 illustrates this process.

Algorithm 1. Active learning procedure

Input: A small, labelled dataset L and an unlabeled data pool U
Output: The final labelled dataset L
while stopping criteria not met:
   M = train (L, U) # train the Model using L for training
   x = Predict (M, U)
   I = Query(X) # Select the optimal candidates I from U
   O = annotate(I)   # Get the label for x from the oracle
   L.append(O) # Add O to the labelled dataset
   U.remove(O)
   Return L,M

The query strategy is a fundamental component of active learning, wherein a selective approach chooses the samples to be labelled by an oracle or domain expert [7]. This involves selecting instances that are expected to have the greatest impact on the model's learning process, typically by focusing on samples that are difficult to correctly identify, or marginal to, or representative of the underlying data distribution. As a result, the model can effectively learn from a limited labeled dataset and achieve high performance with minimal human annotation effort.

The query strategy should aim to choose either (or both) the most informative or (and) the most representative samples [8]. A strategy that identifies *informative* samples ensures that the selected instances provide insights into the margins or edges of the underlying data distribution, addressing areas of uncertainty and improving the model's overall understanding of its data boundaries. There are several query strategies that fall into this category. *Uncertainty sampling* involves selecting instances that the model is most uncertain about [9]. *Disagreement sampling* entails choosing instances based on disagreements among multiple models [10]. *Gradient* information measures informativeness based on the norm of the gradient [11]. *Performance prediction* selects instances that would result in the greatest reduction in error if labelled and added to the training set [12].

A strategy that selects *representative* samples ensures that the chosen instances adequately cover the main patterns and characteristics of the dataset, allowing the model to generalize well to similar unseen data. *Density-based* strategies work by clustering the unlabelled data so that instances can be identified from clusters to represent the main features of the unlabelled data [13]. *Discriminative* methods select instances that are different from already labelled instances. This can be done by finding instances that are misclassified by the current model [14]. These methods aim to ensure batch diversity, so that instances that have diverse characteristics are selected. This can be done by iteratively selecting instances that are different from the previously selected instances [15], or by using clustering-based approaches [16]. Many studies have explored hybrid query strategies, considering both the uncertainty and diversity of query samples [17–19].

Active learning has found applications in various clinical tasks, such as concept extraction [20], medical records deidentification [21], text classification [22, 23], and named entity recognition [24].

Emergency department triage notes are collected at the start of a patient's visit and are used to prioritize their care. They can also be used to track trends in patient visits in near real time. The content of these notes is concise and lacks grammatical structure, is fragmented, and comprises short phrases, measurements, and abbreviations

[25]. This characteristic gives rise to a distinctive language specific to triage notes. Developing robust supervised machine learning (ML) classifiers for these notes relies on obtaining high-quality manual annotations. However, this process is expensive as it demands clinical expertise from healthcare professionals who are familiar with the language used in emergency care.

Some research studies have used coding against pre-existing data as a substitute for labeling triage notes [26]. On the other hand, some researchers have manually annotated their datasets [27, 28]. It is important to note that using codes as the gold standard method has well-documented limitations. These codes may not always accurately reflect the actual reason for a patient's visit. For example, codes could be assigned to identify the underlying cause of a condition, or for purposes such as financial incentives [29, 30]. Often, only a single code is used, which might not correlate to the surveillance condition of interest that is available in the ED text. For example, asthma related to a chest infection could be coded with either an 'asthma' or an 'infection' code. Most importantly in the case of training for real-time classification of ED notes, annotation needs to conform to what is discernable in the limited text available at triage time, whereas diagnostic codes may reference subsequent textual information.

In this work, we aim to assess the effectiveness of active learning in developing a high-performing model for a clinical text classification task from emergency department notes, specifically focusing on syndrome detection of presentations of asthma, a leading cause of disease burden in Australia [31]. The research aims to explore pool-based active learning: different query strategies will be investigated to identify the most valuable instances for labelling, ultimately enhancing the model's ability to identify syndromes from emergency department triage notes.

2 Data

The data comes from the SynSurv near real-time syndromic surveillance system, which primarily collects data from the emergency departments of 26 of the 39 public hospitals in the Australian State of Victoria, which has a population of 6.5 million. It includes texts written by ED nurses at the triage desk after their initial assessment of each patient. These have a unique structure mainly composed of abbreviations (such as "SOB" for "shortness of breath" or "BIBA" for "Brought in by ambulance") and short phrases (such as "asthma flare up unable to manage with ventolin at home"). Some of the abbreviations may differ across hospitals. The texts typically encompass a presenting complaint, medical history, relevant negatives, and the nurse's observations of the patient. The length of the text varies, ranging from detailed narratives of how a patient presents to the triage nurse to concise mentions of a presumed diagnosis and relevant observations. The notes can contain misspellings and the quality of the text can vary widely. ED records also include an age field, which is considered when assessing a record for labelling, and so we included into the data by prepending it to the text.

The unlabelled text pool consisted of 733,638 ED triage texts – all the unique texts with 3 or more characters, received by SynSurv from 1 January 2022 to 22 May 2023. The average length of text was 34.3 words, the maximum text length was 367 words, and the minimum was 4 words.

Labelling criteria were designed by JBL, an author who is a medical doctor with experience in the emergency department (ED). The criteria were used to identify a cohort of asthma cases for monitoring for any sudden increase in acute asthma reporting in a local area that might indicate an unfolding health event, particularly for "thunderstorm asthma". A detected increase in asthma presentations can trigger the allocation of extra resources to help hospitals experiencing an influx to cope with the event. A record is labeled as an asthma presentation if it mentions: asthma symptoms plus a relevant history; use of asthma medications without an accompanying non-asthma indication (such as anaphylaxis or chronic obstructive pulmonary disease); or if a trained health worker (nurse, doctor, or ambulance paramedic) seems to have considered the diagnosis likely. A record is not labeled as asthma if the patient is under 2 years old or over 70 years unless the nurse explicitly states that the patient has asthma or there is a clear history of asthma. Mentions of symptoms (e.g., wheeze, dyspnea, or increased work of breathing) are not labelled as asthma unless accompanied by other features as above.

Detection of acute or exacerbated asthma in triage notes is a very challenging task. Asthma can have symptoms in common with other health conditions, such as cardiac, allergic reactions, or lung trauma. In some cases, it can be very difficult for even a domain expert to decide if an acute asthma attack is the reason for a presentation. For example, a patient who was 40 years old and presented with shortness of breath and chest pain described as "heart hurting", that started a few hours prior to presentation. The patient had nausea and vomiting, and inspiratory and expiratory wheezing. He was given both asthma and cardiac medications, showing that both heart and lung issues were considered by the healthcare team. The domain expert decided that for the purpose of training a model for detecting acute asthma, this should be considered as a cardiac case, despite the healthcare team's assessment of possible asthma. Table 1 contains examples of asthma (label 1) and non-asthma (label 0) records.

Table 1. Examples of asthma presentations

Asthma	Presentation
Yes	Age: 54- exacerbated asthma. Using Ventolin adn pred, nil improvement. Coughing, tightness in chest. Coughing at triage+++, states feels like breathing very fast, nil wheeze on ausculation. Nil resp distress
Yes	Age: 5Y - cough, vomiting + abdo pain. 3x 6 puffs salbutamol nil effect. OE speaking full sentences, tracheal tug, UL insp wheeze SOB
No	Age: 87 - SOB 3/7 URTI, On arrival resp distress, 89% RA, Crackles 10mg Salbu, 500mcg Atrovent, Phx intubated before, COPD, CABGS, AF, CHOL, HTN
No	Age: 1 year - Croup - woke up struggling to breathe. D2 of croup. Mo gave redipred and went to GP. OE NWOB, nil stridor. Distressed ++
No	Age: 48Y - Unwell with pain, insp and exp wheeze, leg swelling, Febrile tachycardic. EMS Treatment (Lung CA with bone mets, had hydromorph, fenatnyl patch AV 350mcg IN fentanyl 50mg IN ketamine 5/10 post meds pt tachy, febrile, with cellulitis had ventolin puffers pt on 2L oxygen, HR 118, BP 126/74, RR24, T 37.9, BGL 14.6, GCS15, spo2 92%)

3 Methods

The goal of our active learning research was to empirically compare the use of these techniques against traditional approaches for identifying further training data, while using minimal amounts of data to reduce the labelling burden on domain experts. Our approach was to utilize an initially trained asthma classification model's predictions and probabilities, and the vector-space embeddings of the texts in the active learning processes. Therefore, we first trained a classifier on a limited set of training data to give us a discriminator, which we used to select a likely asthma-related cohort and applied active learning to identify the most useful examples for labelling. We implemented a stopping criterion using a metric-based approach, setting a predefined threshold performance of 0.9 on the F1 score. This decision was informed by our experience working with the ED data [32], considering the acceptable performance level, the quality of our test datasets, and the constraints of the limited time available for our experiments.

3.1 Data Preparation for Classification

A pattern matching rule, designed by JBL, was employed on the text pool to identify records for labelling which contained mentions of asthma, asthma medications, or asthma symptoms. The rule searched for records that contained one or more of the following strings: 'asthm', 'wheez', 'salb', 'ventol', and 'tightness' - 56,587 records were identified as potential candidates for labelling. To ensure a representative distribution of text lengths in our labelled data, subsets of 500 records were sampled from the data according to their text lengths and annotated by JBL. Six subsets of 500 labelled records were obtained, resulting in a labelled training dataset containing 2,500 records, with 500 held in reserve.

The labelled 2,500 training dataset contained 647 records with a positive label for an asthma presentation and 1,853 records for a negative presentation. From the 2,500 we sampled out 100 records (55 of negative and 45 of positive labels) for a validation dataset. We also sampled out 150 for a test dataset, and from the reserved dataset of 500 we manually identified 100 more labelled examples for the test set that were likely to challenge the models, resulting in a test dataset of 250 (125 of each label).

These data were approximately 0.3% of the original 733 thousand text pool. Table 2 presents the datasets distributions.

Table 2. The datasets distributions

Dataset	Asthma	Non-Asthma	Total
Pool	–	–	733,638
Asthma pattern matching	–	–	56,587
Training	527	1,723	2,250
Validation	45	55	100
Test	125	125	250

Because of the data imbalance of 527 positive to 1723 negative labels, for training our models we used under-sampling to combine the 527 positive labels and half of the negative labels of the training dataset into two smaller training datasets of 1,389, each containing 527 positive and 862 negative instances (one negative example was repeated to balance the two training sets).

3.2 Classification Model

We used the RoBERTa-large-PM-M3-Voc [33] model, published by Facebook, which has been shown to be superior to other models for classifying biomedical and clinical texts. We fine-tuned a classifier on each dataset, obtaining two models. We selected the best performing classifier checkpoints from each training run by evaluating F1 scores on the validation dataset, then further assessed the models by loading these checkpoints and predicting on the test dataset. Test scores are shown in Table 3.

Table 3. Baseline models' test scores

Model	TN	FN	FP	TP	Precision	Recall	F1
Baseline Model One	100	19	25	106	**0.8092**	0.8480	**0.8281**
Baseline Model Two	78	11	47	114	0.7081	**0.9120**	0.7972

3.3 Predictions on the Pool

The base classifier trained on the first training data subset had a better F1 score on the test data because of its greater precision, however the model trained on the second subset had a far higher recall (0.91 vs. 0.85), though its lack of precision resulted in a lower F1 score. We estimated that their combined positive predictions on the text pool would indicate a likely asthma-related subset of the text pool, and that their overlapping positive predictions would mean that negative predictions would be included, thereby including examples of model disagreement or uncertainty.

After applying the models to the entire 700 thousand record text pool (including labelled records), we obtained 14,101 positive predictions from the first base classifier, and 18,513 from the second base classifier. Their combined positive predictions were 19,073, resulting from 13,541 identical positive predictions from each model and 5,532 positive predictions made by only one of the models. The difference therefore contained negative predictions from one of the classifiers. By constraining our view of the text pool to these data we effectively filtered out records that were unlikely to be asthma-related but still allowed for some uncertainty in the predictions. We refer to these records as the "positive predictions" data.

3.4 Query Strategies

We imported the text pool into a ChromaDB vector database, with Sentence-BERT embeddings [34], using the all-MiniLM-L6-v2 model, which is trained on a large dataset of sentence pairs and maps sentences to a 384-dimensional dense vector space. We used the default Euclidian distance algorithm for mapping the vector embeddings relationships. We treated an entire record as a single text unit—this was appropriate for both the brevity of the ED notes and the type of text found in them. For metadata we included the record identifier, the two base models' predictions and probabilities, and the label, where one existed. Importing into a vector database allowed us to preserve the embeddings information, which we used for clustering the likely asthma-related records, and additionally allowed us to perform ad hoc document similarity analysis via the inbuilt query mechanism of ChromaDB.

We then performed three-dimensional k-means clustering of the group of either labelled or positively predicted text, using UMAP and HBDSCAN over the vector embeddings, again based on Euclidean distance. Forty-one clusters were identified, and we estimated their centroids by calculating the average of their XYZ axes. The clusters and their centroids were the basis for identifying our active-learning records. Since there were 41 clusters, we decided to obtain 4 examples per cluster ($4 \times 41 = 164$) from our active learning methods, and 164 records when evaluating other sampling approaches. We took all the samples of 164 each from the predicted cohort, by using the three most common families of active learning strategies: random sampling, uncertainty sampling, and diversity sampling [35]. We also created a baseline by sampling 164 of the records that had been identified by pattern matching. The sampling is explained in detail in the following sections.

Pattern Matching

In the baseline pattern matching approach, candidate records were selected from the available pool based on the presence of predetermined search strings. The search strings included 'asthm', 'wheez', 'salb', 'ventol', and 'tightness', and any record that exhibited one or more of these search strings was included. These were subsequently labelled in batches of 500. A random sample of 164 was taken from the remaining 400 records of the initial pattern-matched dataset, that had not been used in the initial training. It contained 60 positive and 104 negative labels for asthma.

Random Sampling

This active learning strategy ensures that any record that had been identified as potentially asthma by the initial classification process has an equal opportunity to be chosen. We randomly selected 164 records from the "positive predictions" group, which excluding any that had already been labelled. Random sampling used the same sampling technique as was used to select from pattern-matched records, the difference was the data source. That is, we sampled from the records identified by the two baseline classifiers, rather than records identified via pattern matching. Because of the overlapping predictions of the two baseline classifiers, there were records that one or other of the classifiers said were *not* asthma. According to the best of the baseline classifiers there were 103 positive predictions and 61 negative predictions, but the subsequent labelling showed these as 84 positive and 80 negative labelled asthma records.

Diversity Sampling

Diversity sampling, also referred to as *representativeness sampling,* focuses on assembling a diverse collection of records that represent the data, which in our case are the group of positive predictions records. Using the 41 clusters obtained by k-means clustering, we identified the 4 closest unlabeled records to the centroid of each of the clusters, based on Euclidian distance—and so obtained 164 records that represented the range of topics in the data. The best baseline classifier predicted that there were 93 positive and 71 negative asthma records in this sample, but labelling showed that there were 56 positive and 108 negative labels.

Uncertainty Sampling

The uncertainty-based sampling approach selects data about which the classifier model lacks confidence, aiming to identify data points that can offer the most information about subject (decision) boundaries. This is also referred to as *informativeness* sampling, as labelling such low-confidence data provides more information to the model compared to repeated instances of already well-known data. For each of the 41 clusters we chose the 4 records where the most accurate base classifier (base classifier 1) had the lowest probability for its predictions, which were ranging from around 0.75 to 0.51. The baseline classifier had predicted 94 positive and 70 negative asthma records in the sample, but labelling had these as 39 positive and 125 negative labels.

3.5   Training on Active Learning Enhanced Data

After the samples had been labelled, they were added to the training subsets that had been used in the first training for the two base classifiers. This is depicted in Table 4.

Table 4. Active learning contributions to training datasets

Dataset	New Pos	Pos	Total Pos	New Neg	Neg	Total Neg	Total
Uncertainty	39	527	566	125	862	987	1553
Diversity	56	527	583	108	862	970	1553
Sampling	84	527	611	80	862	942	1553
Pattern Matched	60	527	587	104	862	966	1553
Baseline		527	527		862	862	1389

New RoBERTa classifiers were trained over the updated datasets with identical parameters to those used in the first round of training and used the same test dataset for evaluation. This meant that the different strategies were all evaluated under the same conditions as the initial training and as each other, the only difference being the added information in the data obtained by the various active learning (or pattern matched) strategies, which was labelled and added to the training datasets. When assessing random sampling, in order to accommodate the uncertainties associated with the random approach, we generated three random samples, trained a new model for each sample, and utilized the scores from the median model during the evaluation process.

3.6 Training with Hybrid Query Strategies

We carried out another round of training by adding various query strategy results together: Uncertainty + Diversity, Uncertainty + Random Sampling, Diversity + Random Sampling, and even Random Sampling + Random Sampling (by sampling a further 164 from the combined records from the best and worst of the random sampling datasets). Finally, we evaluated a dataset of Uncertainty + Diversity + Random Sampling strategy. As before, these were added to the initial 2,500 record training dataset used for training the baseline model. A few duplicates existed because of overlaps in the sampling techniques, these were removed.

4 Results

When evaluating random sampling we used the scores of the median model. For the uncertainty and diversity datasets we used the scores of the best model. The models' performance was assessed using the standard evaluation metrics of precision, recall, and F1-score, which are presented in Table 5.

Table 5. Scores of baseline models vs. models including active learning data in training

Model	TN	FN	FP	TP	Precision	Recall	F1
Uncertainty [3]	118	16	7	109	**0.9397**	0.8720	**0.9046**
Sampling [3, 4]	108	10	17	115	0.8712	**0.9200**	0.8949
Diversity [3]	107	13	18	112	0.8615	0.8960	0.8784
Pattern Matched [2]	113	24	12	101	0.8938	0.8080	0.8487
Baseline 1 [1]	100	19	25	106	0.8092	0.8480	0.8281
Baseline 2 [1]	78	11	47	114	0.7081	0.9120	0.7972

1.Baseline is the initial model trained on 2,500 labelled records, identified via pattern matching.
2.Pattern Matched is a model trained on the initial 2,500 plus 164 additional records identified via pattern matching.
3.Uncertainty, Diversity, and Sampling are models trained on the initial 2,500 plus 164 records which were sampled from the baseline models' predictions, according to their respective sampling process (Uncertainty, Diversity or Sampling), and labelled.
4.Sampling was repeated 3 times, to allow for its unpredictability, the median scores are used.

The results of using hybrid query strategies datasets are shown in Table 6, which includes all prior tests for comparison.

Table 6. Scores after combining active learning data for training

Model	TN	FN	FP	TP	Precision	Recall	F1
Uncertainty + Sampling	112	9	13	116	0.8992	**0.9280**	**0.9134**
Uncertainty	118	16	7	109	**0.9397**	0.8720	0.9046
Sampling	108	10	17	115	0.8712	0.9200	0.8949
Uncertainty + Diversity + Sampling	108	10	17	115	0.8712	0.9200	0.8949
Uncertainty + Diversity	116	17	9	108	0.9231	0.8640	0.8926
Sampling + Sampling	110	14	15	111	0.8810	0.8880	0.8845
Diversity + Sampling	104	10	21	115	0.8456	0.9200	0.8812
Diversity	107	13	18	112	0.8615	0.8960	0.8784
Pattern Matched	113	24	12	101	0.8938	0.8080	0.8487
Baseline 1	100	19	25	106	0.8092	0.8480	0.8281
Baseline 2	78	11	47	114	0.7081	0.9120	0.7972

5 Discussions

In this study, we investigated the impact of active learning strategies on selecting data for labelling to train models for classifying emergency department acute asthma presentations. We evaluated three commonly used active learning approaches: uncertainty sampling, diversity sampling, and random sampling.

Our findings confirmed the viability of active learning methods for medical text classification, offering a promising approach to reduce annotation costs. We observed that selecting records that the initially trained model was *uncertain* about provided a more informative training set, as evidenced by the better scores obtained by subsequent models. This aligns with the conclusions drawn from other research [20, 24]. When 164 records identified using uncertainty were added to the baseline of 2500 training records, and a new model was trained, the F1-score improved by 7.6 percentage points compared to the baseline model's score, resulting in a score of 0.905.

The diversity method also elicited a significant difference, but not as large; its F1-score of 0.878 was 5.0 percentage points better than baseline. Our hypothesis is that the initially labeled data sufficiently represented the space, thus limiting the advantage for the representative approach of the diversity method. Other researchers who have found diversity sampling not as effective as uncertainty in the medical domain argue that this could be due to the more restricted sublanguage of the medical domain, having a limited vocabulary [23].

Models trained with the additional records obtained through random sampling from the predicted cohort, on average, exhibited superior performance compared to those trained using diversity sampling. The median F1-score of 0.895 was notably 6.7 points higher than the baseline, surpassing the diversity-trained model's score by 1.6 points. The score was only 1 percentage point less than the uncertainty trained model. The random sampling performance reinforces the observation that we were able to train an accurate model on the initially labelled data, and so it was effective to randomly sample from records it had identified as likely asthma. Also, random sampling allowed for a variety of asthma examples and a more balanced set of labels, which had the potential to work favorably compared with the diversity sampling approach that targeted records near to each k-means cluster center.

Records sampled from pattern matched examples did not however perform as well - training that included a batch of these only raised the F1-score by 2.1 percentage points. This indicates that all the active learning approaches outperformed an approach of merely adding extra records identified by pattern matching.

Interesting results were obtained when the various active learning datasets were combined. The hybrid of the uncertainty and random sampled records improved the model over the previous best score (from uncertainty sampling) by 0.9 percentage points, to achieve an F1-score of 0.913, but all other combinations either made no difference or decreased the model performance to below that of both the uncertainty and random sampling models. Ordinarily, adding more data improves a model's performance, but this result suggested that additional data could be detrimental, compared with the contribution of a single set of extra training records obtained by an active learning method.

We performed error analysis on the uncertainty trained model's false predictions. Out of 250 records, there were 16 instances of false negative predictions. All these records had mentions of other health issues, such as infection, use of antibiotics, fever, or croup. This led the model to incorrectly predict that the patient did not have asthma. Of the 7 false positive predictions, all the patients were over the age of 70 and had a history of asthma. However, they also had strong indications of other health issues, which could have been the likely reason for their presentation (instead of asthma).

The potential utility and efficiency of this active learning approach assists syndromic surveillance systems to be rapidly prepared and implemented to address emerging public health crises. By increasing the scope of AL trained surveillance of health events of interest, early warning of dynamic changes in presentation patterns can improve care and outcomes. In the case of thunderstorm asthma events, health systems could respond more rapidly and specifically to allocate resources and inform providers and the community to be alert for symptoms, maximise preventive therapies and stock up on relieving medicines.

Our study has limitations. We were not able to explore the effect of iterative rounds of active learning, because we reached close to a maximum possible performance in our classifiers after just the initial application of active learning. Ideally, we should have started with a smaller number of training records, and consequently our initial classifier would have not been so performant, and we could have observed at least two rounds of improvement due to the application of repeated active learning.

In the future, we plan to initiate the task with a smaller set of labeled data and extend our investigation into the effects of active learning, aiming to gain a more comprehensive understanding of its potential benefits, and investigate other active learning approaches.

6 Conclusion

Our study highlighted the impact of active learning strategies on enhancing the efficiency and efficacy of medical text classification models, specifically focusing on emergency department asthma presentations. Our findings underscored the superiority of uncertainty sampling, wherein selecting records that initially perplexed the model yielded substantial improvements in subsequent model training. Notably, the effectiveness of random sampling highlighted the model's ability to glean valuable insights from its own predictions. The use of active learning in syndromic surveillance allows development of detection models in timely way enabling public health systems' responsiveness, preparedness, and safety, ultimately improving patient care and outcomes.

References

1. Picard, C., Kleib, M., Norris, C., O'Rourke, H.M., Montgomery, C., Douma, M.: The use and structure of emergency nurses' triage narrative data: scoping review. JMIR Nurs. **6**, e41331 (2023)
2. Schröder, C., Niekler, A.: A survey of active learning for text classification using deep neural networks (2020)
3. Olsson, F.: A literature survey of active machine learning in the context of natural language processing (2009)
4. Thien, F., et al.: The Melbourne epidemic thunderstorm asthma event 2016: an investigation of environmental triggers, effect on health services, and patient risk factors. Lancet Planet. Heal. **2**, e255–e263 (2018)
5. Settles, B.: Computer Sciences Active Learning Literature Survey (2009)
6. Tharwat, A., Schenck, W.: A survey on active learning: state-of-the-art, practical challenges and research directions. Mathematics **11**, 820 (2023). https://doi.org/10.3390/math11040820
7. Kumar, P., Gupta, A.: Active learning query strategies for classification, regression, and clustering: a survey. J. Comput. Sci. Technol. **35**, 913–945 (2020). https://doi.org/10.1007/s11390-020-9487-4
8. Zhan, X., Wang, Q., Huang, K., Xiong, H., Dou, D., Chan, A.B.: A Comparative Survey of Deep Active Learning (2022)
9. Lewis, D.D.: A sequential algorithm for training text classifiers: Corrigendum and additional data. In: Acm Sigir Forum. ACM New York, NY, USA, pp. 13–19 (1995)
10. Seung, H.S., Opper, M., Sompolinsky, H.: Query by committee. In: Proceedings of the Fifth Annual Workshop on Computational learning theory, pp. 287–294 (1992)
11. Settles, B., Craven, M., Ray, S.: Multiple-instance active learning. In: Advances in Neural Information Processing Systems, vol. 20 (2007)
12. Roy, N., McCallum, A.: Toward optimal active learning through sampling estimation of error reduction. Int. Conf. Mach. Learn. 441–448 (2001)
13. Settles, B., Craven, M.: An analysis of active learning strategies for sequence labeling tasks. In: Proceedings of the 2008 Conference on Empirical Methods in Natural Language Processing, pp. 1070–1079 (2008)

14. Gissin, D., Shalev-shwartz, S.: Discriminative active learning, pp. 1–11 (2013)
15. Brinker, K.: Incorporating diversity in active learning with support vector machines. In: Proceedings of the 20th International Conference on Machine Learning (ICML-03), pp. 59–66 (2003)
16. Dasgupta, S., Hsu, D.: Hierarchical sampling for active learning. In: Proceedings of the 25th International Conference on Machine learning, pp. 208–215 (2008)
17. Ash, J.T., Zhang, C., Krishnamurthy, A., Langford, J., Agarwal, A.: Deep batch active learning by diverse, uncertain gradient lower bounds. arXiv Preprint arXiv:1906.03671 (2019)
18. Shui, C., Zhou, F., Gagné, C., Wang, B.: Deep active learning: unified and principled method for query and training. In: International Conference on Artificial Intelligence and Statistics, pp. 1308–1318. PMLR (2020)
19. Kim, Y.: Deep active learning for sequence labeling based on diversity and uncertainty in gradient. arXiv Preprint arXiv:2011.13570 (2020)
20. Kholghi, M., Sitbon, L., Zuccon, G., Nguyen, A.: Active learning: a step towards automating medical concept extraction. J. Am. Med. Inform. Assoc. **23**, 289–296 (2016)
21. Li, M., Scaiano, M., El Emam, K., Malin, B.A.: Efficient active learning for electronic medical record de-identification. AMIA Summits Transl. Sci. Proc. **2019**, 462 (2019)
22. Mottaghi, A., Sarma, P.K., Amatriain, X., Yeung, S., Kannan, A.: Medical symptom recognition from patient text: an active learning approach for long-tailed multilabel distributions, pp. 1–14 (2020)
23. Figueroa, R.L., Zeng-Treitler, Q., Ngo, L.H., Goryachev, S., Wiechmann, E.P.: Active learning for clinical text classification: is it better than random sampling? J. Am. Med. Inform. Assoc. **19**, 809–816 (2012)
24. Chen, Y., Lasko, T.A., Mei, Q., Denny, J.C., Xu, H.: A study of active learning methods for named entity recognition in clinical text. J. Biomed. Inform. **58**, 11–18 (2015)
25. Horng, S., Greenbaum, N.R., Nathanson, L.A., McClay, J.C., Goss, F.R., Nielson, J.A.: Consensus development of a modern ontology of emergency department presenting problems-the hierarchical presenting problem ontology (HaPPy). Appl. Clin. Inform. **10**, 409–420 (2019). https://doi.org/10.1055/s-0039-1691842
26. Lee, S.H., Levin, D., Finley, P.D., Heilig, C.M.: Chief complaint classification with recurrent neural networks. J. Biomed. Inform. **93**, 103158 (2019)
27. Rozova, V., Witt, K., Robinson, J., Li, Y., Verspoor, K.: Detection of self-harm and suicidal ideation in emergency department triage notes. J. Am. Med. Inform. Assoc. **29**, 472–480 (2022). https://doi.org/10.1093/jamia/ocab261
28. Chapman, A.B., et al.: Development and evaluation of an interoperable natural language processing system for identifying pneumonia across clinical settings of care and institutions. JAMIA Open. **5**, 1 (2022). https://doi.org/10.1093/jamiaopen/ooac114
29. Ryan, J.: Comparison of presenting complaint vs discharge diagnosis for identifying "Nonemergency" emergency department visits. J. Emerg. Med. **45**, 152–153 (2013). https://doi.org/10.1016/j.jemermed.2013.05.036
30. Singleton, J., Li, C., Akpunonu, P.D., Abner, E.L., Kucharska-Newton, A.M.: Using natural language processing to identify opioid use disorder in electronic health record data. Int. J. Med. Inform. **170**, 104963 (2023). https://doi.org/10.1016/j.ijmedinf.2022.104963
31. AIHW: Asthma. https://www.aihw.gov.au/reports/chronic-respiratory-conditions/asthma-1
32. Khademi, S., Palmer, C., Dimaguila, G.L., Javed, M., Buttery, J., Black, J.: Data augmentation to improve syndromic detection from emergency department notes. In: ACM International Conference Proceeding Series, pp. 198–205 (2023)
33. Lewis, P., Ott, M., Du, J., Stoyanov, V.: Pretrained Language Models for Biomedical and Clinical Tasks: Understanding and Extending the State-of-the-Art (2020)

34. Reimers, N., Gurevych, I.: Sentence-bert: sentence embeddings using siamese bert-networks. arXiv Preprint arXiv:1908.10084 (2019)
35. Zhang, Z., Strubell, E., Hovy, E.: A survey of active learning for natural language processing. In: Proceedings of the 2022 Conference on Empirical Methods in Natural Language Processing. EMNLP 2022, pp. 6166–6190 (2022)

Author Index

© The Editor(s) (if applicable) and The Author(s), under exclusive license
to Springer Nature Singapore Pte Ltd. 2024
D. Benavides-Prado et al. (Eds.): AusDM 2023, CCIS 1943, pp. 299–300, 2024.
https://doi.org/10.1007/978-981-99-8696-5

Printed in the United States
by Baker & Taylor Publisher Services

Printed in the United States
by Baker & Taylor Publisher Services